Mechanical Behavior of Advanced Engineering Materials

Mechanical Behavior of Advanced Engineering Materials

Guest Editors

Zhubin He
Haiming Zhang

MDPI

Basel • Beijing • Wuhan • Barcelona • Belgrade • Novi Sad • Cluj • Manchester

Guest Editors

Zhubin He
Department of Mechanical Engineering
Dalian University of Technology
Dalian
China

Haiming Zhang
School of Materials Science and Engineering
Shanghai Jiao Tong University
Shanghai
China

Editorial Office
MDPI AG
Grosspeteranlage 5
4052 Basel, Switzerland

This is a reprint of the Special Issue, published open access by the journal *Materials* (ISSN 1996-1944), freely accessible at: https://www.mdpi.com/journal/materials/special_issues/mech_behav_adv_eng_mater.

For citation purposes, cite each article independently as indicated on the article page online and as indicated below:

Lastname, A.A.; Lastname, B.B. Article Title. *Journal Name* **Year**, *Volume Number*, Page Range.

ISBN 978-3-7258-7903-8 (Hbk)
ISBN 978-3-7258-7904-5 (PDF)
https://doi.org/10.3390/books978-3-7258-7904-5

Contents

About the Editors

Zhubin He

Zhubin He is a professor at the School of Mechanical Engineering, Dalian University of Technology in China. His research interests include constitutive modeling, materials characterization, new forming technology, lightweight manufacturing and hot fluid forming.

Haiming Zhang

Haiming Zhang is a professor at the School of Materials Science and Engineering, Shanghai Jiao Tong University in China. His research interests include constitutive model, multiscale modeling, metal forming, plasticity and damage, materials characterization, lightweight materials and deformation heterogeneity.

Preface

In recent years, remarkable advancements have been achieved in the research of advanced engineering materials (AEMs), encompassing lightweight and high-strength metals and alloys, intermetallics, composites, shape-memory alloys, metallic glasses, and other related materials. These materials have found widespread applications across diverse industries, such as automotive, aerospace and astronautics, electronics, medical devices, and sports equipment, underscoring their pivotal role in driving technological innovation and industrial progress.

The mechanical properties of AEMs—including elastoplasticity, anisotropy, formability, fracture mechanics, defect interactions, and strengthening–toughening mechanisms—exert a decisive influence on both their manufacturing/forming processes and in-service performance. A deeper understanding of the mechanical behavior of these materials is therefore indispensable for their effective development and application. This has prompted the rapid advancement of cutting-edge experimental techniques for multiscale mechanical testing and microstructural characterization, as well as state-of-the-art computational and theoretical approaches integrated with advanced constitutive models.

The Special Issue "The Mechanical Behaviors of Advanced Engineering Materials" is dedicated to showcasing high-quality original research findings that span the broad spectrum of AEMs' mechanical properties. The topics covered herein will be of great interest to scientists, researchers, and professionals affiliated with universities, research institutes, laboratories and industries, who are engaged in the design, optimization, and practical application of advanced engineering materials.

Zhubin He and Haiming Zhang
Guest Editors

materials

MDPI

Article

Hot Deformation Behavior and Microstructure Evolution Mechanisms of Ti6Al4V Alloy under Hot Stamping Conditions

Mingjia Qu [1,2], Zhengwei Gu [1,2], Xin Li [1,2,*], Lingling Yi [1,2], Yi Li [1,2], Ge Yu [1,2] and Yafu Zhao [3,*]

[1] State Key Laboratory of Automobile Materials, Jilin University, Changchun 130025, China; qumj21@mails.jlu.edu.cn (M.Q.); guzw@jlu.edu.cn (Z.G.); yill20@mails.jlu.edu.cn (L.Y.); henrylee@jlu.edu.cn (Y.L.); yuge@jlu.edu.cn (G.Y.)
[2] Department of Materials Science and Engineering, Jilin University, Changchun 130025, China
[3] CRRC Changchun Railway Vehicles Co., Ltd., Changchun 130022, China
* Correspondence: li_xin@jlu.edu.cn (X.L.); zhaoyafu6265@163.com (Y.Z.)

Abstract: Through the study of the thermal rheological behavior of Ti6Al4V alloy at different temperatures (500 °C, 600 °C, 700 °C, and 800 °C) and different strain rates (0.1 s^{-1}, 0.05 s^{-1}, 0.01 s^{-1}, and 0.005 s^{-1}), a constitutive model was developed for Ti6Al4V alloy across a wide temperature range in the hot stamping process. The model's correlation coefficient reached 0.9847, indicating its high predictive accuracy. Hot processing maps suitable for the hot stamping process of Ti6Al4V alloy were developed, demonstrating the significant impact of the strain rate on the hot formability of Ti6Al4V alloy. At higher strain rates (>0.05 s^{-1}), the hot processing of Ti6Al4V alloy is less prone to instability. Combining hot processing maps with hot stamping experiments, it was found that the forming quality and thickness uniformity of parts improved significantly with the increase in stamping speed. The phase composition and microstructures of the forming parts under different heating temperature conditions have been investigated using SEM, EBSD, XRD, and TEM, and the maximum heating temperature of hot stamping forming was determined to be 875 °C. The recrystallization mechanism in hot stamping of Ti6Al4V alloys was proposed based on EBSD tests on different sections of a hot stamping formed box-shaped component. With increasing deformation, the effect of dynamic recrystallization (DRX) was enhanced. When the thinning rate reached 15%, DRX surpassed dynamic recovery (DRV) as the dominant softening mechanism. DRX grains at different thinning rates were formed through both discontinuous dynamic recrystallization (DDRX) and continuous dynamic recrystallization (CDRX), with CDRX always being the dominant mechanism.

Keywords: titanium alloy; processing map; hot stamping; microstructure evolution; dynamic recrystallization

1. Introduction

In recent years, titanium alloys have been widely used in the aerospace and transportation industries, attributed to their high specific strength and corrosion resistance [1,2]. The application of titanium alloy products in major transportation vehicles like airplanes and trains reduces both self-weight and the frequency of component replacement [3,4], driving an increasing demand for titanium alloy machining processes characterized by low manufacturing costs, short machining cycles, and high forming quality. However, the poor plasticity and significant springback of titanium alloy at room temperature make its forming challenging [5]. Consequently, plastic processing of titanium alloys is typically performed at high temperatures. Conventional hot working processes for titanium alloy sheets involve superplastic forming [6] and isothermal hot press forming. However, superplastic forming which requires blow molding of the sheet at high temperatures, imposes stringent requirements on forming equipment. Although isothermal hot press forming uses relatively simple forming equipment, it necessitates heating the forming tools, leading to short die life and considerable energy consumption. Additionally, both superplastic forming and isothermal hot press forming are relatively time consuming, with dies requiring several hours of preheating before use, which diminishes production efficiency. These factors are among the reasons that hinder the widespread application of titanium alloy components. Therefore, identifying environmentally friendly, efficient, and high-quality processing methods has become a current research focus.

The hot stamping process involves heating sheet metal to a predetermined temperature before quickly transfer to dies with an internal cooling system. This allows for simultaneous stamping and quenching of the sheet [7]. As hot stamping eliminates the need for die heating, it lowers the quality requirements for the dies are reduces heating and forming times. This significantly shortens the processing cycle, saves energy and protects the environment. Therefore, researching and exploring the hot stamping process for titanium alloys is highly beneficial [8]. Current research on the hot stamping of titanium alloys primarily focuses on Ti6Al4V alloy, which is the most widely used titanium alloy globally. To characterize its high-temperature flow behavior, extensive constitutive models have been developed in previous studies. Current research on the hot stamping of titanium alloys primarily focuses on Ti6Al4V alloy, the most used titanium alloy in the world today. To describe its high-temperature flow behavior, generous constitutive models have already been developed in previous studies. Gao et al. [9] constructed six constitutive models to describe the high-temperature flow behavior of Ti6Al4V alloy sheet under conditions ranging from 650 to 750 °C and discussed their accuracy. The results showed that the Hensel–Spittel (HS) model has the best prediction result while the Johnson–Cook (JC) model has the worst prediction result. Chao et al. [10] formulated the constitutive equations based on the stress–strain curves of Ti6Al4V alloy between 700 and 800 °C. The strain-compensated Arrhenius constitutive model showed high predictive capability under most experimental conditions. Yang et al. [11] developed a viscoplastic constitutive model for Ti6Al4V alloy, incorporating damage and softening mechanisms, and incorporating predicted the evolution of damage and fracture processes in hot tensile specimens. While constitutive models can predict the thermal flow behavior of materials, they are not directly applicable to guide hot forming process. In contrast, hot processing maps can characterize the processing performance of materials under various conditions, allowing for a more direct discussion of specific processing parameters. Qiu et al. [12] constructed hot processing maps for SP700 titanium alloy, verified microstructures, and identified optimal processing parameter for the alloy. Luo et al. [13] analyzed hot processing maps of Ti6Al4V alloy, noting significant variations in the instability parameter across different strain rates and deformation temperatures when the alloy was in $\alpha + \beta$ phase region. These studies demonstrate that constitutive models and hot processing maps are widely applied in isothermal hot forming of titanium alloys. However, few studies have specifically focused on the hot stamping process of titanium alloys under conditions with a wide temperature range.

Recent research has extensively explored the microstructural evolution during hot forming process of titanium alloys. Kopec et al. [14] studied the microstructural evolution of Ti6Al4V alloy during hot stamping, revealing phase transformations during heating and the formation of transformed β phase during forming. While the phase transition in hot stamping of titanium is well documented, the deformation mechanisms remain underexplored. Dynamic recrystallization (DRX) is a crucial method for enhancing the microstructure, plasticity, and processability of metallic materials during hot forming. Previous studies indicate that the softening behavior of titanium alloys during hot forming typically results from both DRX and dynamic recovery (DRV) [15,16]. Ning et al. [17] investigated the dynamic softening behavior of TC18 titanium alloy, finding that DRV and DRX dominate under different deformation conditions. Li et al. [18] found that during hot deformation of Ti-6Cr-5Mo-5V-4Al alloy in the $\alpha + \beta$ phase region, the DRX volume fraction is mainly influenced by strain and shows little sensitivity to deformation temperature and strain rate changes. The primary mechanisms of DRX include discontinuous dynamic recrystallization (DDRX) and continuous dynamic recrystallization (CDRX). An et al. [19] investigated the deformation mechanism of Ti-2.5Zr-2Al alloy in $\alpha + \beta$ phase region, finding that the CDRX mechanism-dominated DRX mechanism intensified with increased strain. As Ti6Al4V is the most widely used titanium alloy in the world today, the study of its hot mechanical properties has high application value. In addition, Ti6Al4V alloy, as a two-phase titanium alloy, has excellent mechanical properties, so the study of its hot stamping process is of great significance to the current lightweighting research. Despite extensive literature on softening mechanisms during hot deformation of titanium alloys, research on deformation mechanisms in Ti6Al4V alloy under hot stamping conditions remains incomplete, necessitating further study.

In order to achieve this objective, this paper investigated the flow behavior of Ti6Al4V alloy within temperatures of 500–800 °C and strain rates of 0.1 s^{-1}–0.005 s^{-1}. The strain-compensated Arrhenius constitutive model and a hot processing map are established. Based on hot processing map, the effects of processing parameters such as stamping speed, heating temperature and deformation amount on the quality in hot stamping of Ti6Al4V alloys were investigated, and the microstructural evolution of Ti6Al4V alloys under different deformation conditions was analyzed. And the relationship between the DRX mechanism and deformation amount in hot stamping forming of Ti6Al4V alloy was revealed, which provides theoretical support for precise control of microstructure and mechanical performance in Ti6Al4V alloy hot stamping during industrial production.

2. Materials and Experimental Details

2.1. Materials

The material used in this paper is an industrial Ti6Al4V alloy (Baoji Titanium Industry Co., Ltd., Baoji, China) sheet with a thickness of 2 mm. Table 1 displays its chemical composition. Figure 1a illustrates the initial microstructure, consisting of equiaxed α grains with a small amount of β phase, constituting approximately 6.35% of the structure. Figure 1b shows the room temperature tensile curve of the sheet. The Ti6Al4V alloy sheet was found to have a tensile strength of 1083.19 MPa and an elongation of 10.23%.

Table 1. Chemical composition of Ti6Al4V alloy (%).

Element	Ti	Al	V	Fe	Si
wt%	Bal	6.85	4.00	0.25	0.22

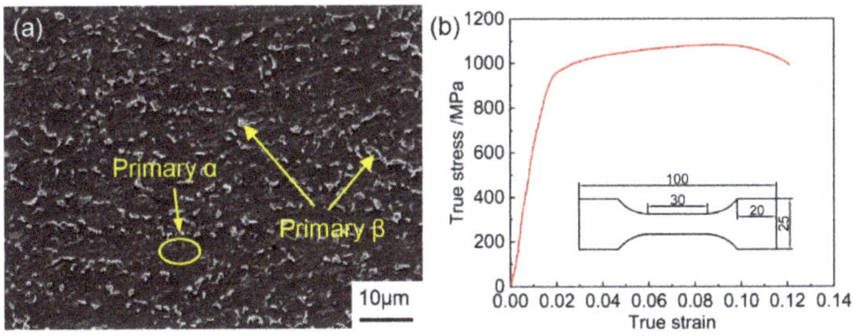

Figure 1. Initial microstructure (**a**) and room temperature tensile curve (**b**) of Ti6Al4V titanium alloy.

2.2. Experimental Details

2.2.1. The Hot Tensile Test

In order to explore the hot deformation behavior of a Ti6Al4V alloy sheet, the hot tensile test was conducted. The schematic diagram of the hot tensile test for sheet material is shown in Figure 2. Based on the existing research, in the hot stamping forming process of Ti6Al4V alloy, the deformation starting temperature is usually between 700 and 800 °C, and in current studies of hot stamping of Ti6Al4V alloy, the temperature range chosen is usually covered by 500–800 °C [14,20], so the deformation temperatures were set to be 500 °C, 600 °C, 700 °C and 800 °C, and according to the real hot stamping process, the strain rates were set to be 0.005 s^{-1}, 0.01 s^{-1}, 0.05 s^{-1} and 0.1 s^{-1}. The specimens were subjected to 800 °C for 2 min air cooling heat treatment before the hot tensile test. In order to ensure even temperature division, the specimens were held for 2 min after heating to the target temperature.

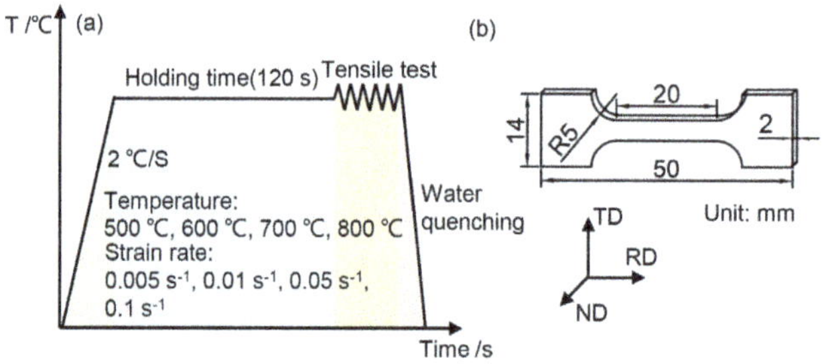

Figure 2. Schematic diagrams of the (**a**) hot tensile test process and (**b**) size of tensile specimen.

2.2.2. The Hot Stamping Test

The hot stamping experiment of Ti6Al4V alloy was divided into several stages. The first stage involved heating. To achieve a higher heating rate during deformation, this study utilized a heating furnace equipped with molybdenum–silicon rods, enabling rapid sheet heating to the target temperature. During heating, the sheet's average heating rate was 14 °C/s, and the maximum rate was 60 °C/s. After heating to the target temperature, the sheet was held in the furnace for 2 min to ensure uniform temperature distribution. The second stage involved the forming stage. The sheet was then removed from the furnace and quickly transferred to room temperature, box-shaped part dies for stamping. The third stage involved holding pressure and quenching. After closing the dies, pressure was held for 1 min to reduce part springback. Before forming, the surface of the sheet was sprayed with boron nitride lubricant, and the top and bottom dies were coated with boron nitride grease to reduce friction.

The whole forming process and the dies profile are shown in Figure 3. The sheet size was 210 × 170 × 2 mm. To reduce wrinkling, it is necessary to cut quarter-circle notches with a radius of 38 mm at four corners of the rectangular sheet. The sheet pressing force was set at approximately 10 kN. The formed part was a box-shaped component including the flange. To determine the optimal forming temperature, titanium alloy specimens with a size of 170 × 20 × 2 mm were previously subjected to hot stamping experiment. Holes were drilled in the sheet's side, and thermocouples were inserted to measure the precise temperature changes during hot stamping experiment.

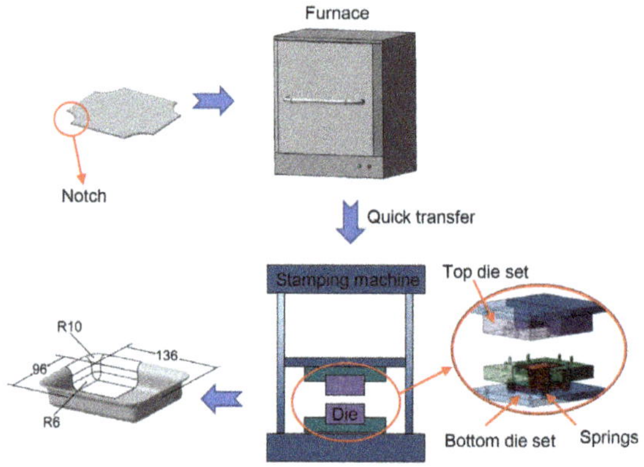

Figure 3. Demonstration of hot stamping formation process.

3. Results and Discussion

3.1. Thermal Rheological Behavior

The true stress–strain curves of Ti6Al4V alloy sheet were obtained through hot tensile tests, as shown in Figure 4. As observed in Figure 4a–d, with increasing temperature and decreasing strain rate, the tensile strength of Ti6Al4V alloy decreases and the elongation increases, indicating enhanced material plasticity. During the initial loading stage, the flow stress rapidly increases, dislocation density significantly rises, and work hardening predominates. With increasing deformation, the continuous accumulation of dislocations leads to the activation of softening mechanisms such as DRV and DRX and their gradual enhancement [21,22]. Under different strain rates and temperature conditions, the materials display markedly different softening behaviors. The flow characteristics of hot deformation depend significantly on the competition between work hardening, DRV and DRX [23].

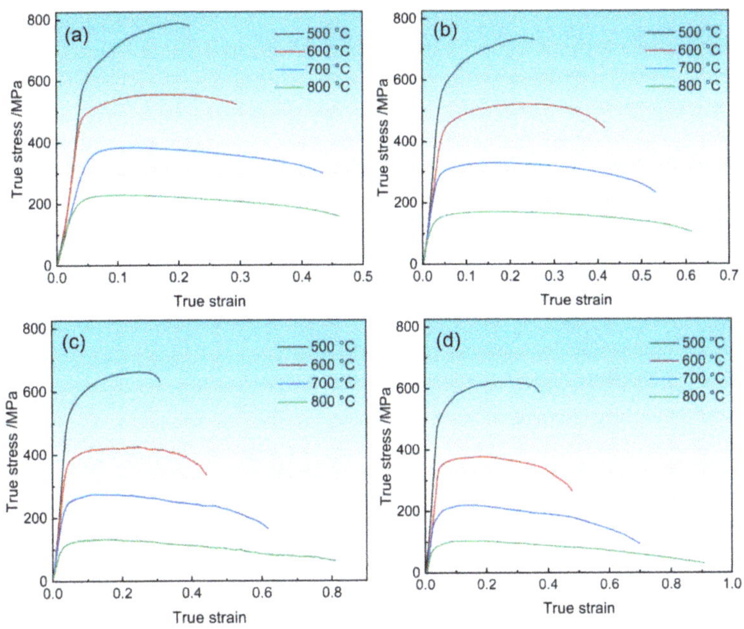

Figure 4. True stress–strain curves at strain rates of (**a**) 0.1 s^{-1}, (**b**) 0.05 s^{-1}, (**c**) 0.01 s^{-1}, and (**d**) 0.005 s^{-1}.

Taking Figure 4a as an example, under the same strain rate, as the deformation temperature increases, the softening mechanisms are enhanced. When the heating temperature reached 600 °C, a balance between softening mechanisms and work hardening is achieved, leaving the flow stress unchanged with increasing strain. At a heating temperature of 700 °C, the softening mechanisms continue to strengthen, surpassing work hardening and resulting in material softening. To further investigate the causes of softening at elevated deformation temperatures, electron backscatter diffraction (EBSD) tests were conducted on the fractured sections of tensile specimens under a strain rate of 0.1 s^{-1} and heating temperatures between 600 and 700 °C in Figure 5. When the orientation difference between adjacent grain boundaries is less than 2°, grains are marked in white to indicate deformation. As hot processing proceeds, DRV action causes subgrain boundaries to gradually appear within deformed grains. Grains are marked in yellow when the orientation difference between adjacent grain boundaries ranges from 2 to 15°, indicating substructure grains. During deformation, low angle grain boundaries (LAGBs) continue to rotate, forming high angle grain boundaries (HAGBs). Grains are marked in green when the orientation difference between adjacent grain boundaries exceeds 15°, indicating recrystallized grains. At both 600 °C and 700 °C, a small number of fine DRX grains are produced, and the proportion of DRV grains significantly exceeds that of DRX, highlighting DRV's dominant role in softening behavior [24]. However, at 700 °C, the percentage of both DRX and DRV grains is higher than at 600 °C, indicating that the material's softening is due to the concurrent growth of DRX and DRV at elevated temperatures.

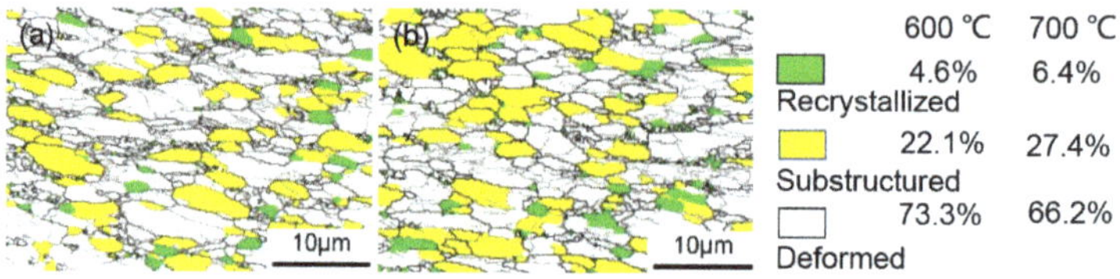

Figure 5. Crystal distribution map at temperatures of (**a**) 600 °C and (**b**) 700 °C.

At a deformation temperature of 500 °C, significantly different softening behaviors were observed across various strain rates. The true stress–strain curves in Figure 4c,d show a brief equilibrium stage absent in Figure 4a,b. This occurs because lower strain rates extend deformation cycles, enabling DRX and DRV to significantly reduce dislocation density and enhanced dynamic softening effects. Conversely, higher strain rates result in greater dislocation densities, leading to enhanced work hardening effect [25]. Additionally, at lower temperatures, the resistance to dislocation motion increases, resulting in more pronounced work hardening [26]. This amplifies the enhancement of the softening effect as the strain rate decreases, leading to a significant difference in softening behavior under varying strain rates at lower temperatures.

3.2. Establishing the Constitutive Model

To describe the flow behavior of materials at high temperatures, Zener et al. [27] investigated the effects of temperature and strain rate on deformation to obtain Equation (1), and a further calculation able to produce Equation (2). In recent studies, the Arrhenius constitutive model has been widely used to describe the flow behavior of Ti6Al4V alloys at high temperatures. The expression can be expressed as:

$$Z = \dot{\varepsilon} \exp(Q/RT) \tag{1}$$

$$\ln Z = \ln A + n \ln[\sinh(\alpha\sigma)] \tag{2}$$

where Z is the Zener–Hollomon parameter, $\dot{\varepsilon}$ is the strain rate, σ is flow stress, Q is hot deformation activation energy, R is molar gas constant, T is the forming temperature, A is material constant, α is stress-level coefficient and n is stress exponent. Based on above parameters, the Arrhenius constitutive equation for Ti6Al4V alloy with strain of 0.06 can be expressed as:

$$\dot{\varepsilon} = 4.827 \times 10^{10}[\sinh(0.00203\sigma)]^{12.0325} \exp[(-207859/RT)] \tag{3}$$

However, the previous equation can not consider the influence of strain on the parameters. To improve the accuracy of the constitutive model, the values of α, n, Q, and $\ln A$ are calculated for strains of 0.06, 0.08, 0.1, 0.12, 0.04, 0.06, 0.08, and 0.2. The resulting strain-compensated Arrhenius constitutive model is shown in Table 2. The stress values were calculated according to the constitutive model and compared with the test values, as shown in Figure 6. The constitutive model produced an average error of only 8.3% in the stress–strain curve, and the correlation coefficient value R is as high as 0.9847 obtained after linear fitting, indicating that the constitutive model has a good prediction ability for the flow stress of Ti6Al4V alloy under a large temperature span.

Table 2. Polynomial fitting results of α, *n*, In*A* and *Q*.

	α	*n*	*Q*	In*A*
k_0	−0.0103	15.96865	676.94558	87.48507
k_1	0.76216	−507.81285	−22,345.4023	−3034.32718
k_2	−18.23749	9887.6117	419,169.9129	58,534.95292
k_3	222.92064	−101,148.339	−4,010,947.11	−582,395.4021
k_4	−1474.3765	560,983.548	20,473,000	3,120,280.7897
k_5	5015.36434	−159,233	−52,513,400	−8,506,609.959
k_6	−6865.6785	1,809,399	52,517,400	9,217,898.9467

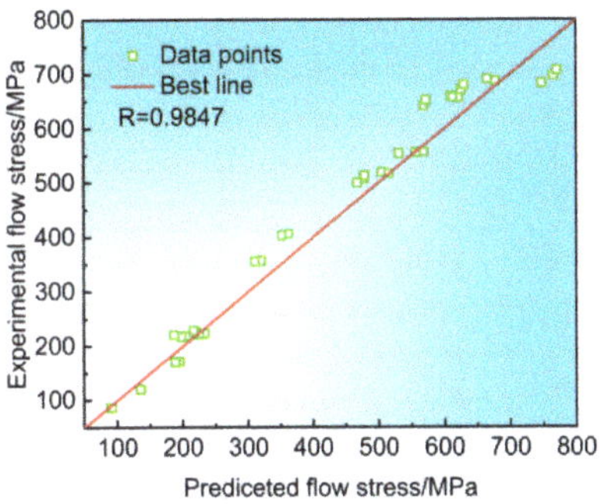

Figure 6. Ti6Al4V alloy predictor stress and experimental stress correlation.

3.3. Hot Processing Maps

Hot processing maps reveal the formability of materials under various deformation conditions and assist in optimizing hot processing parameters. The hot processing map consists of a power dissipation diagram based on the power dissipation efficiency factor η and the rheological instability diagram, based on the instability parameter ξ [28]. Rheological stresses with strain variables of 0.1 and 0.2 were selected to determine the power dissipation efficiency and instability coefficient, respectively. The strain rate sensitivity coefficient m is first calculated based on the stress level at specific temperature and strain rate conditions, and is defined by the following equation:

$$m = \frac{\partial \ln \sigma}{\partial \ln \dot{\varepsilon}} = b + 2c\ln\dot{\varepsilon} + 3d(\ln\dot{\varepsilon})^2 \tag{4}$$

where σ is flow stress, $\dot{\varepsilon}$ is the strain rate, and b, c and d are material constants. The m value obtained from the solution is shown in Table 3. The value of m can determine the plasticity of the material. A larger m value indicates better plasticity of the material [29]. A negative of m suggests that the material is difficult to deform plastically. From Table 3, it can be observed that when strain is 0.1 and temperature is 500 °C, the m value is negative for all strain rate conditions, so the material is difficult to be formed under this deformation condition. A higher m value is obtained at higher temperatures and strain rates, indicating that the plasticity of the material improves with the increase in temperature and strain.

Table 3. Values of m calculated at different hot tensile condition.

ε	0.1				0.2			
T/°C	500 °C	600 °C	700 °C	800 °C	500 °C	600 °C	700 °C	800 °C
$0.1\ s^{-1}$	−0.19	0.21	0.39	0.61	0.057	0.14	0.29	0.49
$0.05\ s^{-1}$	−0.21	0.078	0.13	0.31	0.033	0.066	0.10	0.28
$0.01\ s^{-1}$	−0.20	0.052	0.26	0.30	0.049	0.078	0.28	0.318
$0.005\ s^{-1}$	−0.17	0.17	0.63	0.60	0.088	0.16	0.61	0.56

The power dissipation efficiency η is a parameter that describes the dissipative characteristics of the material, which can be expressed as:

$$\eta = 2m/(m+1) \tag{5}$$

The expression of instability parameter ξ is as follows:

$$\xi(\dot{\varepsilon}) = \frac{\partial \ln[m/(m+1)]}{\partial \ln \dot{\varepsilon}} + m \tag{6}$$

When the instability coefficient ξ is negative, the material exhibits a higher tendency destabilization. Figure 7 shows the hot processing maps of Ti6Al4V alloy at different strain rates. In Figure 7, the gray area indicates the instability region, and the white area denotes the safe region. In general, at higher strain rates (>0.05 s^{-1}), the material is less prone to instability, and the processing temperature range of the material is increased as the deformation amount increases. Within the temperature range of 500–550 °C, when the strain is 0.1, flow instability is likely to occur. When the strain increases to 0.2, the flow instability disappears in the low temperature region, enabling good processing performance even at lower temperatures.

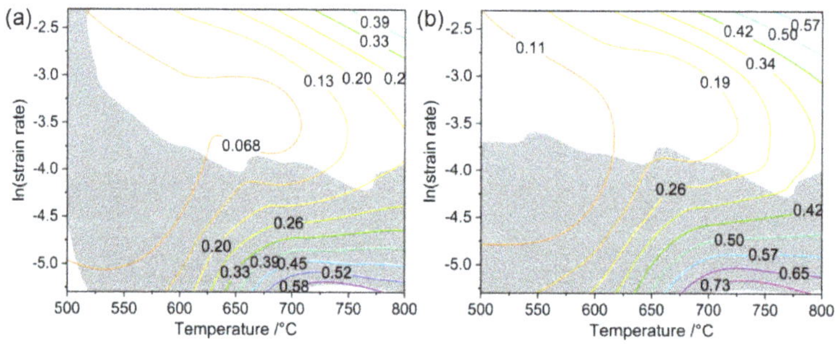

Figure 7. Processing maps of Ti6Al4V alloy at strains of (**a**) ε = 0.1 and (**b**) ε = 0.2.

3.4. *Hot Stamping of Titanium Alloy*

Based on the hot processing maps of Ti6Al4V alloy, a series of hot stamping experiments were conducted using custom-made box-shaped part dies to further investigate the influence of process parameters such as stamping speed, deformation temperature and deformation amount on the microstructure and forming quality of Ti6Al4V alloy.

3.4.1. Effect of Stamping Speed on Hot Stamping

According to hot processing maps, it is observed that the Ti6Al4V alloy exhibits less instability tendency at higher deformation rates. To select an appropriate stamping speed, under the heating temperature of 850 °C [14], the stamping speed of press machine was adjusted to 40 mm/s, 20 mm/s, 10 mm/s, and 5 mm/s, respectively. The forming quality is shown in Figure 8. The temperature evolution of the sheet at stamping speeds of 40 mm/s and 5 mm/s is illustrated in Figure 9. At a stamping speed of 5 mm/s, the formed component fractures. Combined observations from Figure 9, the fracture is attributed to excessive temperature loss at a stamping speed of 5 mm/s, resulting in a forming temperature range of 200–500 °C, unsuitable for forming. At stamping speeds of 10 mm/s and 20 mm/s, while the components can be shaped, cracks appear on the rounded surface. At a stamping speed of 40 mm/s, the U-shaped part is formed perfectly. A study on part thickness shows that as stamping speed increases, the variation in sheet thickness decreases. This occurs because at higher stamping speeds, strain rate strengthening delays local deformation, leading to more uniform deformation of the sheet [20]. Therefore, it is difficult to obtain high-quality titanium alloy parts at lower stamping speeds, which is consistent with the hot processing maps provided in this paper. Therefore, in the hot stamping experiment of Ti6Al4V alloy, the stamping speed of the top die should not be lower than 40 mm/s.

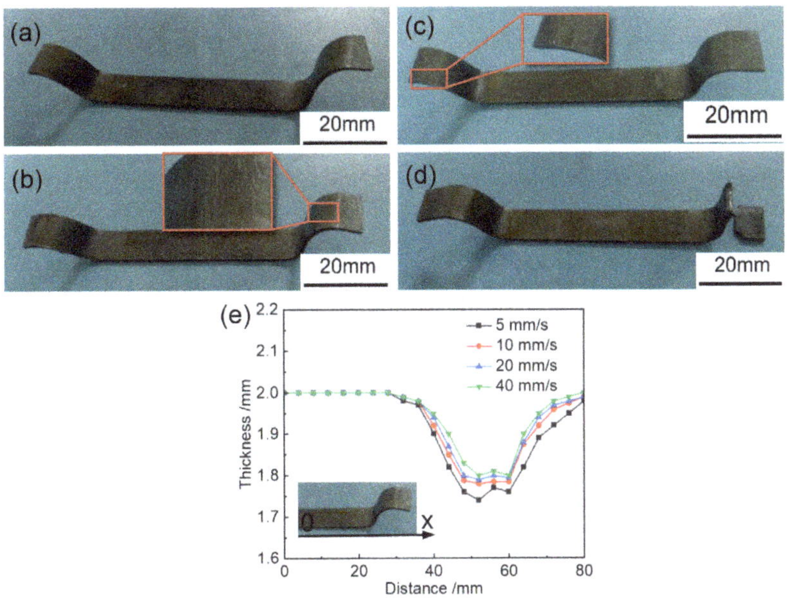

Figure 8. View of part formed at (**a**) 40 mm/s, (**b**) 20 mm/s, (**c**) 10 mm/s, and (**d**) 5 mm/s, and (**e**) variation in thickness.

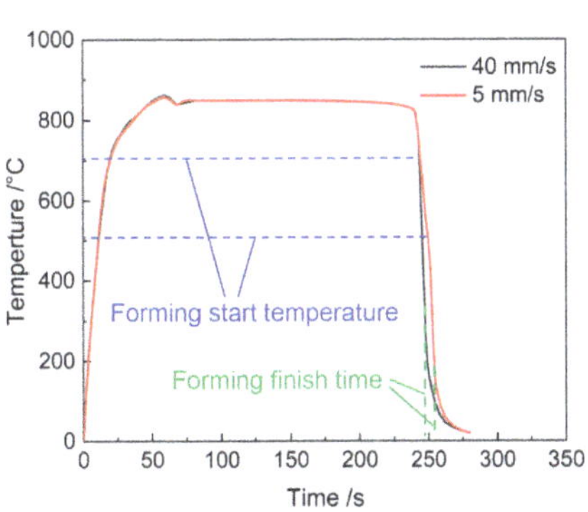

Figure 9. Evolution of temperature during forming process.

3.4.2. Effect of Temperature on Hot Stamping

Figure 9 shows that at a stamping speed of 40 mm/s, the temperature during the forming stage ranges from 693 to 384 °C. The hot processing maps indicate that instability is likely when the strain rate is between 0.1 s^{-1} and 0.05 s^{-1}, and the deformation temperature is below approximately 525 °C. Therefore, to minimize the risk of instability at low temperatures, the heating temperature of the sheet was increased. Specimens were heated to 825 °C, 850 °C, 875 °C and 900 °C, respectively, before being stamped into components. It was found that components formed well of 825 °C, 850 °C, and 875 °C, with cracking occurring only at 900 °C.

The microstructure of the formed components is shown in Figure 10. As the heating temperature increases, the content of equiaxial α phase decreases, while β phase grains grow and their percentage increases. At heating temperatures of 825 °C and 850 °C, protuberances extending from the β phase grains can be observed. This occurs because the high cooling rate in cold die hot stamping forming leads to incomplete diffusion of alloying elements. V elements accumulate along α phase grain boundaries, causing morphological instability of some primary α phases at the phase interfaces [30]. At a heating temperature of 875 °C, the β phase grains grow and the protuberances extending are replaced by elongated β grains. The element distribution at 875 °C shows that the V element content is highest at the α-β grain boundaries, followed by within β grains. This indicates that with the increase in heating temperature, the atomic long-range mobility is enhanced, resulting in a higher diffusion capability. Consequently, a larger number of V atoms diffuse to α phase grain boundaries, expand β phase region in an enriched manner [31]. As shown in Figure 10d, it can be observed in the microstructure that there is already a small amount of α platelets has already precipitated in part of β matrix, which implies that the transformation of the β phase to the secondary α phase has already occurred in the hot stamping organization of Ti6Al4V alloy when the heating temperature reaches 900 °C [30]

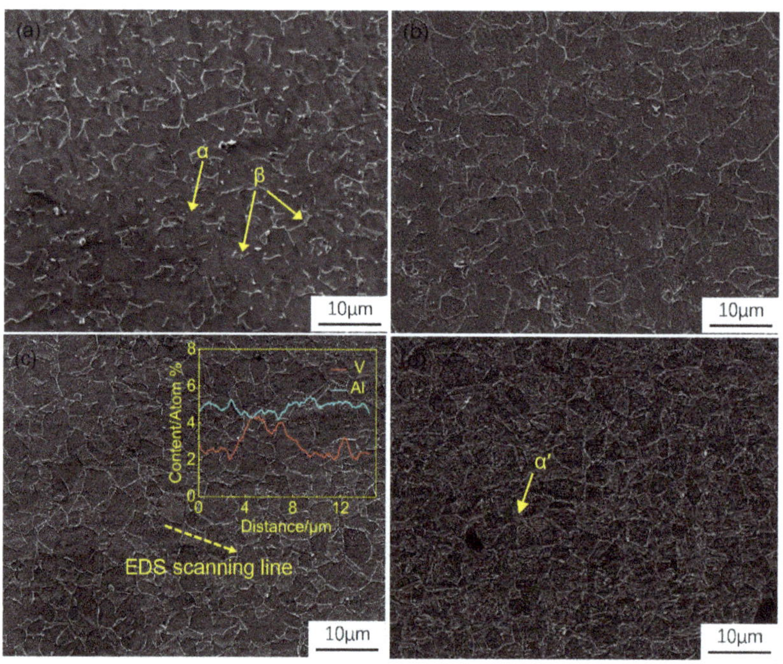

Figure 10. Microstructure images of part formed at (**a**) 825 °C, (**b**) 850 °C, (**c**) 875 °C, and (**d**) 900 °C.

The hardness of the formed parts was tested and the results are shown in Figure 11. It can be found that the hardness value of the formed parts increases from 347 HV at 825 °C to 387 HV at 900 °C. The hardness increases slowly when the heating temperature is between 825 °C and 875 °C, while the hardness increases significantly when the heating temperature is increased to 900 °C. Combined with the microstructure evolution in Figure 10, it indicates that phase transformation is the reason for the hardness enhancement. In order to further investigate the principle of hardness enhancement and probe the upper temperature limit of Ti6Al4V alloy hot stamping, X-ray diffraction patterns tests were carried out on the formed parts obtained under heating temperature of 875 °C and 900 °C, and the results are shown in Figure 12. The results show that only the diffraction peaks of α and β phases can be observed in Figure 12. Since the crystal structure of α' phase and α phase are both hexagonal close-packed structure, the lattice constants of the two phases are very close to each other, so there is no difference in the position of the diffraction peak between α' phase and α phase. The intensity of the diffraction peak of β phase at 875 °C is obviously higher than 900 °C, indicating that the transition from β phase to α' phase is enhanced with the increase in the heating temperature.

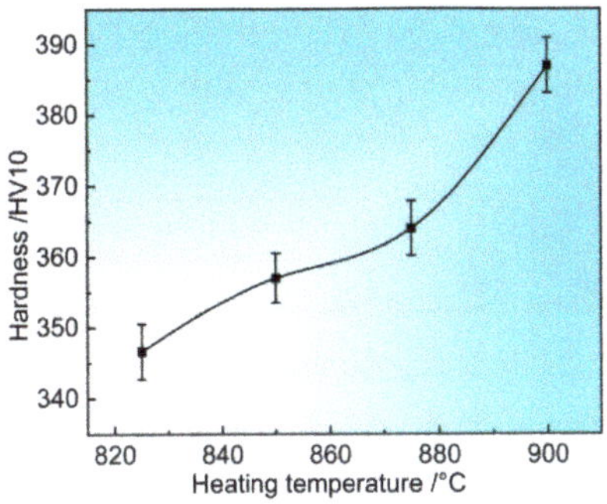

Figure 11. Post-form strength of components formed at different heating temperatures.

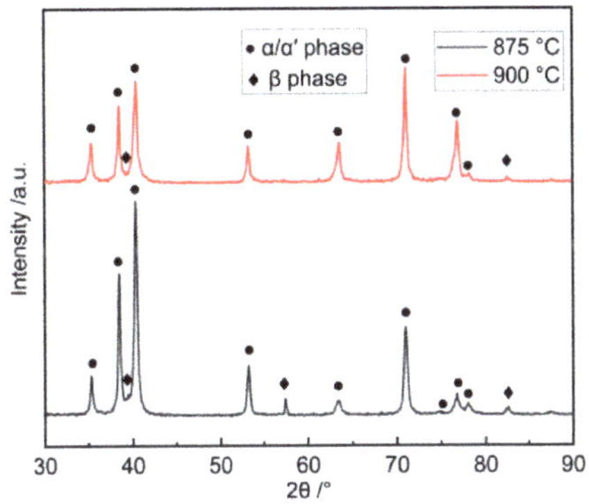

Figure 12. X-ray diffraction patterns for Ti6Al4V alloy under different heating temperatures.

In Figure 13, further transmission electron microscopy (TEM) tests of hot stamping organization at heating temperatures of 875 °C and 900 °C revealed that at a heating temperature of 875 °C, the β-transformed tissues obtained showed nano-sized martensite internally, with twinned α and β phases. However, when the heating temperature reaches 900 °C, the microstructure transforms into coarse slaty α' phase, resulting in a significant increase in material hardness, which is the main cause of fracture. In Figure 13, the hardness increases more slowly within the heating temperature of 825–875 °C compared to 900 °C. Therefore, in order to achieve a higher initial forming temperature of the material, 875 °C is chosen as heating temperature for hot stamping.

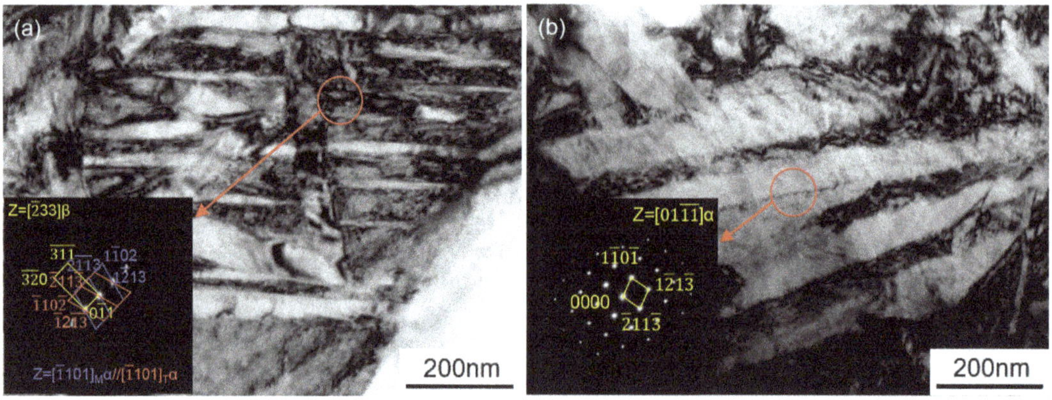

Figure 13. TEM micrograph of Ti6Al4V alloy after hot stamping at (**a**) 875 °C and (**b**) 900 °C.

3.4.3. Effect of Deformation Amount on Hot Stamping

Hot stamping experiments were conducted to explore the effects of deformation amount on microstructure. Figure 14 shows the box-shaped component formed by stamping under the conditions of a heating temperature of 875 °C and a speed of 40 mm/s. The formed box-shaped component shows no cracking and high forming quality.

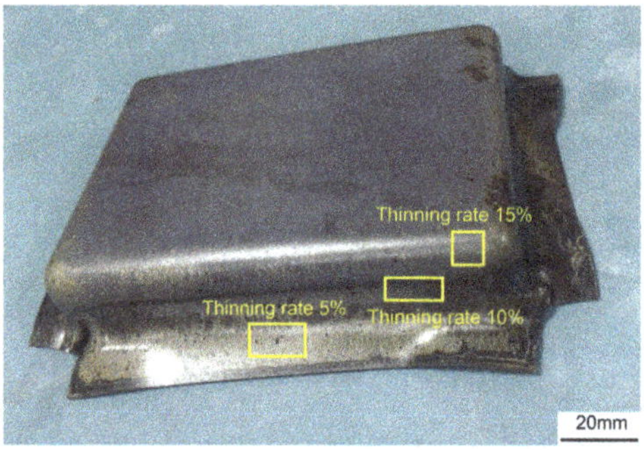

Figure 14. Forming parts and EBSD sample collection location.

Due to the difficulty in quantifying true strain experimentally, deformation amount is often measured using thinning rate. The thinning rate is the ratio of the difference in thickness before and after stamping to the original thickness of the sheet. Larger deformations correspond to higher thinning rates. Sampling was performed at positions with thinning rates of 5%, 10%, and 15% in the box-shaped components, as shown in Figure 14. The EBSD technique was employed to visually observe the distribution and morphology of substructures and characterize the microstructure of samples at different thinning rates.

Figure 15a–c depict inverse pole figure (IPF) of Ti6Al4V alloy at different thinning rates. A comparison between Figure 15a and c reveals that the thinning rate increases with the increase in the amount of deformation of the sheet. Figure 15d–f represent the grain size distribution at different thinning rates. The average grain sizes with thinning rates of 5%, 10%, and 15% were measured to be 1.73 μm, 1.68 μm and 1.57 μm, respectively. This indicates that during the hot stamping process of Ti6Al4V alloy, as the thinning rate increases, the average grain size decreases.

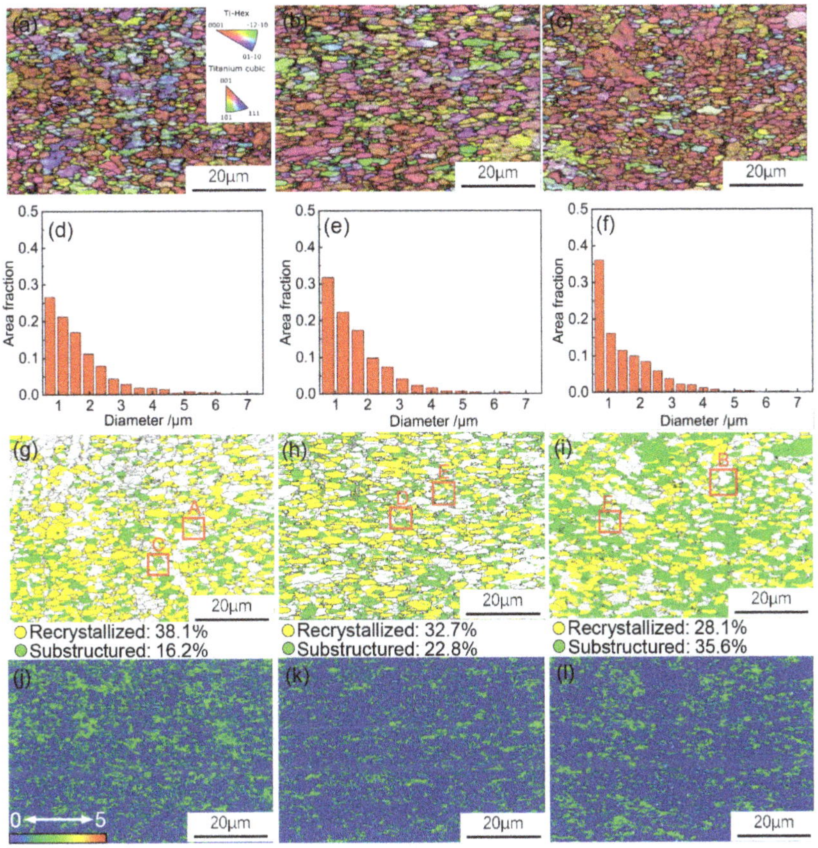

Figure 15. EBSD results of Ti6Al4V alloy for deflation of (**a**,**d**,**g**,**j**) 5%, (**b**,**e**,**h**,**k**) 10%, and (**c**,**f**,**i**,**l**) 15%.

Figure 15g–i display the evolution of the microstructure with the level of thinning rate. As the thinning rate increases, the proportion of DRV grains remains relatively unchanged, and the proportion of DRX grains significantly increases. This indicates that DRX is the main cause of grain refinement. At thinning rate of 5% and 10%, the proportion of DRV grains is higher. This indicates that DRV is the primary softening mechanism at lower deformation amount. However, at a thinning rate of 15%, the proportion of DRX grains surpasses that of DRV grains. This suggests that at higher deformation amount, DRX surpasses DRV as the primary softening mechanism.

Figure 15j–l show the Kernel Average Misorientation (KAM) maps of Ti6Al4V alloy at different thinning rates. The KAM maps characterize the substructure density of the material, with higher values indicating greater defect concentration [32]. At a thinning rate of 5%, the substructure density is higher, indicating numerous defects like dislocations generated during deformation. As the thinning rate increases to 10% and 15%, the substructure density significantly decreases, indicating that many substructures generated by deformation are consumed This also confirms that the deformation mechanism in the hot stamping process is influenced by the amount of deformation. As deformation increases, the DRX mechanism is activated, leading to a decrease in dislocation density and grain refinement.

3.5. Analysis of the Dynamic Recrystallization Mechanism

In Ti6Al4V alloy, both DDRX and CDRX are common in deformation processes. Due to the significant impact of different DRX modes on practical production, it is necessary to analyze their mechanisms further. Figure 16 illustrates several typical DRX grain intercepted from Figure 15g–i. At area A and area B, DRX grains are generated at grain boundaries or triple intersections, which suggests that these DRX grains undergo nucleation and growth processes, a phenomenon that typically occurs in the DDRX mode [33,34]. In zones C and D, there are DRX grains within the original grains, bordering the original grains surrounded by fragments of LAGBs. This is due to the fact that DRX grains are formed by LAGBs and dislocation cells after continuous dislocation uptake without a nucleation process, known as CDRX [35]. At area E and area F, DRX grains are formed within the grains. This results from numerous dislocations into the interior under the influence of DRV, leading to dislocation cell formation within the original grains and subsequent transformation into CDRX grains [36]. The analysis above shows that both DDRX and CDRX exist in the microstructures across various deformation amounts.

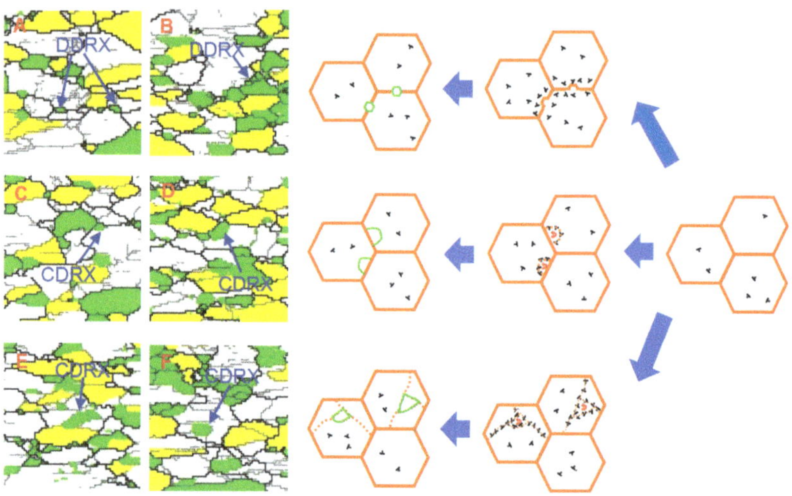

Figure 16. Schematic diagram of dynamic recrystallization formation: (**A**,**B**) DDRX grains and (**C**–**F**) CDRX grains.

Figure 17a–c present the quantitative evolution in fractions of LAGBs and HAGBs with different thinning rates. It can be observed that as the thinning rate increases, the fraction of LAGBs decreases, while the fraction of HAGBs exhibits the opposite trend. This is due to the fact that lots of deformed substructures containing LAGBs are consumed during the formation of DDRX nucleation, resulting in a significant decrease in percentage of LAGBs [37]. It indicates that the DDRX mechanism is enhanced with the increase in thinning rate. Figure 17d–f illustrate the crystallographic orientation changes in DRX grains (mark in red lines) and remained grains (mark in blue lines) in {0001} pole figure at different thinning rates. There is little difference in texture between recrystallized and retained grains as the thinning rate increases. Given that DDRX has a distinct texture from the deformation texture, and CDRX's texture is almost identical to the deformation texture [38], the similar between recrystallized and retained grain textures implies that CDRX consistently dominates this deformation mode. Additionally, DDRX processes often involve the growth of dynamically recrystallized grains, resulting in recrystallized grains of varying sizes and an uneven grain size distribution. Conversely, CDRX involves the continuous absorption of dislocations at subgrain interfaces, directly leading to the formation of recrystallized grains. Therefore, this process involves minimal growth of recrystallized grains, resulting in a more uniform and finer microstructure post-DRX completion. Statistical analysis of Figure 17d–f reveals that the average grain sizes of hot stamping components at thinning rates of 5%, 10%, and 15% are 1.73 μm, 1.68 μm, and

1.57 μm, respectively. The percentage of grains with diameters smaller than 1 μm are 26.62%, 31.74%, and 36.05%, respectively. This indicates that deformation increases, so does the proportion of fine grains, further supporting the dominance of CDRX during deformation, which intensifies with increased deformation.

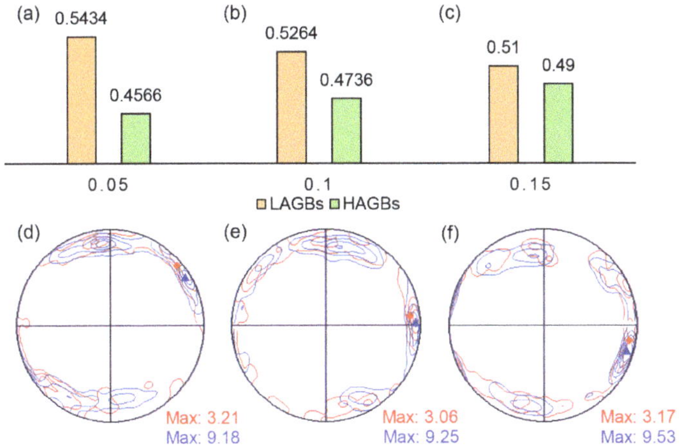

Figure 17. Evolution of recrystallized weaves at thinning rate of (**a**,**d**) 0.05, (**b**,**e**) 0.1, and (**c**,**f**) 0.15.

In summary, both CDRX and DDRX mechanisms coexist in hot stamping of Ti6Al4V alloys. With increased deformation in the hot stamping process, the DDRX mechanism is enhanced to some extent, but the CDRX mechanism consistently dominates. Therefore, designing a rational deformation path and mode for the hot stamping of Ti6Al4V alloy components is crucial to achieving a finer and more uniform microstructure, which leads to enhanced mechanical properties.

4. Conclusions

Based on the stress–strain curve, a constitutive model for strain compensation and a hot processing map for Ti6Al4V alloy under a large temperature span in the hot stamping process were constructed. The hot stamping rheological behavior and the deformation mechanism of Ti6Al4V alloy were systematically studied. The main conclusions are as follows:

1. The stress–strain curve indicated that with the increase in temperature and the decrease in the strain rate, the rheological softening effect was enhanced, and the material exhibited temperature and strain rate sensitivity. A strain-compensated Arrhenius constitutive model suitable for hot stamping of Ti6Al4V alloy was established, which showed good agreement with experimental values. The average error of the stress–strain curve calculated using the constitutive model was only 8.3%.
2. A hot processing map suitable for the hot stamping process of Ti6Al4V alloy was established. The instability region mainly concentrates in the low strain rate region. Ti6Al4V alloy was less prone to instability when the strain rate was higher than 0.05 s^{-1}, and the processing temperature range of the material was increased as the deformation amount increases. Based on the hot processing map, in hot stamping experiments, Ti6Al4V alloy parts can be successfully formed at stamping speeds greater than 40 mm/s and heating temperatures in the range of 825–875 °C.
3. The study of recrystallization mechanisms in Ti6Al4V alloy during hot stamping revealed that increased deformation has an exciting effect on DRX. When the thinning rate of Ti6Al4V alloy during hot stamping was increased from 5% to 15%, DRX was significantly enhanced, dislocation density was reduced, and grain refinement occurred. The characteristics of substructure and texture evolution indicated that both CDRX and DDRX coexist throughout the hot stamping of titanium alloy. With

increasing deformation during hot stamping, the DDRX mechanism was strengthened, but the CDRX mechanism remained dominant.

Author Contributions: M.Q. conceptualized the research, data analysis, and paper writing. Z.G. helped in paper editing. X.L. performed the experiments. G.Y. and L.Y. contributed to supervision and revision of this paper. Y.L. and Y.Z. performed paper editing, project administration and funding acquisition. All authors have read and agreed to the published version of the manuscript.

Funding: This work was supported by the National Natural Science Foundation of China (No. 51101072), the Technology Development Program of Jilin Province (No. 20210201038GX) and the Program for Jilin University Science and Technology Innovative Research Team (No. 20230508039RC).

Institutional Review Board Statement: Not applicable.

Informed Consent Statement: Not applicable.

Data Availability Statement: Datasets generated and/or analyzed during the current study are available from the corresponding author on request.

Conflicts of Interest: Author Yafu Zhao was employed by the company CRRC Changchun Railway Vehicles Co., Ltd. The remaining authors declare that the research was conducted in the absence of any commercial or financial relationships that could be construed as a potential conflict of interest.

References

1. Quadros, F.D.; Kuroda, P.A.B.; Sousa, K.D.J.; Donato, T.A.G.; Grandini, C.R. Preparation, structural and microstructural characterization of Ti-25Ta-10Zr alloy for biomedical applications. *J. Mater. Res. Technol.* **2019**, *8*, 4108–4114. [CrossRef]
2. Jin, B.Q.; Wang, Q.; Zhao, L.Z.; Pan, A.J.; Ding, X.F.; Gao, W.; Song, Y.F.; Zhang, X.F. A Review of Additive Manufacturing Techniques and Post-Processing for High-Temperature Titanium Alloys. *Metals* **2023**, *13*, 1327. [CrossRef]
3. Kumar, N.; Kumar, M.; Kumar, V.; Sharma, A. Microstructural Heterogeneity and Anisotropy Control of Additive Manufactured Ti-6Al-4V Alloy for Aircraft Components. *JOM* **2023**, *75*, 1695–1709. [CrossRef]
4. Tshephe, T.S.; Akinwamide, S.O.; Olevsky, E.; Olubambi, P.A. Additive manufacturing of titanium-based alloys—A review of methods, properties, challenges, and prospects. *Heliyon* **2022**, *8*, e09041. [CrossRef]
5. Li, H.; Chen, S.F.; Zhang, S.H.; Xu, Y.; Song, H.W. Deformation Characteristics, Formability and Springback Control of Titanium Alloy Sheet at Room Temperature: A Review. *Materials* **2022**, *15*, 5586. [CrossRef] [PubMed]
6. Akula, S.P.; Ojha, M.; Rao, K.L.; Gupta, A.K. A review on superplastic forming of Ti-6Al-4V and other titanium alloys. *Mater. Today Commun.* **2023**, *34*, 105343. [CrossRef]
7. Wróbel, I.; Skowronek, A.; Grajcar, A. A Review on Hot Stamping of Advanced High-Strength Steels: Technological-Metallurgical Aspects and Numerical Simulation. *Symmetry* **2022**, *14*, 969. [CrossRef]
8. Chen, Y.; Li, S.H.; Li, Y.F.; Wang, Y.Q.; Li, Z.Q.; Lin, Z.Q. Constitutive modeling of TA15 alloy sheet coupling phase transformation in non-isothermal hot stamping process. *J. Mater. Res. Technol.* **2021**, *12*, 629–642. [CrossRef]
9. Gao, S.; Sang, Y.; Li, Q.H.; Sun, Y.L.; Wu, Y.H.; Wang, H.R. Constitutive modeling and microstructure research on the deformation mechanism of Ti-6Al-4V alloy under hot forming condition. *J. Alloys Compd.* **2022**, *892*, 162128. [CrossRef]
10. Zhang, C.; Li, X.Q.; Li, D.S.; Jin, C.H.; Xiao, J.J. Modelization and comparison of Norton-Hoff and Arrhenius constitutive laws to predict hot tensile behavior of Ti-6Al-4V alloy. *Trans. Nonferrous. Met. Soc. China* **2012**, *22*, S457–S464. [CrossRef]
11. Yang, X.M.; Wang, B.Y.; Zhou, J.; Xiao, W.C.; Feng, P.N. Constitutive modeling of softening mechanism and damage for Ti-6Al-4V alloy and its application in hot tensile simulation process. *Arch. Civ. Mech. Eng.* **2021**, *21*, 69. [CrossRef]
12. Qiu, Q.; Wang, K.L.; Li, X.; Wang, J.; Gao, X.; Zhang, K.M. Hot deformation behavior and processing parameters optimization of SP700 titanium alloy. *J. Mater. Res. Technol.* **2021**, *15*, 3078–3087. [CrossRef]
13. Luo, J.; Li, M.Q.; Yu, W.X.; Li, H. Effect of the strain on processing maps of titanium alloys in isothermal compression. *Mater. Sci. Eng. A* **2009**, *504*, 90–98. [CrossRef]
14. Kopec, M.; Wang, K.H.; Politis, D.J.; Wang, Y.Q.; Wang, L.L.; Lin, J.G. Formability and microstructure evolution mechanisms of Ti6Al4V alloy during a novel hot stamping process. *Mater. Sci. Eng. A* **2018**, *719*, 72–81. [CrossRef]
15. Zheng, Z.Z.; Chen, Y.Y.; Kong, F.T.; Wang, X.P. The interrupted hot compression of an as-forged titanium matrix composite: Flow softening behavior and deformation mechanism. *J. Mater. Res. Technol.* **2022**, *20*, 821–832. [CrossRef]
16. Li, C.H.; Cui, Y.M.; Zheng, W.W.; Song, L.; Wu, Y.L.; Wu, Y.D.; Yu, C.Y.; Hui, X.D. Unveiling the dynamic softening mechanism via micromechanical behavior for a near-β titanium alloy deformed at a high strain rate. *J. Mater. Res. Technol.* **2023**, *26*, 9392–9405. [CrossRef]
17. Ning, Y.Q.; Xie, B.C.; Liang, H.Q.; Li, H.; Yang, X.M.; Guo, H.Z. Dynamic softening behavior of TC18 titanium alloy during hot deformation. *Mater. Des.* **2015**, *71*, 68–77. [CrossRef]
18. Li, C.M.; Huang, L.; Zhao, M.J.; Guo, S.Q.; Su, Y.; Li, J.J. Systematic analysis of the softening mechanism and texture evolution of Ti-6Cr-5Mo-5V-4Al alloy during hot compression in a + b phase region. *Mater. Sci. Eng. A* **2022**, *850*, 143571. [CrossRef]

19. Deng, Y.C.; An, Y.X.; Xiao, Y.Z.H.; Zhan, X.D.; Zhang, X.Y.; Wang, B.F. Deformation mechanism diagram and deformation instability of a Ti-5Al-5Mo-5V-1Cr-1Fe titanium alloy during the hot compression. *J. Alloys Compd.* **2023**, *966*, 171446. [CrossRef]
20. Yang, X.M.; Wang, B.Y.; Zhou, J. Numerical and experimental study on formability of TC4 alloy in a novel multi-layer sheet hot stamping process. *Int. J. Adv. Manuf. Technol.* **2020**, *110*, 1233–1247. [CrossRef]
21. Zhang, J.Q.; Xu, X.Y.; Xue, J.Y.; Liu, S.N.; Deng, Q.H.; Li, F.; Ding, J.; Wang, H.; Chang, H. Hot deformation characteristics and mechanism understanding of Ti-6Al-2Sn-4Zr-6Mo titanium alloy. *J. Mater. Res. Technol.* **2022**, *20*, 2591–2610. [CrossRef]
22. Yu, J.X.; Li, Z.J.; Qian, C.; Huang, S.; Xiao, H. Investigation of deformation behavior, microstructure evolution, and hot processing map of a new near-α Ti alloy. *J. Mater. Res. Technol.* **2023**, *23*, 2275–2287. [CrossRef]
23. Ning, Y.Q.; Luo, X.; Liang, H.Q.; Guo, H.Z.; Zhang, J.L.; Tan, K. Competition between dynamic recovery and recrystallization during hot deformation for TC18 titanium alloy. *Mater. Sci. Eng. A* **2015**, *635*, 77–85. [CrossRef]
24. Li, L.; Liu, J.X.; Ding, N.; Li, M.Q. Substructure evolution in two phases based constitutive model for hot deformation of TC18 in α + β phase region. *Chin. J. Aeronaut.* **2023**, *36*, 573–588. [CrossRef]
25. Wang, K.; Sun, D.L.; Wu, Y.L.; Xin, R.L.; Wu, M.Y.; Liu, Q. Correlation between Flow Behavior and Microstructure Evolution during α/β Deformation of TA19 Titanium Alloy. *Adv. Eng. Mater.* **2016**, *18*, 1808–1815. [CrossRef]
26. Lin, Y.C.; Huang, J.; He, D.G.; Zhang, X.Y.; Wu, Q.; Wang, L.H.; Chen, C.; Zhou, K.C. Phase transformation and dynamic recrystallization behaviors in a Ti55511 titanium alloy during hot compression. *J. Alloys Compd.* **2019**, *795*, 471–482. [CrossRef]
27. Zener, C.; Hollomon, J.H. Effect of strain rate upon plastic flow of steel. *J. Appl. Phys.* **1944**, *15*, 22–32. [CrossRef]
28. Cho, C.H.; Kim, D.O.; Son, K.; Park, H.S. Relationship between hot workability and texture evolution in an Al-Zn-Mg-Cu alloy under hot compressive stress mode. *J. Mater. Sci.* **2023**, *58*, 16537–16549. [CrossRef]
29. Xu, W.X.; Zhao, Z.L.; Wang, Y.B.; Liu, N.; Li, S.; Huang, X.Y. Correlation between the strain rate sensitivity exponent m and microstructure evolution during hot deformation of an as-cast TiAl alloy. *Mater. Charact.* **2023**, *203*, 113050. [CrossRef]
30. Zhu, S.; Yang, H.; Guo, L.G.; Fan, X.G. Effect of cooling rate on microstructure evolution during α/β heat treatment of TA15 titanium alloy. *Mater. Charact.* **2012**, *70*, 101–110. [CrossRef]
31. Wang, K.H.; Kopec, M.; Chang, S.P.; Qu, B.; Liu, J.; Politis, D.J.; Wang, L.L.; Liu, G. Enhanced formability and forming efficiency for two-phase titanium alloys by Fast light Alloys Stamping Technology (FAST). *Mater. Des.* **2020**, *194*, 108948. [CrossRef]
32. Sundar, A.S.; Mugada, K.K.; Kumar, A. Microstructural Evolution, Intermetallic Formation, and Mechanical Performance of Dissimilar Al6061-Ti6Al4V Static Shoulder Friction Stir Welds. *Adv. Eng. Mater.* **2023**, *25*, 2300973. [CrossRef]
33. Zhu, W.G.; Pu, J.; Zhang, C.H.; Wang, Q.B.; Wang, Y.L.; Shu, T.C.; Wang, X.X.; Wang, J. Microstructure Evolution and Anisotropy Effect During Low Temperature Cross Rolling and Annealing of Ti6Al4V. *Metall. Mater. Trans. A* **2023**, *54*, 3550–3562. [CrossRef]
34. Li, X.; Guan, B.; Wang, Y.L.; Wei, Y.L.; Li, B. Ascertaining the microstructural evolution and strengthening mechanisms of the gradient nanostructured pure titanium fabricated by ultrasonic surface rolling process. *Surf. Coat. Technol.* **2023**, *473*, 130047. [CrossRef]
35. Wu, Y.; Fan, R.L.; Chen, M.H.; Wang, K.H.; Zhao, J.; Xiao, W.C. High-temperature anisotropic behaviors and microstructure evolution mechanisms of a near-α Ti-alloy sheet. *Mater. Sci. Eng. A* **2021**, *820*, 141560. [CrossRef]
36. Liu, G.X.; Zhang, D.H.; Yao, C.F. Investigation of the grain refinement mechanism in machining Ti-6Al-4V: Experiments and simulations. *J. Manuf. Process* **2023**, *94*, 479–496. [CrossRef]
37. Fu, Q.; Feng, J.; Xiang, W.; Yuan, W.H. Microstructure Characteristic and Texture Evolution of TB18 Titanium Alloy during Hot Compression in the β Phase Zone. *Mater. Trans.* **2022**, *63*, 1416–1424. [CrossRef]
38. Wang, X.X.; Zhan, M.; Gao, P.F.; Ma, P.Y.; Yang, K.; Lei, Y.D.; Li, Z.X. Deformation mode dependent mechanism and kinetics of dynamic recrystallization in hot working of titanium alloy. *Mater. Sci. Eng. A* **2020**, *772*, 138804. [CrossRef]

materials

MDPI

Article

Thermal Vibration of Thick FGM Conical Shells by Using Third-Order Shear Deformation Theory

Chih-Chiang Hong

Department of Mechanical Engineering, Hsiuping University of Science and Technology, Taichung 412-406, Taiwan; cchong@mail.hust.edu.tw

Abstract: A time-dependent third-order shear deformation theory (TSDT) approach on the displacements of thick functionally graded material (FGM) conical shells under dynamic thermal vibration is studied. Dynamic equations of motion with TSDT for thick FGM conical shells are applied directly with the partial derivative of variable $R^*\theta$ in the curve coordinates (x, θ, z) instead of y in the Cartesian coordinates (x, y, z) for thick FGM plates, where R^* is the middle-surface radius at any point on conical shells. The generalized differential quadrature (GDQ) numerical method is used to solve the dynamic differential equations in equilibrium matrix forms under thermal loads. It is the novelty of the current study to identify the parametric effects of shear correction coefficient, environment temperature, TSDT model, and FGM power law index on the displacements and stresses in the thick conical shells only subjected to sinusoidal heating loads. The physical parts with values on the length-to-thickness ratio equals 5, and 10 FGMs can be used in an area of an airplane engine that usually operates near more than 1000 K of temperatures when the thermal stress is considered and affected. The important findings of the presented study are listed as follows. The values of normal stress are in decreasing tendencies with time in cases when the coefficient c_1 equals $0.925925/\text{mm}^2$ in TSDT and length-to-thickness ratio equals 5. The shear stress values in x plane z direction on the minor middle-surface radius (r) equals the major middle-surface radius (R) over 8 and length-to-thickness ratio equals to 5 can withstand $T = 1000$ K of pressure.

Keywords: thick FGM; TSDT; thermal vibration; nonlinear coefficient

1. Introduction

The importance of the essential background study is to find the temperature effect of sinusoidal thermal loading on the displacement and stress. In the thick materials, the third-order shear deformation theory (TSDT) of displacements is used for more shear effect than the first-order shear deformation theory (FSDT) mode. The functionally graded material (FGM) can operate usually with less displacement and stress values under higher temperature, e.g., a 1000 K environment.

Many recent works in the literature are devoted to studying the mechanical behavior of conical shells made of functionally graded materials (FGMs). Soureshjani et al. [1] studied the numerical solutions of the generalized differential quadrature (GDQ) method for critical buckling pressure by using FSDT. Ansari et al. [2] used the von Karman geometric nonlinearity and higher-order shear deformation theory (HSDT) to study the numerical solutions of the variational differential quadrature (VDQ) method for maximum deflection. The VDQ method was a weak formulation form for governing equations and then was discretized directly by using time differential operators and the arc-length continuation scheme. Das et al. [3] used the modified Hertzian contact theory to study the time response of numerical results in the finite element method (FEM) for power law exponent materials Ti–6Al–4V-ZrO2. Shakouri [4] presented the Donnell's FSDT to investigate the vibration behavior in the thermal environment. Sofiyev [5] presented a literature review in the realms of the vibration and buckling engineering of nuclear, space, marine, electronics, and biomedical. Taraghi et al. [6] used analytical models and experimental methods to study the buckling under external pressure. Yang et al. [7] used the FSDT to study the vibration results under aerodynamic load. Javed [8] investigated the free vibration by using TSDT and spline approximation. Mouli et al. [9] used the FSDT and commercial finite element tool ANSYS to study the free vibration. It would be a novelty of this study to consider both the two directional frequency of vibrations and nonlinear coefficient terms in the TSDT among the thermal vibration studies for thick FGM conical shells.

Some GDQ studies for thick FGM spherical shells, circular cylindrical shells, plates, and plates-cylindrical shells employed the parameters effect of TSDT, heating loads, and environment temperature. The investigations by using the TSDT model approach for thick material were more complicated than the FSDT model approach used for thin material. Hong [10] studied the GDQ solutions of thick plates by including the TSDT model. The material properties of power-law function for thick plates were used. The displacements and thermal stresses of thick plates were also presented. Hong [11] presented the GDQ results of thick circular cylindrical shells by using the TSDT model. The von Karman type of strain-displacement equations and partial derivatives of displacements with respect to thickness direction are assumed. Parametric effect studies of power law index and environment temperature on the circular cylindrical shells were presented.

In this study, the time responses of displacements and stresses are investigated by using the TSDT model and effect of computed shear coefficient in laminated conical shells with four simply supported edges. It is the objective of the current study to identify the parametric effects of TSDT model, environment temperature, and the power law index of FGMs on the displacements and stresses in the thick conical shells under heating loads. The aim of the study is to provide the GDQ results in the thermal vibration analysis for FGM thick conical shells.

2. Procedures of Formulations

A point (x, θ, z) of curve coordinates in two constituent materials-laminated FGM conical shells under heating loads with temperature difference ΔT is shown in Figure 1 with the thickness h_1 of the inner layer constituent material 1 and thickness h_2 of the outer layer constituent material 2. L is the length of conical shells. h^* is the total thickness of conical shells. r is the minor middle-surface radius at x = 0, and R is the major middle-surface radius at x = L. β is the semi-vertex angle. The power-law function with power law index R_n is used and expressed in material properties (E_{fgm}, ν_{fgm}) by Hong [11], as follows.

$$E_{fgm} = (E_2 - E_1)\left[(z + h^*/2)/h^*\right]^{R_n} + E_1, \tag{1}$$

$$\nu_{fgm} = (\nu_1 + \nu_2)/2, \tag{2}$$

in which E_2 and E_1 are Young's modulus, ν_2 and ν_1 are Poisson's ratios of the constituent material 2 and constituent material 1, respectively.

The individual constituent material properties (e.g., for E_2 and E_1) are used and expressed in functions of environment temperature (T) by Chi and Chung [12] as follows.

$$P_i = P_0\left(P_{-1}T^{-1} + 1 + P_1T + P_2T^2 + P_3T^3\right), \tag{3}$$

in which P_0, P_{-1}, P_1, P_2, and P_3 are coefficients of temperature.

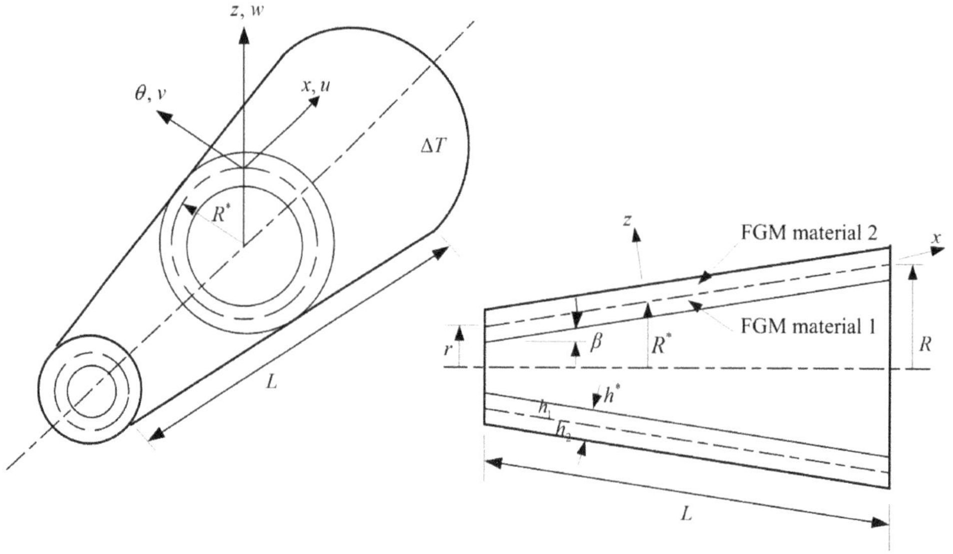

Figure 1. Two constituent materials-laminated thick FGM conical shells under thermal loads.

At any point (x, θ, z) of thick FGM conical shells, the time dependence of nonlinear displacements u, v, and w equations are applied to the coefficient c_1 term of TSDT by Lee et al. [13] with the parameter $R^* = r + x\sin\beta$ instead of the R, as follows:

$$u = u_0(x,\theta,t) + z\phi_x(x,\theta,t) - c_1 z^3\left(\phi_x + \frac{\partial w}{\partial x}\right), \tag{4}$$

$$v = v_0(x,\theta,t) + z\phi_\theta(x,\theta,t) - c_1 z^3\left(\phi_\theta + \frac{\partial w}{R^*\partial\theta}\right), \tag{5}$$

$$w = w(x,\theta,t), \tag{6}$$

where u_0, v_0, and w are the x, θ axes tangential and z axis transverse displacements, respectively, on the middle-plane of conical shells. ϕ_x and ϕ_θ are the shear rotations. R^* is the middle-surface radius of the conical shells at (x, θ, z). When the value $\beta = 0$ is used, it becomes a case of circular cylindrical shell. t is the time, and $c_1 = 4/\left(3h^{*2}\right)$ expression is used for the TSDT model in nonlinear vs. z^3. When $c_1 = 0$ is used in Equations (4) and (5) for the thin material, then the displacement expression became the more simpler FSDT model in linear vs. z.

Normal stress denotations with $\sigma_x = \sigma_{xx}$, $\sigma_\theta = \sigma_{\theta\theta}$, and shear stress denotations with $\sigma_{x\theta}$, $\sigma_{\theta z}$, and σ_{xz} on the (k)th layer of conical shells can be expressed in functions of stiffness, strains (ε_x, ε_θ, $\varepsilon_{x\theta}$, $\varepsilon_{\theta z}$, ε_{xz}), and ΔT, as in Lee and Reddy [14] and Whitney [15]. The expression of ΔT between conical shells and curing area with linear and uncouple effects vs. z is given in the equation as follows:

$$\Delta T = \frac{z}{h^*} T_1(x,\theta,t), \tag{7}$$

in which T_1 is temperature parameter. For no heat generation, the equation of heat conduction in simplified form is applied by Hong [10] with the variable $R^2\partial\theta^2$ instead of the ∂y^2, as follows:

$$K\left(\frac{\partial^2\Delta T}{\partial x^2} + \frac{\partial^2\Delta T}{R^2\partial\theta^2} + \frac{\partial^2\Delta T}{\partial z^2}\right) = \frac{\partial\Delta T}{\partial t}, \tag{8}$$

where $K = K_{fgm}/\left(\rho_{fgm}C_{v\,fgm}\right)$ in which K_{fgm} is thermal conductivity, ρ_{fgm} is density, and $C_{v\,fgm}$ is specific heat. It can be reduced to the following equation to obtain the frequency γ of sinusoidal heat flux under heating loads:

$$\frac{L^2}{K\pi^2[1+(L/R)^2]}\gamma\cos(\gamma t)+\sin(\gamma t)=0 \tag{9}$$

For Equation (9), there is more detail to provide a step-by-step derivation in Appendix A.

Dynamic equations of motion for thick FGM conical shells with TSDT are applied by Hong [10] and by Reddy [16] with the variables $R^*\partial\theta$ and $R^{*2}\partial\theta^2$ instead of the ∂y and ∂y^2, respectively, as follows:

$$\frac{\partial N_{xx}}{\partial x}+\frac{1}{R^*}\frac{\partial N_{x\theta}}{\partial\theta}=I_0\frac{\partial^2 u_0}{\partial t^2}+J_1\frac{\partial^2\phi_x}{\partial t^2}-c_1 I_3\frac{\partial^2}{\partial t^2}\left(\frac{\partial w}{\partial x}\right), \tag{10}$$

$$\frac{\partial N_{x\theta}}{\partial x}+\frac{1}{R^*}\frac{\partial N_{\theta\theta}}{\partial\theta}=I_0\frac{\partial^2 v_0}{\partial t^2}+J_1\frac{\partial^2\phi_\theta}{\partial t^2}-c_1 I_3\frac{\partial^2}{\partial t^2}\left(\frac{\partial w}{R^*\partial\theta}\right), \tag{11}$$

$$\begin{aligned}&\frac{\partial\overline{Q}_x}{\partial x}+\frac{1}{R^*}\frac{\partial\overline{Q}_\theta}{\partial\theta}+c_1\left(\frac{\partial^2 P_{xx}}{\partial x^2}+\frac{2}{R^*}\frac{\partial^2 P_{x\theta}}{\partial x\partial\theta}+\frac{1}{R^{*2}}\frac{\partial^2 P_{\theta\theta}}{\partial\theta^2}\right)+q=I_0\frac{\partial^2 w}{\partial t^2}-{c_1}^2 I_6\frac{\partial^2}{\partial t^2}\left(\frac{\partial^2 w}{\partial x^2}+\right.\\&\left.\frac{1}{R^{*2}}\frac{\partial^2 w}{\partial\theta^2}\right)+c_1\left[I_3\frac{\partial^2}{\partial t^2}\left(\frac{\partial u_0}{\partial x}+\frac{1}{R^*}\frac{\partial v_0}{\partial\theta}\right)+J_4\frac{\partial^2}{\partial t^2}\left(\frac{\partial\phi_x}{\partial x}+\frac{1}{R^*}\frac{\partial\phi_\theta}{\partial\theta}\right)\right],\end{aligned} \tag{12}$$

$$\frac{\partial\overline{M}_{xx}}{\partial x}+\frac{1}{R^*}\frac{\partial\overline{M}_{x\theta}}{\partial\theta}-\overline{Q}_x=\frac{\partial^2}{\partial t^2}\left(J_1 u_0+K_2\phi_x-c_1 J_4\frac{\partial w}{\partial x}\right), \tag{13}$$

$$\frac{\partial\overline{M}_{x\theta}}{\partial x}+\frac{1}{R^*}\frac{\partial\overline{M}_{\theta\theta}}{\partial\theta}-\overline{Q}_\theta=\frac{\partial^2}{\partial t^2}\left(J_1 v_0+K_2\phi_\theta-c_1 J_4\frac{1}{R^*}\frac{\partial w}{\partial\theta}\right), \tag{14}$$

where $\overline{M}_{\alpha\beta}=M_{\alpha\beta}-c_1P_{\alpha\beta}$, $\overline{Q}_\alpha=Q_\alpha-3c_1R_\alpha$, $(\alpha,\beta=x,\theta)$,

$$\begin{Bmatrix}N_{xx}\\N_{\theta\theta}\\N_{x\theta}\end{Bmatrix}=\int_{-\frac{h^*}{2}}^{\frac{h^*}{2}}\begin{Bmatrix}\sigma_{xx}\\\sigma_{\theta\theta}\\\sigma_{x\theta}\end{Bmatrix}dz,\ \begin{Bmatrix}M_{xx}\\M_{\theta\theta}\\M_{x\theta}\end{Bmatrix}=\int_{-\frac{h^*}{2}}^{\frac{h^*}{2}}\begin{Bmatrix}\sigma_{xx}\\\sigma_{\theta\theta}\\\sigma_{x\theta}\end{Bmatrix}zdz,\ \begin{Bmatrix}P_{xx}\\P_{\theta\theta}\\P_{x\theta}\end{Bmatrix}=\int_{-\frac{h^*}{2}}^{\frac{h^*}{2}}\begin{Bmatrix}\sigma_{xx}\\\sigma_{\theta\theta}\\\sigma_{x\theta}\end{Bmatrix}z^3dz,$$

$$\begin{Bmatrix}R_\theta\\R_x\end{Bmatrix}=\int_{-\frac{h^*}{2}}^{\frac{h^*}{2}}\begin{Bmatrix}\sigma_{\theta z}\\\sigma_{xz}\end{Bmatrix}z^2dz,\ \begin{Bmatrix}Q_\theta\\Q_x\end{Bmatrix}=\int_{-\frac{h^*}{2}}^{\frac{h^*}{2}}\begin{Bmatrix}\sigma_{\theta z}\\\sigma_{xz}\end{Bmatrix}dz,\ I_i=\sum_{k=1}^{N^*}\int_k^{k+1}\rho^{(k)}z^i dz,\ (i=0,1,2,\cdots,6),$$

where N^* is total number of constituent layers and $\rho^{(k)}$ is the density of (k)th constituent ply. $J_i=I_i-c_1I_{i+2}$, $(i=1,4)$ and $K_2=I_2-2c_1I_4+{c_1}^2I_6$.

Assuming $\frac{\partial v_0}{\partial z}=-\frac{v_0}{R^*}$, $\frac{\partial u_0}{\partial z}=-\frac{u_0}{R^*}$, $\frac{\partial w}{\partial z}=\frac{\partial\phi_x}{\partial z}=\frac{\partial\phi_\theta}{\partial z}=0$, terms of $\frac{1}{2}\left(\frac{\partial w}{\partial x}\right)^2$, $\frac{\partial w}{\partial x}\frac{\partial w}{R^*\partial\theta}$, $\frac{1}{2}\left(\frac{\partial w}{R^*\partial\theta}\right)^2$ are in constant value and can be applied by Hong [11] with the parameter R^* instead of the R. Thus, dynamic differential equations with TSDT expressed in equilibrium matrix forms can be obtained for thick FGM conical shells under heating loads. Also, the varied values of shear correction coefficient k_α are used for the integrals of shear stiffness $\overline{Q}_{i^*j^*}$ in which subscripts i^*, j^* = 4, 5.

The GDQ numerical method can be used to solve the dynamic differential equations in equilibrium matrix forms. In the method history for differential quadrature (DQ) was issued by Bert et al. [17]. After then, the GDQ method was presented by Shu and Du [18]. Usually, the GDQ method approximated the function $f(x,\theta)$ derivatives at an arbitrary grid point (x_i,θ_j) in which subscripts i = 1,2,. . .,N and j = 1,2,. . .,M for two-dimensional grid points N and M. The oscillations of time sinusoidal displacement with frequency ω_{mn}, in which subscripts m and n are the mode shape numbers, shear rotations, environment temperature, and sinusoidal heat flux with frequency γ due to ΔT, are applied for the thermal vibrations by Hong [10]. And the simple vibration of ΔT for thermal loads can be given in the following sinusoidal expression:

$$\Delta T = \frac{z}{h^*}\overline{T}_1 \sin(\pi x/L)\sin(\pi\theta/R)\sin(\gamma t), \tag{15}$$

in which γ is calculated directly from Equation (9) and $\overline{T}_1$ is the temperature amplitude.

3. Numerical Results and Discussions

The Visual Studio 2015 Lahey–Fujitsu Fortran high level language can be used and implemented as the computational tool. The author has written the software program based on Lahey–Fujitsu Fortran to perform the simulation. Also, the current study with the method can be used from previously published work for the circular cylindrical shells by the same author Hong [11]. This current manuscript can be applied with the same methodology and materials to obtain the results for the thick conical shells by referring the previous publication works by the same author Hong [19]. The curve coordinates x_i and θ_j for the conical shells are used as follows to calculate the GDQ solutions with constituent layers in FGMs under sinusoidal displacement vibration and heating loads only.

$$x_i = 0.5\left[1 - \cos\left(\frac{i-1}{N-1}\pi\right)\right]L \tag{16}$$

$$\theta_j = 0.5\left[1 - \cos\left(\frac{j-1}{M-1}\pi\right)\right]R \tag{17}$$

Usually, the linear k_α in varied values are functions of h^*, R_n, and T used in the preliminary study of thick FGM conical shells. The dynamic thermal vibration for thick FGM plates has been presented by the same author Hong [10]. Currently, it is very interesting and getting further study in the field of thick FGM conical shells. There are many material combinations for functionally graded (FG) mode. Usually, the FGM constituent material in general used metal material are SUS304, Ti–6Al–4V, Al, and Ni; ceramic material are Si_3N_4, ZrO_2, and Al_2O_3. Also, the constituent material can be in the intelligent form for its particular function used, e.g., piezoelectric material $BaTiO_3$ and PZT, magnetostrictive material $CoFe_2O_4$, and Terfenol-D. For the choosing that Si_3N_4/SUS304 used in the interesting present study, especially this type of composite materials possess temperature-dependent properties. The constituent material 1 on inner position is SUS304, and the FGM constituent material 2 on outer position is Si_3N_4 applied for the GDQ calculations. In thermal vibration of TSDT computation, the thermal loads in terms of ΔT in Equation (15) are only used for the external loads. Thermal vibration in the thermal loads are coupled with the temperature difference in frequency of sinusoidal heat flux Equation (9) and integrals of stiffness.

3.1. Dynamic Convergence

The iteration flow algorithm for the dynamic convergence studies is listed in Appendix B. The maximum value of displacement w can be founded particularly at center position $(x,\theta) = (L/2, 2\pi/2)$. The center displacement $w(L/2, 2\pi/2)$(mm) vs. $N \times M$ of dynamic convergence studies both in nonlinear TSDT with c_1 = 0.925925/mm^2 and in linear with c_1 = 0/mm^2 are investigated. For thick L/h^* = 10, γ = 0.2618004/s and L/h^* = 5, γ = 0.2618019/s under sinusoidal heat flux at t = 6 s, L/R = 1, h^* = 1.2 mm, $h_1 = h_2$ = 0.6 mm, T = 100 K, $\overline{T}_1$ = 1 K vs. $r = 3R/4$ are shown in Table 1. Also, the varied k_α and ω_{11} for R_n values (0.5, 1, and 2) are considered in the sinusoidal oscillation. For example, in the L/h^* = 5 of c_1 = 0.925925/mm^2 case, when R_n = 0.5, the values k_α = 0.111874 and ω_{11} = 0.001730/s are used; when R_n = 1, the values k_α = 0.149001 and ω_{11} = 0.001730/s are used; when R_n = 2, the values k_α = 0.231364 and ω_{11} = 0.001730/s are used. In the L/h^* = 5 of c_1 = 0/mm^2 case, when R_n = 0.5, the values k_α = 0.111874 and ω_{11} = 0.004731/s are used; when R_n = 1, the values k_α = 0.149001 and ω_{11} = 0.004191/s are used; when R_n = 2, the values k_α = 0.231364 and ω_{11} = 0.003575/s are used. The 8.8×10^{-5} error accuracy can be found for nonlinear $w(L/2, 2\pi/2)$ of typical c_1 = 0.925925/mm^2, R_n = 2 and L/h^* = 10 for $N \times M$ = 15 × 15 and 17 × 17 listed in Table 1. The 8.8 × 10^{-5} error accuracy can be considered acceptable in this presented preliminary study for conical shells. Table 1 also contains the comparison between the results obtained by using TSDT and FSDT ($c_1 = 0$). Consider Table 1, which reports the $w(L/2, 2\pi/2)$ after 6 s, e.g., at t = 12 s. Table 2 presents for the convergence of TSDT FGM conical shells with γ = 0.1309014/s, ω_{11} = 0.003457/s, r = 3R/4, and R_n = 0.5 at t = 12 s. It can be shown that the smaller value of the displacement is decreasing as the number of grid points $N \times M$ increases. Thus, grid points $N \times M$ = 17 × 17 can be considered in a good convergence and used in the GDQ calculations of time responses for thick FGM conical shells. Also, the σ_x, σ_θ, $\sigma_{x\theta}$, $\sigma_{\theta z}$, and σ_{xz} on the FGMs can be calculated. There are no other researcher results in this area of research. Without additional evidence for the validation of the results, it would be premature to definitively claim sole authorship within this research area. The calculated similar values of concern cylindrical shells were studied by Brischetto and Torre [20], although there are some available similar but different analysis models used for the interesting cases.

Table 1. Convergence of TSDT FGM conical shells with r = 3R/4.

c_1(1/mm^2)	L/h^*	GDQ Method	Displacement $w(L/2, 2\pi/2)$(mm) at t = 6 s		
		$N \times M$	R_n = 0.5	R_n = 1	R_n = 2
0.925925	10	7 × 7	0.011270	0.011308	0.011308
		9 × 9	0.011274	0.011275	0.011276
		11 × 11	0.011274	0.011275	0.011276
		13 × 13	0.011272	0.011272	0.011274
		15 × 15	0.011270	0.011272	0.011273
		17 × 17	0.011263	0.011267	0.011272
	5	7 × 7	0.000852	0.000851	0.000851
		9 × 9	0.000846	0.000846	0.000847
		11 × 11	0.000845	0.000846	0.000845
		13 × 13	0.000836	0.000838	0.000839
		15 × 15	0.000834	0.000836	0.000837
		17 × 17	0.000816	0.000822	0.000825
0	10	7 × 7	−0.045997	−0.044647	−0.077112
		9 × 9	0.028134	0.030911	0.034508
		11 × 11	0.028229	0.030891	0.034972
		13 × 13	0.028143	0.030926	0.034519
		15 × 15	0.028710	0.030805	0.034919
		17 × 17	0.028140	0.030926	0.034534
	5	7 × 7	0.002116	0.002409	0.002833
		9 × 9	0.002199	0.002478	0.002898
		11 × 11	0.002167	0.002443	0.002858
		13 × 13	0.002201	0.002480	0.002900
		15 × 15	0.002183	0.002461	0.002877
		17 × 17	0.002201	0.002480	0.002739

Table 2. Convergence of TSDT FGM conical shells with r = 3R/4 at t = 12 s.

c_1(1/mm^2)	L/h^*	$N \times M$	$w(L/2, 2\pi/2)$(mm) for R_n = 0.5
0.925925	5	7 × 7	0.000071
		9 × 9	0.000064
		11 × 11	0.000063
		13 × 13	0.000056
		15 × 15	0.000053
		17 × 17	0.000050

3.2. Time Response

The time responses of $w(L/2, 2\pi/2)$ are computed with γ of sinusoidal heat flux and ω_{11} of sinusoidal displacement vibration. Figure 2 shows the responses of $w(L/2, 2\pi/2)$(mm) versus t(s) for $c_1 = 0.925925/\text{mm}^2$ and $c_1 = 0/\text{mm}^2$, respectively, with $L/h^* = 5$ and 10, $L/R = 1$, $R_n = 1$, $k_\alpha = 0.120708$, $T = 100$ K, $\overline{T}_1 = 1$ K, and $r = 3R/4$. The $w(L/2, 2\pi/2)$ maximum value 0.149393 mm is found at $t = 0.1$ s for $L/h^* = 5$ and $c_1 = 0/\text{mm}^2$, as shown in Figure 2a. The $w(L/2, 2\pi/2)$ maximum value 1.863387 mm is found at $t = 0.1$ s for $L/h^* = 10$ and $c_1 = 0/\text{mm}^2$, as shown in Figure 2b. The $w(L/2, 2\pi/2)$ values are in decreasing tendencies for both $c_1 = 0.925925/\text{mm}^2$, $c_1 = 0/\text{mm}^2$, and $L/h^* = 5$, $L/h^* = 10$. The $w(L/2, 2\pi/2)$ values in $c_1 = 0/\text{mm}^2$ case are greater than that in $c_1 = 0.925925/\text{mm}^2$ case. The $w(L/2, 2\pi/2)$ result clearly state the improvement obtained from using this third order derivation TSDT case of $c_1 = 0.925925/\text{mm}^2$, as opposed to the first or second order derivation case of $c_1 = 0/\text{mm}^2$.

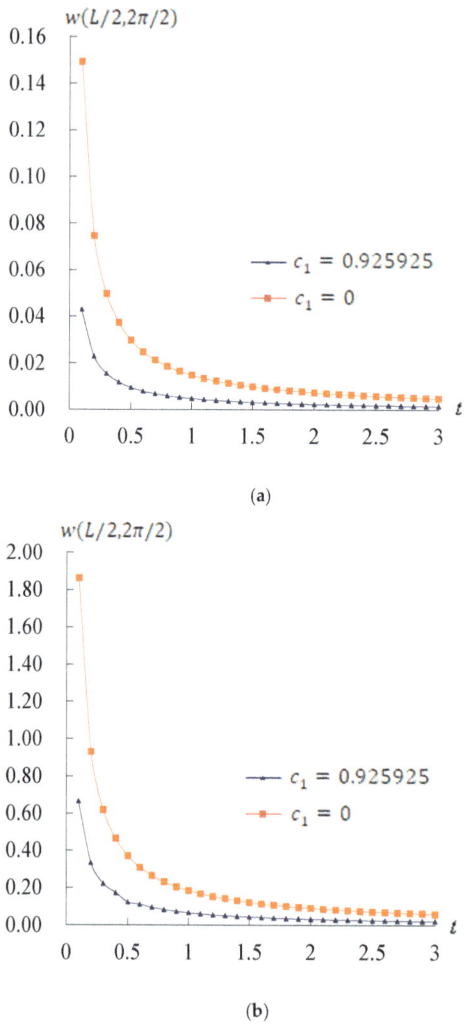

Figure 2. $w(L/2, 2\pi/2)$(mm) vs. t(s) with $r = 3R/4$: (**a**) for $L/h^* = 5$; (**b**) for $L/h^* = 10$.

Figure 3 shows the time responses of σ_x, $\sigma_{x\theta}$, and σ_{xz} on $x = L/2$, $\theta = 2\pi/2$ for the conical shells. Figure 3a shows the σ_x(Pa) versus z/h^*. Figure 3b,c show the $\sigma_{x\theta}$ and σ_{xz}(Pa) versus z/h^*, respectively, at t = 3.0 s for a/h^* = 10, c_1 = 0.925925/mm^2, R_n = 1, $L/R = 1$, T = 100 K, $\overline{T}_1$ = 1 K, and r = 3R/4. The σ_{xz} maximum value -1.8715×10^5 Pa is found on z = 0.0h^* for L/h^* = 10. The σ_{xz} absolute value for -1.8715×10^5 Pa is found on z = 0.0h^* and in greater value than σ_x value 1.3210×10^4 Pa on $z = -0.5h^*$. Figure 3d,e show the σ_x(Pa) time responses on inner surface $z = -0.5h^*$ for R_n = 1, c_1 = 0.925925/mm^2, L/h^* = 5 and 10, respectively. The values of σ_x are in decreasing tendencies for c_1 = 0.925925/mm^2 and L/h^* = 5. The σ_x maximum value 1.3225×10^4 Pa is found at t = 0.2 s and σ_x values can be considered in order of 1.32×10^4 Pa for L/h^* = 10. Figure 3f,g show the time responses of the σ_{xz}(Pa) on middle surface z = 0.0h^* for R_n = 1, c_1 = 0.925925/mm^2, L/h^* = 5 and 10, respectively. The σ_{xz} maximum absolute value for -4.3348×10^5 Pa is found at t = 0.2 s for L/h^* = 10.

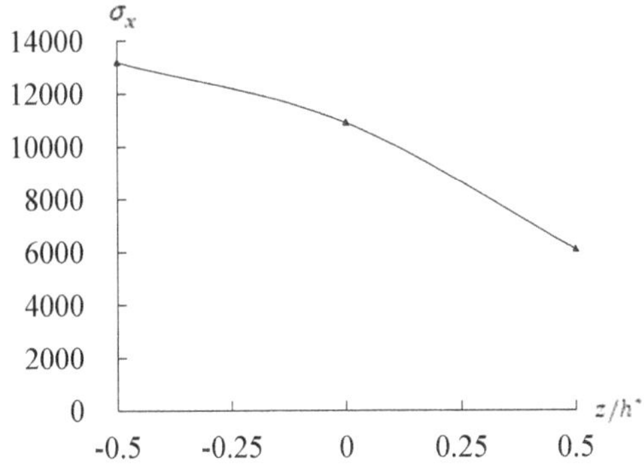

(**a**)

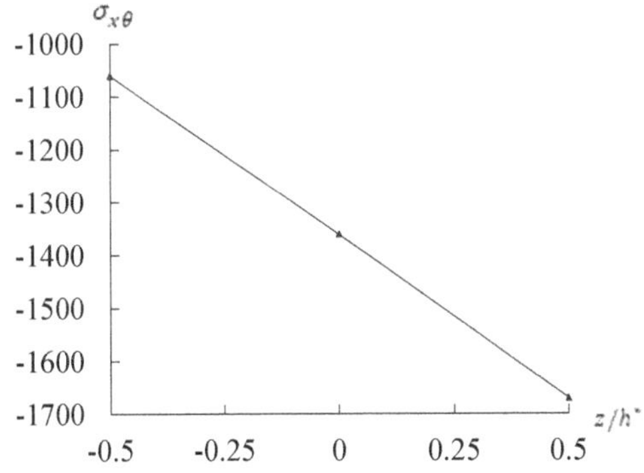

(**b**)

Figure 3. *Cont.*

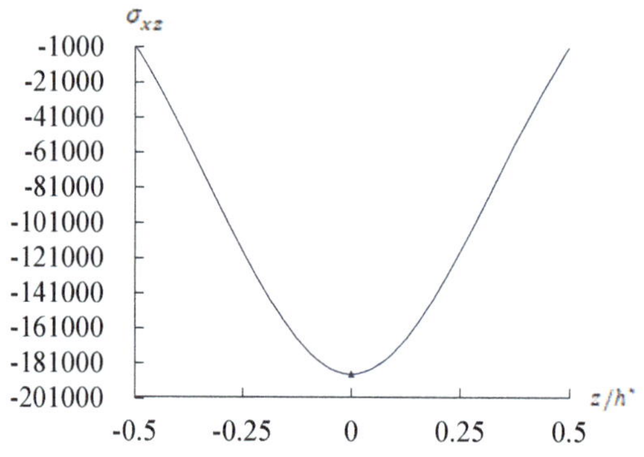

(**c**)

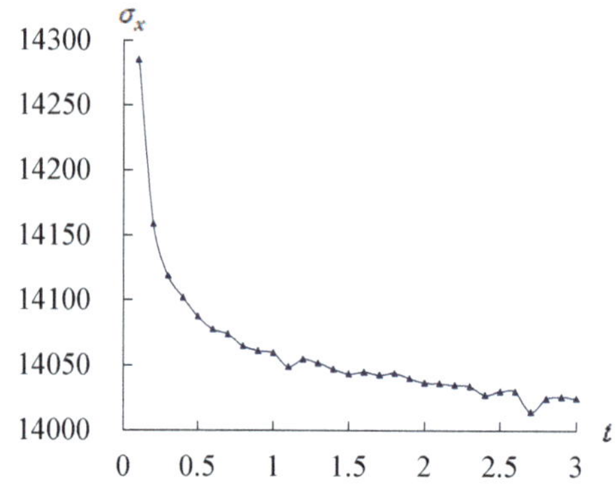

(**d**)

Figure 3. *Cont.*

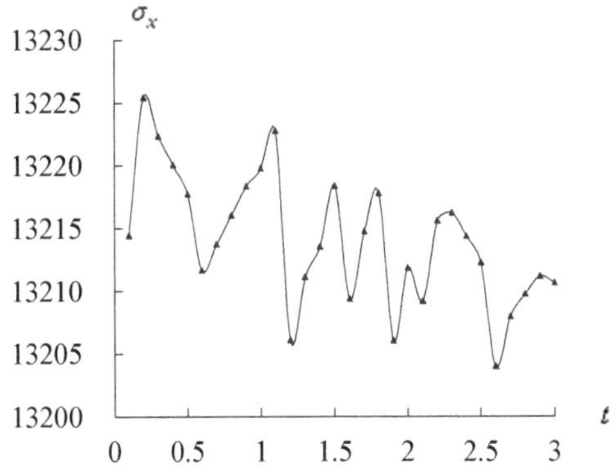

(e)

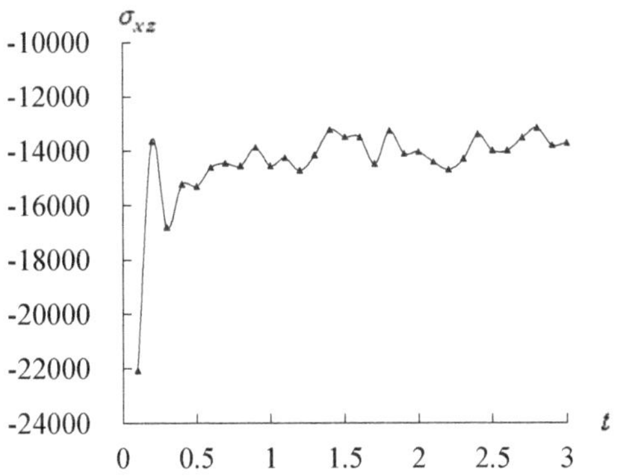

(f)

Figure 3. *Cont.*

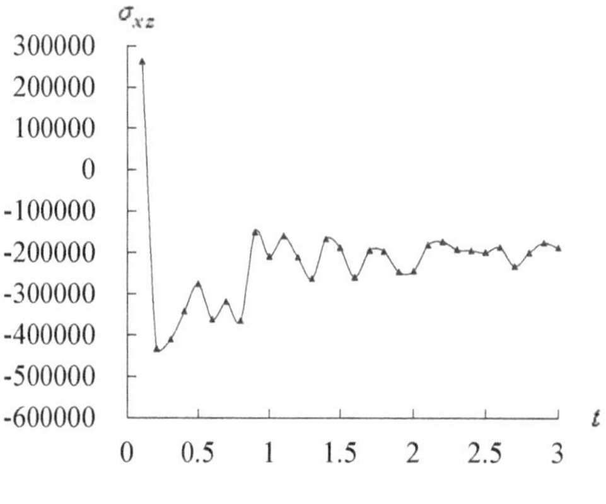

(g)

Figure 3. Stresses vs. z/h^* and t with $r = 3R/4$: (**a**) σ_x(Pa) vs. z/h^* for $L/h^* = 10$; (**b**) $\sigma_{x\theta}$(Pa) vs. z/h^* for $L/h^* = 10$; (**c**) σ_{xz}(Pa) vs. z/h^* for $L/h^* = 10$; (**d**) σ_x(Pa) vs. t(s) for $L/h^* = 5$; (**e**) σ_x(Pa) vs. t(s) for $L/h^* = 10$; (**f**) σ_{xz}(Pa) vs. t(s) for $L/h^* = 5$; (**g**) σ_{xz}(Pa) vs. t(s) for $L/h^* = 10$.

3.3. Parameters Effect

Figure 4 shows the $w(L/2, 2\pi/2)$(mm) response values vs. T(100 K, 600 K and 1000 K) and R_n(from 0.1 to 10) at $t = 0.1$ s under $c_1 = 0.925925/\text{mm}^2$ for conical shells $L/h^* = 5$ and 10, respectively, with $L/R = 1$, $\overline{T}_1 = 1$ K and $r = 3R/4$. Figure 4a shows the $w(L/2, 2\pi/2)$ curves vs. T and R_n for $L/h^* = 5$ case, the $w(L/2, 2\pi/2)$ maximum value 0.089998 mm is found on $T = 1000$ K and $R_n = 10$. The values of $w(L/2, 2\pi/2)$ are all in increasing tendencies for T from $T = 100$ K to $T = 1000$ K and all of R_n. The $w(L/2, 2\pi/2)$ amplitude for $L/h^* = 5$ cannot withstand on $T = 1000$ K of environment. Figure 4b shows the $w(L/2, 2\pi/2)$ curves vs. T and R_n for $L/h^* = 10$ case, they are almost located in the same curves for all value of R_n. The $w(L/2, 2\pi/2)$ maximum value 1.223374 mm is found on $T = 1000$ K and $R_n = 0.2$. The $w(L/2, 2\pi/2)$ values are all in increasing tendencies for T and all of R_n, the $w(L/2, 2\pi/2)$ amplitude for $L/h^* = 10$ also cannot withstand on $T = 1000$ K of environment.

More interpretation for the stresses can be calculated from the following equations when the displacements and shear rotations are already computed. Thus, a peculiarly maximum values occur at $T = 600$ K for displacements and stresses.

$$
\begin{aligned}
\sigma_x = {} & \{[\sum_{l=1}^{N} A_{i,l}^{(1)} U_{l,j} + \tfrac{z}{L}\sum_{l=1}^{N} A_{i,l}^{(1)} \varphi_{x\ l,j} - c_1 z^3 (\tfrac{1}{L}\sum_{l=1}^{N} A_{i,l}^{(1)} \varphi_{x\ l,j} + \tfrac{h^*}{L^2}\sum_{l=1}^{N} A_{i,l}^{(2)} W_{l,j})] \sin(\omega_{mn} t) \\
& + \tfrac{1}{2}(\tfrac{h^* L}{L^2}\sum_{l=1}^{N} A_{i,l}^{(1)} W_{l,j} \sin(\omega_{mn} t))^2 - \alpha_x \tfrac{z}{h^*}\overline{T_1} \sin(\pi x_i/L)\sin(\pi \theta_j/R)\sin(\gamma t)\}\overline{Q}_{11} \\
& + \{[\sum_{m=1}^{M} B_{j,m}^{(1)} V_{i,m} + \tfrac{z}{R}\sum_{m=1}^{M} B_{j,m}^{(1)} \varphi_{\theta\ i,m} - c_1 z^3 (\tfrac{1}{R}\sum_{m=1}^{M} B_{j,m}^{(1)} \varphi_{\theta\ i,m} + \tfrac{h^*}{L^2}\tfrac{L^2}{R^2}\sum_{m=1}^{M} B_{j,m}^{(2)} W_{i,m})] \sin(\omega_{mn} t) \\
& + \tfrac{1}{2}(\tfrac{h^*}{L^2}\tfrac{L^2}{R}\sum_{m=1}^{M} B_{j,m}^{(1)} W_{i,m} \sin(\omega_{mn} t))^2 - \alpha_\theta \tfrac{z}{h^*}\overline{T_1} \sin(\pi x_i/L)\sin(\pi \theta_j/R)\sin(\gamma t)\}\overline{Q}_{12} \\
& + \{[\tfrac{L}{R}\sum_{m=1}^{M} B_{j,m}^{(1)} U_{i,m} + \tfrac{z}{R}\sum_{m=1}^{M} B_{j,m}^{(1)} \varphi_{x\ i,m} - c_1 z^3 (\tfrac{1}{R}\sum_{m=1}^{M} B_{j,m}^{(1)} \varphi_{x\ i,m} + \tfrac{h^*}{L^2}\tfrac{L}{R}\sum_{l=1}^{N}\sum_{m=1}^{M} A_{i,l}^{(1)} B_{j,m}^{(1)} W_{l,m}) \\
& + \tfrac{R}{L}\sum_{l=1}^{N} A_{i,l}^{(1)} V_{l,j} + \tfrac{z}{L}\sum_{l=1}^{N} A_{i,l}^{(1)} \varphi_{\theta\ l,j} - c_1 z^3 (\tfrac{1}{L}\sum_{l=1}^{N} A_{i,l}^{(1)} \varphi_{\theta\ l,j} + \tfrac{h^*}{L^2}\tfrac{L}{R}\sum_{l=1}^{N}\sum_{m=1}^{M} A_{i,l}^{(1)} B_{j,m}^{(1)} W_{l,m})] \sin(\omega_{mn} t) \\
& + (\tfrac{h^* L}{L^2}\sum_{l=1}^{N} A_{i,l}^{(1)} W_{l,j} \sin(\omega_{mn} t))(\tfrac{h^*}{L^2}\tfrac{L^2}{R}\sum_{m=1}^{M} B_{j,m}^{(1)} W_{l,m} \sin(\omega_{mn} t)) \\
& - \alpha_{x\theta} \tfrac{z}{h^*}\overline{T_1} \sin(\pi x_i/L)\sin(\pi \theta_j/R)\sin(\gamma t)\}\overline{Q}_{16}
\end{aligned}
\quad (18)
$$

where $A_{i,l}^{(m)}$ and $B_{j,m}^{(m)}$ denote the weighting coefficients used in the GDQ calculations for the superscript (m)th-order derivative of functions, e.g., displacements ($U = u_0/L$, $V = v_0/R$ and $W = w/h^*$) and shear rotations ($\varphi_x(x,\theta)$ and $\varphi_\theta(x,\theta)$). α_x and α_θ denoted the thermal expansion coefficients, $\alpha_{x\theta}$ denoted the thermal shear coefficient. And the stiffness $\overline{Q}_{ij}$ can be defined as follows:

$$\overline{Q}_{11} = \overline{Q}_{22} = \frac{E_{fgm}}{1 - {v_{fgm}}^2}, \overline{Q}_{12} = \overline{Q}_{21} = \frac{v_{fgm} E_{fgm}}{(1 + z/R)(1 - {v_{fgm}}^2)} \text{and} \overline{Q}_{16} = \overline{Q}_{26} = \overline{Q}_{45} = 0 \tag{19}$$

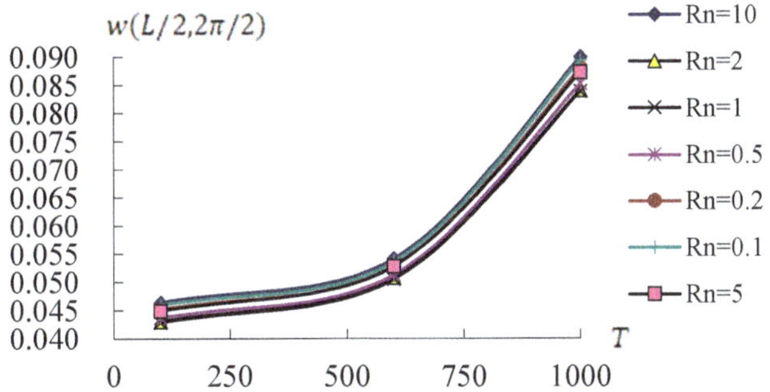

(**a**)

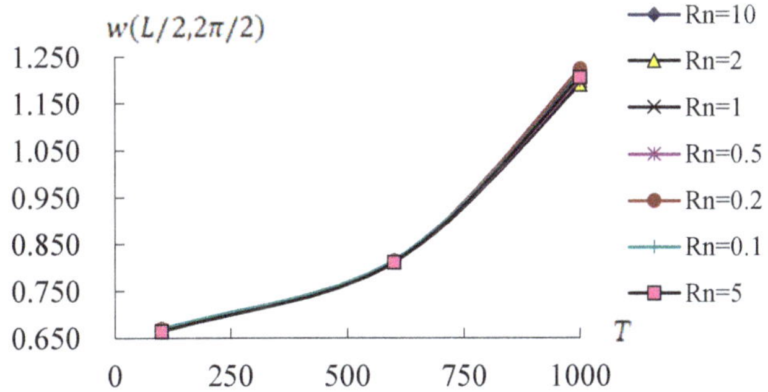

(**b**)

Figure 4. $w(L/2, 2\pi/2)$(mm) versus T(K) with r = $3R/4$: (**a**) for L/h^* = 5 with R_n from 0.1 to 10; (**b**) for L/h^* = 10 with R_n from 0.1 to 10.

To provide physical justification for (18) used in the GDQ computation as follows, the values of stress along x direction (σ_x) are functions of symbolic parameters, e.g., displacements ($U_{l,j}$, $V_{i,m}$, $W_{l,j}$), shear rotations ($\phi_{x\,l,j}$, $\phi_{\theta\,i,m}$), TSDT coefficient term (c_1), thermal loads temperature amplitude ($\overline{T}_1$), environment temperature (T), sinusoidal heat flux frequency (γ), sinusoidal displacement fundamental frequency (ω_{11}), located curve coordinates (x_i, θ_j, z), and at time (t). One of the most important parameters is T to be considered for the stress σ_x of FGM and used in higher temperature of environment. To provide the interpretation of the peculiar behavior of σ_x on T = 600 K as follows, it is considered the increment 100 K used for T (100 K, 200 K, 300 K, 400 K, 500 K, 600 K, 700 K, 800 K, 900 K, and 1000 K) to calculate the values and interpret the behaviors of σ_x. Figure 5 shows the σ_x(Pa) on $x = L/2$, $\theta = 2\pi/2$ of inner surface $z = -0.5h^*$ vs. T and R_n = 0.5 for conical shells L/h^* = 5 with $\overline{T}_1$ = 1 K and $r = 3R/4$ at t = 0.1 s. Thus, the curve provided a maximum interpretation of the peculiar behavior of σ_x around at T = 600 K. Then, the increment for T (100 K, 600 K, and 1000 K) is used to calculate the values and interpret the behaviors of σ_x.

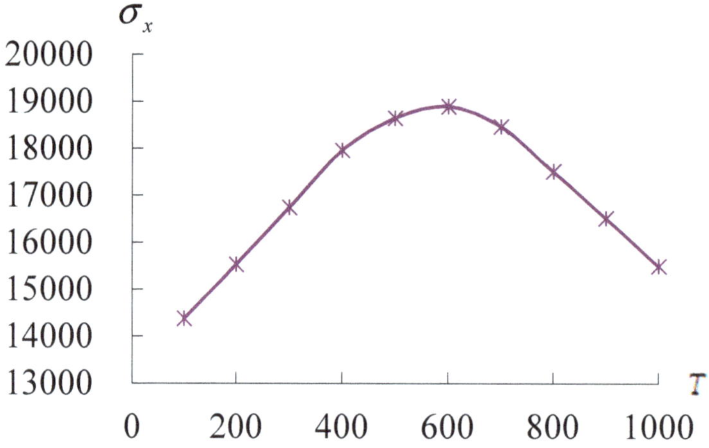

Figure 5. σ_x(Pa) versus T(K).

Figure 6 shows the σ_x(Pa) on $x = L/2$, $\theta = 2\pi/2$ of inner surface $z = -0.5h^*$ vs. T (100 K, 600 K and 1000 K) and R_n(from 0.1 to 10) for conical shells L/h^* = 5 and 10 with $\overline{T}_1$ = 1 K and $r = 3R/4$ at t = 0.1 s. Figure 6a shows the σ_x curves vs. T and R_n for L/h^* = 5, the σ_x values vs. T are all in increasing tendencies from T = 100 K to T = 600 K and then all in decreasing tendencies from T = 600 K to T = 1000 K for all of R_n. The σ_x maximum value 1.9215×10^4 Pa is found on T = 600 K and R_n = 0.1. The σ_x values of L/h^* = 5 can withstand on T = 1000 K of environment. Figure 6b shows the σ_x curves vs. T and R_n for L/h^* = 10, they are almost located in the same curves for all value of R_n. The σ_x values vs. T are all in increasing tendencies from T = 100 K to T = 600 K, and then all in decreasing tendencies from T = 600 K to T = 1000 K for all of R_n. The σ_x maximum value 1.7094×10^4 Pa is found on T = 600 K. The σ_x values of L/h^* = 10 also can withstand on T = 1000 K of environment. Usually, the thermal stress occurred and be stood on the FGMs under thermal loading of extreme high temperatures by Noda [21]. The physical parts on values L/h^* = 5 and 10 of FGMs can be used in area of airplane engine that usually operate near more than 1000 K of temperatures when the thermal stress to be considered and affected. To understand the effect of thermal loading on the thermal stress, the middle-near 600 K of temperatures can be used to study.

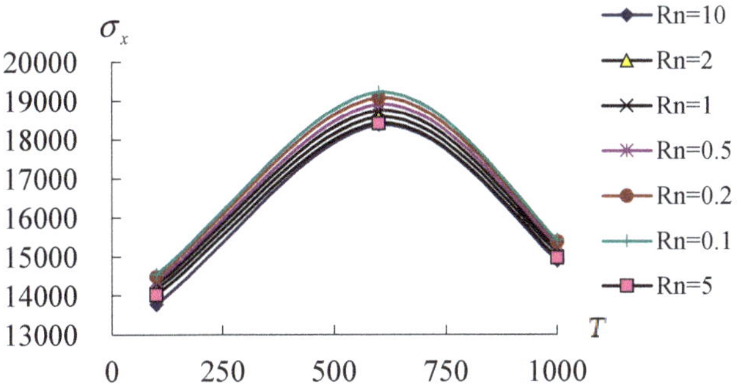

(**a**)

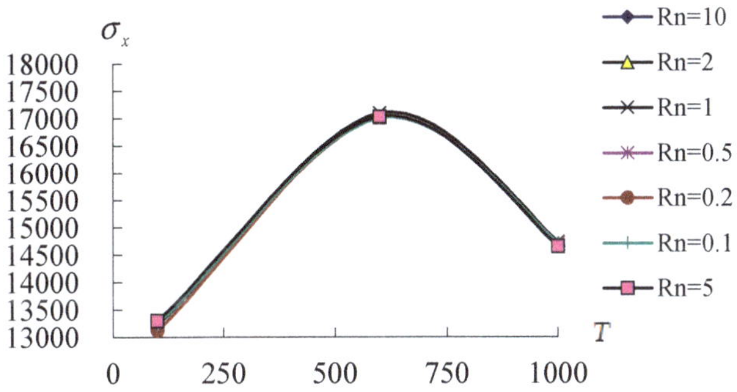

(**b**)

Figure 6. σ_x(Pa) versus T(K) with $r = 3R/4$: (**a**) for $L/h^* = 5$; (**b**) for $L/h^* = 10$.

3.4. Compared Results

Figure 7 shows the compared values of $w(L/2, 2\pi/2)$(mm) vs. T(100 K, 600 K and 1000 K) for r ($R/8$, $R/4$, $R/2$, $3R/4$ and R) of conical shells with $c_1 = 0.925925/\text{mm}^2$, $L/R = 1$ and $\overline{T}_1 = 1$ K at $t = 0.1$ s. Figure 7a shows the $w(L/2, 2\pi/2)$ values vs. T and r for $L/h^* = 5$ and $R_n = 10$. The $w(L/2, 2\pi/2)$ maximum value 0.143956 mm is found on $r = R/8$ and $T = 600$ K. The $w(L/2, 2\pi/2)$ values vs. T are in increasing tendencies from $T = 100$ K to $T = 600$ K and then in decreasing tendencies from $T = 600$ K to $T = 1000$ K for the case $r = R/8$. The $w(L/2, 2\pi/2)$ values for the case $r = R/8$ and $L/h^* = 5$ can withstand on $T = 1000$ K of environment. Thus, physical parts on values $R_n = 10$, $r = R/8$ and $L/h^* = 5$ of FGMs can be used in an area of an airplane engine that usually operates near more than 1000 K of temperatures. To understand the effect of thermal loading on the displacement, the middle-near 600 K of temperatures can be used to study. Figure 7b shows the $w(L/2, 2\pi/2)$ values vs. T and r for $L/h^* = 10$ and $R_n = 0.2$. The $w(L/2, 2\pi/2)$ maximum value 1.265982 mm is found on $r = R/8$ and $T = 1000$ K. The $w(L/2, 2\pi/2)$ values are all in increasing tendencies vs. T for all values of r. The $w(L/2, 2\pi/2)$ values for $L/h^* = 10$ cannot withstand on $T = 1000$ K of environment.

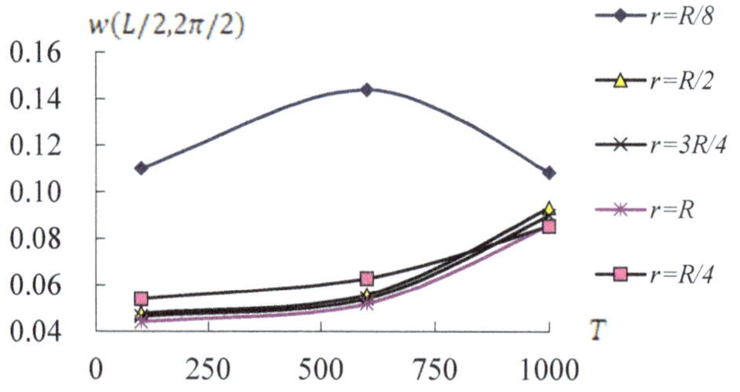

(**a**)

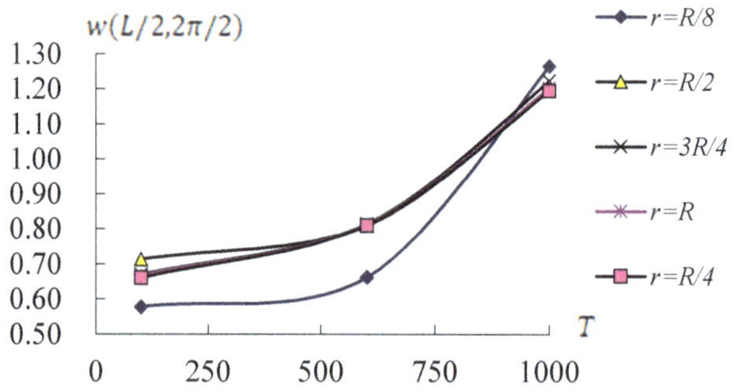

(**b**)

Figure 7. Compared $w(L/2, 2\pi/2)$(mm) vs. T(K) and r: (**a**) for $L/h^* = 5$ with $R_n = 10$; (**b**) for $L/h^* = 10$ with $R_n = 0.2$.

Figure 8 shows the compared values of σ_{xz}(Pa) on center position of middle surface $z = 0.0h^*$ vs. T(100 K, 600 K, and 1000 K) for r($R/8$, $R/4$, $R/2$, $3R/4$, and R) with $c_1 = 0.925925/\text{mm}^2$. Figure 8a shows the values of σ_{xz} vs. T and r in the nonlinear TSDT FGM conical shells for $L/h^* = 5$, $R_n = 10$, $L/R = 1$, and $\overline{T}_1 = 1$ K at $t = 0.1$ s. The σ_{xz} maximum value -7.7207×10^6 Pa is found on $r = R/8$, $T = 600$ K and $L/h^* = 5$. The σ_{xz} values vs. T are in increasing tendencies from $T = 300$ K to $T = 600$ K and then in decreasing tendencies from $T = 600$ K to $T = 1000$ K for the case $r = R/8$. The σ_{xz} values for the case $r = R/8$ and $L/h^* = 5$ can withstand on $T = 1000$ K of environment. Figure 8b shows the σ_{xz} values vs. T and r in the nonlinear TSDT FGM conical shells for $L/h^* = 10$, $R_n = 0.2$, $L/R = 1$ and $\overline{T}_1 = 1$ K at $t = 0.1$ s. The σ_{xz} maximum value -3.4165×10^7 Pa is found on $r = R/8$, $T = 1000$ K and $L/h^* = 10$. The σ_{xz} values are all in increasing tendencies vs. T from $T = 600$ K to $T = 1000$ K for all values of r. The σ_{xz} values of the $L/h^* = 10$ cannot withstand on $T = 1000$ K of environment.

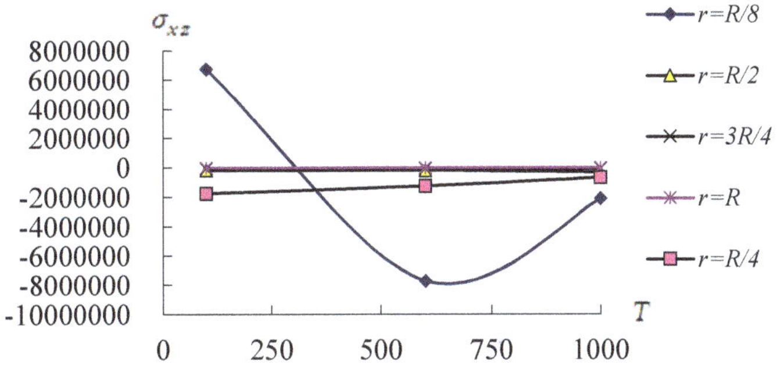

(a)

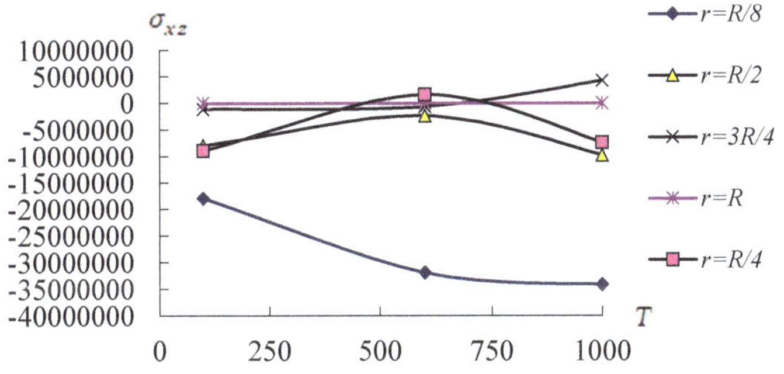

(b)

Figure 8. Compared σ_{xz}(Pa) vs. T(K) and r: (**a**) for L/h^* = 5 with R_n = 10; (**b**) for L/h^* = 10 with R_n = 0.2.

3.5. Limitations

For further study, the imperfection of FGMs can be considered. Also, the more details of imperfection in the material composition of FGMs can be referred to in the literature by Malikan and Eremeyev [22]. Also, the inhomogeneousness of FGMs can be studied in the future. The more details of inhomogeneousness in the material composition of FGMs can be referred in the literature by Dastjerdi and Akgöz [23].

4. Conclusions

Dynamic GDQ thermal vibration results are studied for the displacements and stresses in the of laminated thick FGM conical shells by considering the effects of computed shear coefficient and c_1 term of TSDT. The displacement values in $c_1 = 0/\text{mm}^2$ case are greater than that $c_1 = 0.925925/\text{mm}^2$ case for L/h^* = 5 and L/h^* = 10, respectively. It is over estimated in the center displacement values for the linear case. The values of stress σ_x are in decreasing tendencies with time in $c_1 = 0.925925/\text{mm}^2$ and L/h^* = 5. The σ_{xz} values on $r = R/8$ and L/h^* = 5 can withstand T = 1000 K of the environment. Main highlights included: (a) thermal vibration of conical shells by TSDT are computed (b) with varied shear correction to obtain the displacements and stresses results; (c) FGM power law and environment temperature are considered. Main findings included: (d) result improvement obtained from using the TSDT case better than that FSDT case; and (e) compared values of displacement and stress versus environment temperature (100 K, 600 K, and 1000 K) for minor middle-surface radius ($R/8$, $R/4$, $R/2$, $3R/4$, and R). A limitation would be that too many grid points used in the calculation would cause too much memory space occupied in RAM. Areas of future research in advanced thermal vibration of FGM conical shells would include studies of TDST and the nonlinear varied shear coefficient.

Funding: The funding of this paper is MOST 111-2221-E-164-001 from NSTC, Taiwan.

Institutional Review Board Statement: In this section, the study did not require ethical approval.

Informed Consent Statement: This article does not contain any studies with human participants performed by any of the authors.

Data Availability Statement: In the manuscript completely mentioned the data used to generate the figures and tables. Data are all available on request. The authors declare that all the data are generated by the author, and also that data are openly available.

Acknowledgments: The author expresses his thanks to the people helping with this work, and acknowledge the valuable suggestions from the peer reviewers.

Conflicts of Interest: The author declares no conflicts of interest.

Appendix A

For no heat generation, the heat conduction equation can be given as follows:

$$K\left(\frac{\partial^2 \Delta T}{\partial x^2}+\frac{\partial^2 \Delta T}{R^2 \partial \theta^2}+\frac{\partial^2 \Delta T}{\partial z^2}\right)=\frac{\partial \Delta T}{\partial t}. \quad \text{(A1)}$$

And the sinusoidal expression ΔT can be given simply with linear vs. z, as follows:

$$\Delta T=\frac{z}{h^*}\overline{T}_1 \sin(\pi x/L)\sin(\pi\theta)\sin(\gamma t), \quad \text{(A2)}$$

then (A1) can be obtained as follows:

$$K\left(-\frac{z}{h^*}\frac{\pi^2}{L^2}\overline{T}_1\sin\left(\frac{\pi x}{L}\right)\sin(\pi\theta)\sin(\gamma t)-\frac{1}{R^2}\frac{z}{h^*}\pi^2\overline{T}_1\sin\left(\frac{\pi x}{L}\right)\sin(\pi\theta)\sin(\gamma t)\right)=\gamma\frac{z}{h^*}\overline{T}_1\sin\left(\frac{\pi x}{L}\right)\sin(\pi\theta)\cos(\gamma t), \quad \text{(A3)}$$

thus:

$$K\left(-\frac{\pi^2}{L^2}-\frac{\pi^2}{R^2}\right)\sin(\gamma t)=\gamma\cos(\gamma t), \quad \text{(A4)}$$

also:

$$-\frac{K\pi^2}{L^2}\left(1+\frac{L^2}{R^2}\right)\sin(\gamma t)=\gamma\cos(\gamma t). \quad \text{(A5)}$$

It can be reduced to the following equation:

$$\frac{L^2}{K\pi^2[1+(L/R)^2]}\gamma\cos(\gamma t)+\sin(\gamma t)=0. \quad \text{(A6)}$$

Appendix B

The five-degree polynomial equation can be obtained for the sinusoidal form of free vibration in dynamic equilibrium differential equations, as follows;

$$A(1)\lambda_{mn}{}^5+A(2)\lambda_{mn}{}^4+A(3)\lambda_{mn}{}^3+A(4)\lambda_{mn}{}^2+A(5)\lambda_{mn}+A(6)=0, \quad \text{(A7)}$$

where

$$\lambda_{mn} = I_0 \omega_{mn}{}^2, \tag{A8}$$

$$A(1) = -sd, \tag{A9}$$

$$A(2) = (FH_{11} + FH_{22})sd + sc, \tag{A10}$$

$$A(3) = -[(FH_{11}FH_{22} - FH_{12}FH_{12})\, sd + (FH_{11} + FH_{12})sc + sb], \tag{A11}$$

$$A(4) = (FH_{11}FH_{22} - FH_{12}FH_{12})\, sc + (FH_{11} + FH_{22})sb + sa, \tag{A12}$$

$$A(5) = -[(FH_{11}FH_{22} - FH_{12}FH_{12})\, sb + (FH_{11} + FH_{22})sa], \tag{A13}$$

$$A(6) = (FH_{11}FH_{22} - FH_{12}FH_{12})sa, \tag{A14}$$

in which ω_{mn} denotes the vibration frequency with mode shape numbers m and n in the subscripts,

$$sd = (K_2/I_0)^2, \tag{A15}$$

$$sc = FH_{33}sd + FH_{44}K_2/I_0, \tag{A16}$$

$$sb = (FH_{33}FH_{55} + FH_{44}FH_{55} + FH_{33}FH_{44} - FH_{35}FH_{35} - FH_{34}FH_{34})K_2/I_0 - FH_{45}FH_{45}, \tag{A17}$$

$$sa = FH_{33}FH_{44}FH_{55} + FH_{44}FH_{34}FH_{35} + FH_{35}FH_{34}FH_{45} - FH_{35}FH_{35}FH_{44} - FH_{34}FH_{34}FH_{55} - FH_{45}FH_{45}FH_{33}, \tag{A18}$$

and

$$FH_{11} = A_{11}(m\pi/L)^2 + A_{66}(n\pi/R)^2, \tag{A19}$$

$$FH_{12} = (A_{12} + A_{66})(m\pi/L)(n\pi/R), \tag{A20}$$

$$FH_{22} = A_{66}(m\pi/L)^2 + A_{22}(n\pi/R)^2, \tag{A21}$$

$$\begin{aligned} FH_{33} = {} & A_{55}(m\pi/L)^2 + A_{44}(n\pi/R)^2 + c_1{}^2H_{11}(m\pi/L)^4 + (2c_1{}^2H_{12} + 4c_1{}^2H_{66})(m\pi/L)^2(n\pi/R)^2 \\ & + c_1{}^2H_{22}(n\pi/R)^4 - 3c_1\left(2D_{55} - 3c_1F_{55})(m\pi/L)^2 - 3c_1(2D_{44} - 3c_1F_{44})(n\pi/R)^2, \right. \end{aligned} \tag{A22}$$

$$\begin{aligned} FH_{34} = {} & A_{55}m\pi/L - (c_1F_{11} - c_1{}^2H_{11})\,(m\pi/L)^3 - (2c_1F_{66} - 2c_1{}^2H_{66} + c_1F_{12} - c_1{}^2H_{12})(m\pi/L)\,(n\pi/ \\ & R)^2 - (6c_1D_{55} - 9c_1{}^2F_{55})(m\pi/L), \end{aligned} \tag{A23}$$

$$\begin{aligned} FH_{35} = {} & A_{44}n\pi/R - (c_1F_{22} - c_1^2H_{22})(n\pi/R)^3 - (2c_1F_{66} - 2c_1^2H_{66} + c_1F_{12} - c_1^2H_{12})(m\pi/L)^2(n\pi/R) \\ & - (6c_1D_{44} - 9c_1^2F_{44})(n\pi/R), \end{aligned} \tag{A24}$$

$$FH_{44} = (D_{11} - 2c_1F_{11} + c_1^2H_{11})(m\pi/L)^2 + (D_{66} - 2c_1F_{66} + c_1^2H_{66})(n\pi/R)^2 + A_{55} - 6c_1D_{55} + 9c_1^2F_{55}, \tag{A25}$$

$$FH_{45} = (D_{12} + D_{66} - 2c_1F_{12} + c_1^2H_{12} - 2c_1F_{66} + c_1^2H_{66})(m\pi/L)(n\pi/R), \tag{A26}$$

$$FH_{55} = (D_{66} - 2c_1F_{66} + c_1^2H_{66})(m\pi/L)^2 + (D_{22} - 2c_1F_{22} + c_1^2H_{22})(n\pi/R)^2 + A_{44} - 6c_1D_{44} + 9c_1^2F_{44}, \tag{A27}$$

$$(A_{i^sj^s}, B_{i^sj^s}, D_{i^sj^s}, E_{i^sj^s}, F_{i^sj^s}, H_{i^sj^s}) = \int_{\frac{-h^*}{2}}^{\frac{h^*}{2}} \overline{Q}_{i^sj^s}(1, z, z^2, z^3, z^4, z^6)dz,\ (i^s, j^s = 1, 2, 6), \tag{A28}$$

$$(A_{i^*j^*}, B_{i^*j^*}, D_{i^*j^*}, E_{i^*j^*}, F_{i^*j^*}, H_{i^*j^*}) = \int_{\frac{-h^*}{2}}^{\frac{h^*}{2}} k_\alpha \overline{Q}_{i^*j^*}(1, z, z^2, z^3, z^4, z^5)dz,\ (i^*, j^* = 4, 5), \tag{A29}$$

Then, the iteration flow algorithm of Newton's method [24] can be used to find the real zeros λ_{mn} of polynomial Equation (A7). Firstly, a first gauss of λ_{mn} is chosen and read in, then the iterations of do loop in Newton's method continuously to find the λ_{mn} solution until the error less than 1×10^{-6}.

References

1. Soureshjani, A.H.; Talebitooti, R.; Talebitooti, M. A semi-analytical approach on the effect of external lateral pressure on free vibration of joined sandwich aerospace composite conical-conical shells. *Aerosp. Sci. Technol.* **2020**, *99*, 105559. [CrossRef]
2. Ansari, R.; Hasrati, E.; Torabi, J. Nonlinear vibration response of higher-order shear deformable FG-CNTRC conical shells. *Compos. Struct.* **2019**, *222*, 110906. [CrossRef]
3. Das, A.; Singha, T.D.; Karmakar, A. Low velocity normal impact performance of functionally graded conical shell with simple power law. *Mater. Today Proc.* **2019**, *11*, 729–739. [CrossRef]
4. Shakouri, M. Free vibration analysis of functionally graded rotating conical shells in thermal environment. *Compos. Part B* **2019**, *163*, 574–584. [CrossRef]
5. Sofiyev, A.H. Review of research on the vibration and buckling of the FGM conical shells. *Compos. Struct.* **2019**, *211*, 301–317. [CrossRef]
6. Taraghi, P.; Showkati, H.; Firouzsalari, S.E. The performance of steel conical shells reinforced with CFRP laminates subjected to uniform external pressure. *Constr. Build. Mater.* **2019**, *214*, 484–496. [CrossRef]
7. Yang, S.W.; Zhang, W.; Hao, Y.X.; Niu, Y. Nonlinear vibrations of FGM truncated conical shell under aerodynamics and in-plane force along meridian near internal resonances. *Thin-Walled Struct.* **2019**, *142*, 369–391. [CrossRef]
8. Javed, S. Free vibration characteristic of laminated conical shells based on higher order shear deformation theory. *Compos. Struct.* **2018**, *204*, 80–87. [CrossRef]
9. Mouli, B.C.; Kar, V.R.; Ramji, K.; Rajesh, M. Free vibration of functionally graded conical shell. *Mater. Today Proc.* **2018**, *5*, 14302–14308. [CrossRef]
10. Hong, C.C. GDQ computation for thermal vibration of thick FGM plates by using third-order shear deformation theory. *Mater. Sci. Eng. B* **2023**, *294*, 116208. [CrossRef]
11. Hong, C.C. Thermal vibration of thick FGM circular cylindrical shells by using TSDT. *Mater. Plus* **2022**, *1*, 2–11. [CrossRef]
12. Chi, S.H.; Chung, Y.L. Mechanical behavior of functionally graded material plates under transverse load, part I: Analysis. *Int. J. Solids Struct.* **2006**, *43*, 3657–3674. [CrossRef]
13. Lee, S.J.; Reddy, J.N.; Rostam-Abadi, F. Transient analysis of laminated composite plates with embedded smart-material layers. *Finite Elem. Anal. Des.* **2004**, *40*, 463–483. [CrossRef]
14. Lee, S.J.; Reddy, J.N. Non-linear response of laminated composite plates under thermomechanical loading. *Int. J. Non-Linear Mech.* **2005**, *40*, 971–985. [CrossRef]
15. Whitney, J.M. *Structural Analysis of Laminated Anisotropic Plates*; Technomic Publishing Company, Inc.: Lancaster, PA, USA, 1987.
16. Reddy, J.N. *Energy Principles and Variational Methods in Applied Mechanics*; Wiley: New York, NY, USA, 2002.
17. Bert, C.W.; Jang, S.K.; Striz, A.G. Nonlinear bending analysis of orthotropic rectangular plates by the method of differential quadrature. *Comput. Mech.* **1989**, *5*, 217–226. [CrossRef]
18. Shu, C.; Du, H. Implementation of clamped and simply supported boundary conditions in the GDQ free vibration analyses of beams and plates. *Int. J. Solids Struct.* **1997**, *34*, 819–835. [CrossRef]
19. Hong, C.C. Advanced dynamic thermal vibration of thick FGM plates-cylindrical shells. *Ocean Eng.* **2022**, *266 Pt 1*, 112701. [CrossRef]
20. Brischetto, S.; Torre, R. 3D shell model for the thermo-mechanical analysis of FGM structures via imposed and calculated temperature profiles. *Aerosp. Sci. Technol.* **2019**, *85*, 125–149. [CrossRef]
21. Noda, N. Thermal stresses in functionally graded materials. *J. Therm. Stress.* **1999**, *22*, 477–512. [CrossRef]
22. Malikan, M.; Eremeyev, V.A. A new hyperbolic-polynomial higher-order elasticity theory for mechanics of thick FGM beams with imperfection in the material composition. *Compos. Struct.* **2020**, *249*, 112486. [CrossRef]
23. Dastjerdi, S.; Akgöz, B. New static and dynamic analyses of macro and nano FGM plates using exact three-dimensional elasticity in thermal environment. *Compos. Struct.* **2018**, *192*, 626–641. [CrossRef]
24. Conte, S.D.; de Boor, C. *Elementary Numerical Analysis an Algorithmic Approach*; McGraw-Hill Book Company: New York, NY, USA, 1980.

materials MDPI

Article

A Method for Straightening Distorted Giga-Cast Large Thin-Walled Components

Donghwi Park, Joonhee Park and Naksoo Kim *

Department of Mechanical Engineering, Sogang University, Seoul 04107, Republic of Korea;
pdhwi93@sogang.ac.kr (D.P.); yory11@sogang.ac.kr (J.P.)
* Correspondence: nskim@sogang.ac.kr; Tel.: +82-2-705-8635

Abstract: Giga-casting, a revolutionary approach for manufacturing large, single-piece car body components from aluminium, has emerged as a potential game-changer in the automotive industry. However, these large, thin-walled castings are prone to distortions during solidification and heat treatment processes. Straightening these distortions is crucial to ensure structural integrity, facilitate downstream assembly, and maintain aesthetic qualities. This paper proposes a novel method for straightening giga-cast components using a multi-pin straightening machine. The machine's versatility stems from its ability to adapt to various geometries through multiple strategically controlled straightening pins. This paper introduces the concept of a "straightening stroke decision algorithm" to achieve precise straightening and overcome the challenges of complex shapes. This algorithm determines the stroke length for each pin, combining a polynomial model representing the global stiffness of the component with a machine learning model that captures the stiffness changes arising from the current geometry. The effectiveness of the proposed approach is evaluated through comprehensive numerical experiments using finite element analyses. The straightening performance is assessed for the straightening algorithm with different machine learning models (deep neural network and XGBoost) and compared to a traditional optimisation method. The proposed surrogate models decided the straightening strokes so that the maximum remaining distortion became 0.02% of the largest dimension of each target geometry. The results of the numerical experiment showed that the proposed straightening method is suitable for straightening distortion in large thin-walled components.

Keywords: distortion straightening; thin-walled structures; die-casting; giga-casting; machine learning; finite element analysis

1. Introduction

The automotive industry witnessed a significant shift with the emergence of giga-casting, a revolutionary approach pioneered by Tesla [1]. This technology utilises colossal, high-pressure die-casting (HPDC) machines, aptly named Giga Presses, to manufacture large, single-piece chassis components from aluminium [2,3]. Prior to giga-casting, car bodies were traditionally constructed by welding together numerous smaller parts. By eliminating the need for extensive welding processes, giga-casting offers the possibility of significantly reduced production times and costs [4]. Additionally, using single-piece components can enhance structural integrity and improve vehicle performance metrics like weight and, potentially, fuel efficiency [5]. As the electric vehicle market continues its rapid expansion, giga-casting's ability to optimise production for lighter, more efficient car bodies positions it as a potentially disruptive force [6].

Despite the numerous advantages of giga-casting, the process has its challenges. One significant hurdle lies in the inherent tendency for these large, thin-walled components to warp and distort during solidification or subsequent heat treatment processes [7]. Several factors contribute to this phenomenon. The immense size of the castings creates uneven cooling rates across the part, leading to thermal stresses that can cause warping [8]. Additionally, the high pressures employed during casting and the potential for non-uniform heating during heat treatment can induce residual stresses within the material [9]. These combined effects can result in components that deviate from their intended geometry.

Straightening these distortions is crucial for several reasons. Firstly, deviations from the designed shape can compromise the structural integrity of the component [10]. Uneven load distribution and stress concentrations can arise due to warping, potentially leading to safety hazards. Secondly, distorted components can create challenges during downstream assembly processes. Misaligned components can require additional processing and can even lead to production line disruptions [11]. Finally, deviations from the intended form can negatively impact the aesthetic qualities of the final product.

Traditional straightening methods, which involve mechanical pressing, often rely on custom-designed die blocks that match specific sections of the component. These die blocks are expensive to manufacture and can only be used for a particular component geometry [12]. This dependence on custom tooling renders traditional methods incapable of addressing the general complex geometries encountered in giga-castings. The sheer variety of shapes and sizes within these large components would necessitate an extensive and costly collection of die blocks, making traditional straightening impractical for large-scale production. In addition, most of the straightening process has been traditionally carried out by workers' manual handwork [13], and a systematic approach for the straightening process has yet to be proposed.

Given the limitations of traditional techniques and the growing need for efficient straightening solutions for versatile geometries, this paper proposes a novel approach utilising a straightening machine with multiple straightening pins. This innovative method offers several advantages over conventional methods. Unlike die blocks, which require a custom design for each specific geometry, a straightening machine with multiple straightening pins can readily adapt to a wide range of complex shapes encountered in giga-castings. By strategically pressing the component with each straightening pin, this method allows the distortion to be removed after straightening. It enables the machine to address distortions in specific areas of the giga-casting without introducing new ones elsewhere, making it a highly versatile solution.

Machine learning has emerged as a powerful tool for tackling complex challenges in various mechanical engineering domains [14]. This study utilised two prominent algorithms, deep neural network (DNN) and XGBoost, to address the intricate problem of giga-casting straightening. With their ability to learn complex relationships from vast datasets [15], deep neural networks are well suited for capturing the non-linear interactions of geometric variations in giga-castings. XGBoost, a robust gradient boosting algorithm, excels in handling high-dimensional data [16] and modelling intricate feature interactions [17], making it suitable for predicting stiffness variations within the component. Compared to traditional methods that rely on simpler models or that lack the ability to adapt to complex geometries, incorporating DNN and XGBoost can enhance the straightening performance, leading to more accurate and efficient correction of distortions in these large, complex components.

Numerous researchers have successfully adopted machine learning algorithms for mechanical engineering problems. Salb et al. [18] proposed a method for enhancing IoT network security by combining CNNs for feature extraction with XGBoost for intrusion detection. They further introduced a modified Reptile Search algorithm for hyperparameter optimisation, leading to a more robust defence against emerging threats in IoT security. Park et al. [19] proposed a method for designing patterns in tubular robots that utilises deep neural networks to extract key features and metaheuristics optimisation to achieve desired mechanical properties, surpassing previous designs in performance and efficiency. Ref. [20] developed a system for the intelligent fault diagnosis of rotary machinery using a convolutional neural network (CNN) with automatic hyperparameter optimisation via Bayesian optimisation, achieving accurate fault detection without manual network configuration. Lin et al. [21] combined finite element simulation to generate data on self-piercing riveted joints and utilised the XGBoost algorithm to analyse it, achieving highly accurate predictions of their cross-tension strength with an impressive error rate of only 7.6%, which offered a significant advancement in predicting joint performance, potentially replacing traditional testing methods. Hashemi et al. [22] utilised machine learning to create surrogate finite element models. The surrogate finite element models could efficiently predict the dynamic response of mechanical systems, significantly reducing the time and resources needed compared to traditional full-scale finite element analysis.

This paper introduces the concept of a "straightening stroke decision algorithm" to achieve precise straightening and overcome the challenges associated with complex geometries. This innovative algorithm plays a crucial role in determining the optimal straightening stroke for each individual pin within a straightening machine. The algorithm utilises the following two key components: The first component is a polynomial model representing the global stiffness of the giga-casting component. This model provides a baseline understanding of the component's overall stiffness to the deformation. The second component is a machine learning model that captures the stiffness changes arising from the current geometry of the specific component being straightened. This model accounts for the deviations of the stiffness of the current geometry from the global stiffness. By combining these two models, the straightening stroke decision algorithm calculates the optimal stroke length required for each straightening pin. The following sections will delve deeper into the details of the straightening stroke decision algorithm, explaining the underlying models and their role in achieving efficient and accurate straightening of giga-cast components.

This paper presents the results of comprehensive numerical experiments to evaluate the effectiveness of the proposed straightening approach. Finite element analyses were employed to simulate the straightening process for various target component geometries. These target components included a simple box, a centre spine, and a side member, each representing different levels of geometric complexity encountered in giga-castings.

The straightening performance was meticulously assessed by comparing the following four different approaches: The first and second approaches are the straightening algorithms with the deep neural network (DNN) and the XGBoost [23] models. These models utilise machine learning algorithms to capture the intricate stiffness variations due to the geometry deviations. The third approach is the straightening algorithm with only the polynomial model for the global stiffness, which means that the model does not consider the effect of the geometry deviations. The last approach is the naive L-BFGS-B method. It serves as a benchmark representing a traditional optimisation technique commonly used for solving engineering problems. By comparing the straightening performance of these approaches across various component geometries, this paper aims to demonstrate the effectiveness of the proposed straightening stroke decision algorithm.

2. Distortion Straightening Method

2.1. Straightening Machine Concept and Straightening Process

The design of the straightening machine suggested in this study is described in Figure 1. The straightening machine has upper and lower dies and numerous straightening pin modules attached to the dies. The lower and upper dies are stationary. The number of straightening pin modules varies depending on the shape of the part to be straightened. The feature of the straightening pin module is that the tip of the straightening pin can move, which makes the machine able to handle various distortions. The component is straightened by pressing it using the strokes of the straightening pins. The movement of the straightening pin can be achieved electrically, pneumatically, hydraulically, or through any other mechanism that can manage the height with enough precision; however, the module system should have appropriate stiffness so that the height is not altered while pressing the part.

The shape of the straightening pin's tip can be any shape. It can be cylindrical, hemispherical, the local geometry of the part, or any shape suitable for straightening. It is recommended for the straightening pins to be located at the positions where the part has higher stiffness in the press direction. If a pin is located at a position with lower stiffness in the press direction, unwanted local deformation can be generated to the part during the straightening process.

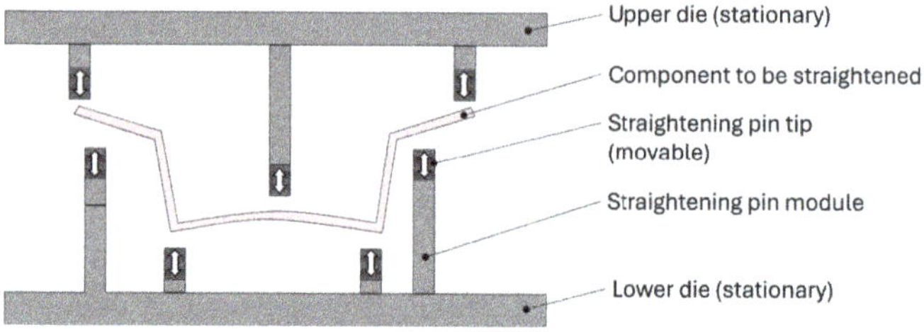

Figure 1. Schematic concept of the straightening machine suggested in this study and its straightening stroke.

The procedure of the straightening process is depicted in Figure 2 in the form of a flow chart. The straightening process aims to make the geometry of the distorted part as close as possible to the target geometry so that the dimensional tolerance is satisfied. The first procedure is to evaluate the distortion of the part by comparing the current geometry of the part with the target geometry. If the distortion satisfies the dimensional tolerance, the straightening does not need to be conducted, which means the process is finished. However, if the distortion is out of the dimensional tolerance, the distortion should be straightened. In this scenario, we should determine how much each pin should press the distorted part based on the measured distortion. Once the strokes of each pin are decided, the straightening stroke is conducted on each pin, and the distortion is straightened by the pins pressing the part. When the straightening stroke is finished, the process moves back to the distortion evaluation phase, and the straightening result is saved in the history database to improve the straightening process.

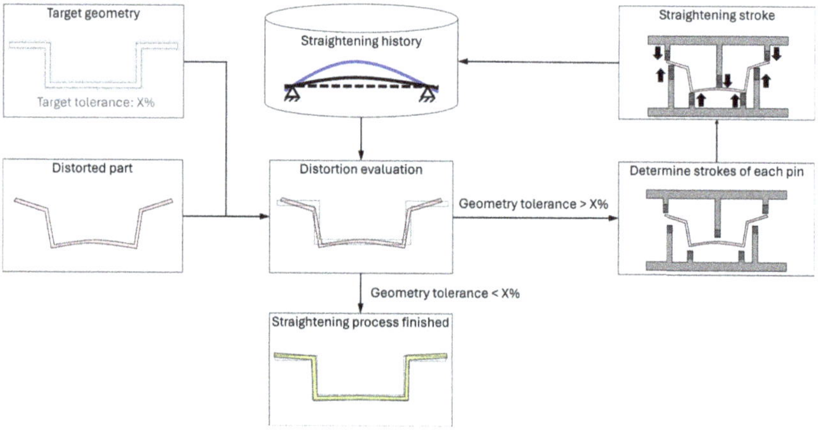

Figure 2. Flow chart of the straightening process.

2.2. *Quantification of Distortion*

The distortion of a component can be defined in various ways. In this study, distortion is defined and quantified in the following manner.

Consider a current geometry Ω and the target geometry Ω_t shown in Figure 3. There is a supporting fixture which prevents the rigid body motion of the geometries. From the measurement origin $\mathbf{O}_m$, there are intersection points $\mathbf{Q}$ and $\mathbf{Q}_t$ with Ω and Ω_t in the direction of measurement **q**. Let the distances from $\mathbf{O}_m$ to **Q** and $\mathbf{Q}_t$ be d and d_t, respectively, and define these distances as the distortion measurements. Based on this conceptualisation, in this paper, the distortion is quantified through the difference of the distortion measurements d and d_t as follows:

$$\delta_t = d_t - d. \tag{1}$$

δ_t represents how the current geometry should be corrected at point **Q** in the direction of **q**, which means that δ_t can represent how the current geometry is distorted from the target geometry.

The suggested distortion measurement can be easily measured in real situations. A tactile probe can be adopted as an example of a contacting measurement, and laser distance or displacement sensors are examples of non-contacting measurements. In addition, laser scanners could also be an example that rapidly measures the distortion over a wide range.

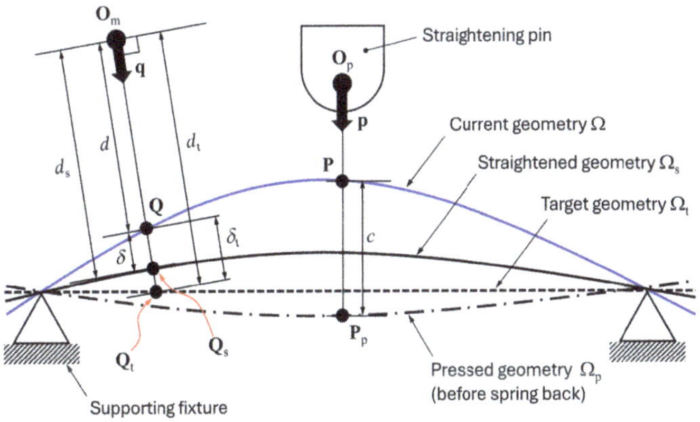

Figure 3. Schematic of a straightening stroke. The current geometry Ω is pressed by a straightening pin in the direction **p**, and the geometry becomes the pressed geometry Ω_p. After unloading, Ω_p becomes the straightened geometry Ω_s due to spring-back.

2.3. *Straightening Stroke Decision Algorithm*

The decision process of the pins' heights is one of the essential elements in the straightening process since the pins' heights determine how the part deforms. The straightening stroke decision algorithm proposed in this study decides how much the part should be pressed at the straightening pin locations. The following is a detailed description of the straightening stroke decision algorithm.

Consider a straightening stroke that a straightening pin presses the current geometry Ω from the origin $\mathbf{O}_p$ with a direction **p** until the geometry becomes the pressed geometry Ω_p as shown in Figure 3. From the initial contact point **P** between the pin and Ω, the pin moves the distance c until the contact point $\mathbf{P}_p$ of Ω_p. The pressed geometry Ω_p will be changed to the straightened geometry Ω_s after unloading due to spring-back. The straightening stroke decision algorithm aims to find the distance c that makes the straightened geometry Ω_s as close as possible to the target geometry Ω_t.

How the current geometry is corrected can be quantified using d_s, which is the distance between the measurement origin $\mathbf{O}_m$ and the intersection point $\mathbf{Q}_s$. δ, the correction amount at the measurement point **Q**, is defined as follows:

$$\delta = d_s - d. \tag{2}$$

δ means how the distortion measurement changed at **Q** after a straightening stroke. It can be said that the component is straightened correctly at the measurement point **Q** if the measured δ is close enough to δ_t, which is defined in Equation (1). We consider the component straightened when all δ are close enough to their corresponding δ_t since there are multiple straightening pins and measurement points. In other words, the algorithm aims to find the set of c values that makes all δ as close as possible to their corresponding δ_t.

The straightening stroke decision algorithm utilises a surrogate model that approximates the correction amount δ for the distance c based on the current geometry. Say that there are n measurement points and m straightening pins. In the surrogate model, at the i-th measurement point, the correction amount δ_i is modelled as an N-th order polynomial based on the assumption that δ_i is a function of the distance c and the current geometry. The N-th order polynomial for δ_i can be expressed as follows:

$$\delta_i = (d_s)_i - d_i = \sum_{j=1}^{N} \left(\sum_{j_1+\cdots+j_m=j} a_{j_1\cdots j_m} c_1^{j_1} \cdots c_m^{j_m} \right), \tag{3}$$

where c_k denotes the k-th pin movement distance, and a denotes a coefficient of the polynomial. The parameter j in Equation (3) starts from one since the constant term is neglected in the model. Equation (3) can be expressed as the following equation using vector notation:

$$\boldsymbol{\delta} = \mathbf{K}\mathbf{C}, \tag{4}$$

where $\boldsymbol{\delta}$ and $\mathbf{C}$ are the vectorised δ_i and polynomial combinations of c_k, respectively, and $\mathbf{K}$ is the matrix representing the polynomial coefficients. $\mathbf{K}$ is called the stiffness matrix in this study since it represents the "stiffness" of the current geometry. $\boldsymbol{\delta}$ has n components, while the number of components of $\mathbf{C}$ varies which depends on the order N. The number of components of $\mathbf{C}$ is calculated as follows:

$$\dim(\mathbf{C}) = \sum_{r=1}^{N} \frac{(m+r-1)!}{(m-1)!r!}, \tag{5}$$

which is the sum of the combinations with repetition of r objects from m objects.

The stiffness matrix $\mathbf{K}$ strongly depends on the current geometry. It is assumed that the stiffness matrix can be split into two terms, as follows, to consider the effect of the current geometry:

$$\mathbf{K} = \mathbf{K}_{\text{global}} + \Delta\mathbf{K}, \tag{6}$$

where $\mathbf{K}_{\text{global}}$ represents the global stiffness of the component, and $\Delta\mathbf{K}$ represents the change in the stiffness from the global stiffness for the current geometry. Figure 4 shows the schematic representation of the global stiffness matrix $\mathbf{K}_{\text{global}}$ and the current geometry's stiffness matrix $\mathbf{K}_{\text{global}} + \Delta\mathbf{K}$. The role of $\mathbf{K}_{\text{global}}$ is to represent the global stiffness of the part regardless of how the current geometry differs from the target geometry. Assuming that the material properties are constant, how the current geometry looks determines the change in the stiffness matrix, which means that the stiffness matrix of the current geometry can be represented as the sum of these two matrices.

The global stiffness matrix $\mathbf{K}_{\text{global}}$ is determined so that the following objective functions $f_1, \cdots, f_n$ are minimised:

$$f_i = \frac{1}{M}\sqrt{\sum_{p=1}^{M}\left[\delta_{i,p}^{\text{measured}} - \left(\mathbf{K}_{\text{global}}\mathbf{C}_p\right)_i\right]^2}, \tag{7}$$

where $i = 1, \cdots, n$ is the distortion measurement number, $p = 1, \cdots, M$ is the observation number, and the superscript "measured" denotes the measured value.

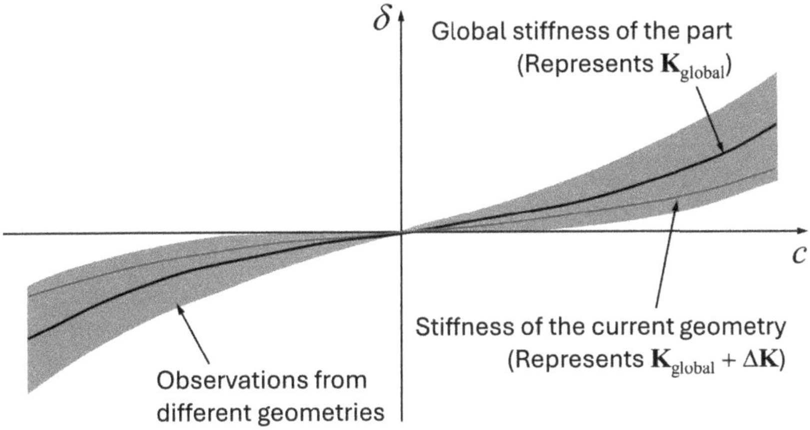

Figure 4. Schematic representation of the global stiffness matrix and the stiffness matrix of the current geometry.

$\Delta\mathbf{K}$, the change in the stiffness matrix, is modelled as a function of the distortion measurements of the current geometry since the distortion measurements can represent the current geometry. We adopted machine learning algorithms for modelling $\Delta\mathbf{K}$ since the correlation between the current geometry and $\Delta\mathbf{K}$ might be highly complex. The objective functions $g_1, \cdots, g_n$ to be minimised for the training of the machine learning models are

$$g_i = \frac{1}{M}\sqrt{\sum_{p=1}^{M}\left[\delta_{i,p}^{\text{measured}} - \left(\left\{\mathbf{K}_{\text{global}} + \Delta\mathbf{K}\right\}\mathbf{C}_p\right)_i\right]^2}, \tag{8}$$

where g_i is the i-th objective function of the i-th machine learning model which is for an i-th row of $\Delta\mathbf{K}$. Note that $\mathbf{K}_{\text{global}}$ is fixed during the training of the machine learning algorithm. Two machine learning algorithms were adopted to model $\Delta\mathbf{K}$, and a detailed description of these machine learning models is in Section 3.3.

The flow chart of the straightening stroke decision algorithm is depicted in Figure 5. The algorithm starts with input $\mathbf{d}$ and the vectorised current distortion measurement d_i. The algorithm determines $\Delta\mathbf{K}$ from $\mathbf{d}$ with the machine learning model and calculates the current stiffness matrix $\mathbf{K}$. $\mathbf{c}_{\text{opt}}$ is the optimum $\mathbf{c}$, where $\mathbf{c}$ is the vectorised pressure distance c_k, and the subscript "opt" denotes the optimum value. $\mathbf{c}_{\text{opt}}$ is determined through an optimisation process with the following objective function f_c:

$$f_c = \max\left(\left|\boldsymbol{\delta}^{\text{pred}} - \boldsymbol{\delta}_{\text{t}}\right|\right) = \max(|\mathbf{KC} - \boldsymbol{\delta}_{\text{t}}|), \tag{9}$$

where $\boldsymbol{\delta}_{\text{t}}$ is vectorised $(\delta_{\text{t}})_i$, and the superscript "pred" denotes the predicted value from the surrogate model. Plenty of optimisation algorithms can be used for the optimisation process of f_c. In this paper, L-BFGS-B [24] is utilised for the optimisation. The straightening stroke is conducted with $\mathbf{c}_{\text{opt}}$, and the distortion measurements of the straightened geometry come into the algorithm. The straightening result is saved to the straightening database. $\mathbf{K}_{\text{global}}$ and the machine learning model for $\Delta\mathbf{K}$ are updated with the updated database, and the algorithm ends. The proposed algorithm can be considered a self-learning algorithm since it can improve its performance by itself.

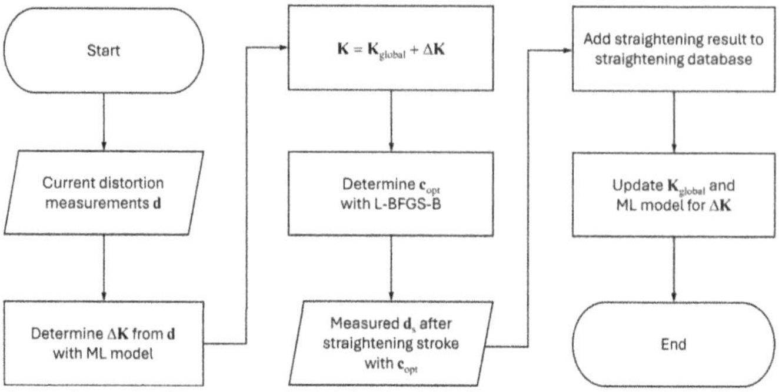

Figure 5. Flow chart of the straightening stroke decision algorithm. ML is an abbreviation of machine learning.

3. Numerical Experiments

3.1. Target Components, Straightening Pins, and Distortion Measurement Locations

Three thin-walled components, shown in Figure 6, were adopted to show the performance of the proposed distortion straightening method. The names of the components are "simple box", "centre spine", and "side member" in the sequence of Figure 6a–c. The simple box is a box shape with ribs, which was chosen to represent the very simple geometry of a thin-walled die-cast component. Meanwhile, the centre spine and the side member geometries are from the currently manufactured car frames using the die-cast process.

As the size of the component becomes larger, the possibility becomes higher that the component has a substantial amount of distortion. As a result, the straightening process is likely to be applied to large thin-walled structures; therefore, the die-cast products with the largest dimension above 1200 mm were chosen as the target components. The simple box shape has the largest dimension of 1200 mm and the most straightforward geometry among the target components, so one would expect it to be the easiest part to straighten. The centre spine has the largest dimension of 1400 mm. The centre spine is expected to have intermediate difficulty since the height difference is less than the side member and has symmetric geometry. The side member could be the most challenging component among the target components because it has no symmetric geometry, and the height difference is higher than the two other geometries.

The locations of the straightening pins and the distortion measurements are also depicted in Figure 6. Blue markers indicate the datum points for the distortion measurement. Red and yellow markers indicate the locations of the straightening pins and distortion measurements, respectively. In this study, we only considered cases in which straightening pin locations and distortion measurement locations were identical. On each pin location, two pins were located on the front and back sides so that the components could be pressed from both sides. The distortion measurement direction $\mathbf{q}$ and the press direction $\mathbf{q}$ were parallel to the direction of view of the diagrams describing the straightening pin and distortion measurement locations in Figure 6 for all components. The locations were chosen at the point of intersection of the ribs, which had higher stiffness, and the distance between the straightening pins was greater than 80 mm so as to not make them too close to each other. The number of locations was 12 for the simple box, 16 for the centre spine, and 18 for the side member.

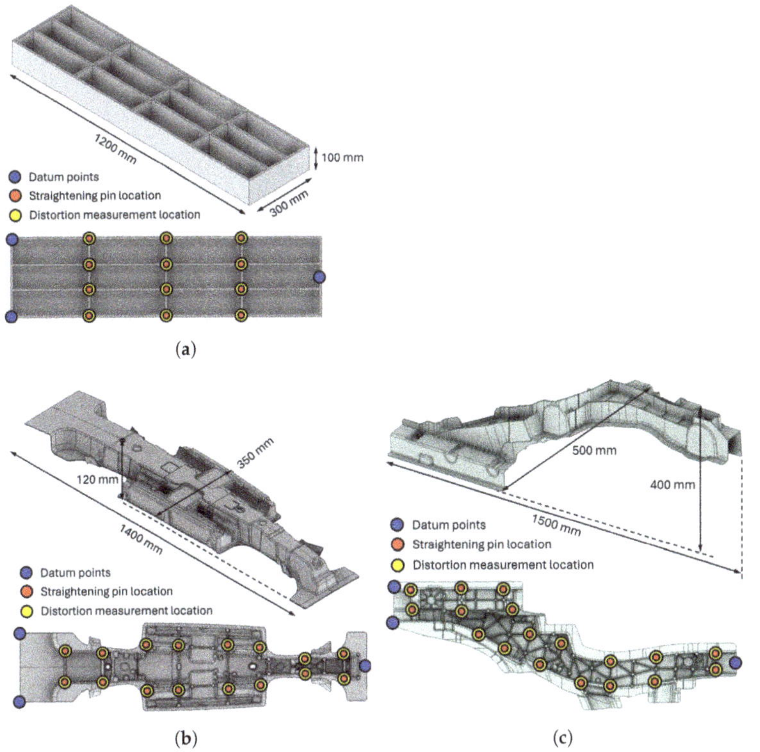

Figure 6. Geometries of the components adopted for the numerical experiments. (**a**) Simple box. (**b**) Centre spine. (**c**) Side member.

3.2. Numerical Experiment Procedure

The numerical experiment procedure is depicted in Figure 7 as a flow chart. The first procedure is to generate a straightening database for the surrogate model. The database has 1500 straightening data points, which consist of 15 random straightening strokes of each of 100 randomly distorted geometries. The random distortion of the geometry was generated by pressing the part randomly with the straightening pins. Each randomly distorted geometry's maximum absolute distortion measurement ranges from 6 mm to 9 mm. Once the database is constructed, $\mathbf{K}_{\text{global}}$ is fitted, and the machine learning model for $\Delta\mathbf{K}$ is trained with the fitted $\mathbf{K}_{\text{global}}$. The polynomial order for $\mathbf{K}_{\text{global}}$ was chosen as $N = 3$. The number of data points in the database was chosen as 1500 since the minimum required number of data points for 18 straightening pins with a third-order polynomial model is 1329 according to Equation (5).

The performance of the constructed surrogate model was evaluated by comparing the straightening results with naive L-BFGS-B as shown in the right side of the flow chart in Figure 7. The L-BFGS-B algorithm is a quasi-Newton algorithm that aims to minimise a scalar function f. For a given $\mathbf{x}_k$, the position at the k-th iteration, the quantities are defined as follows:

$$\mathbf{s}_k = \mathbf{x}_{k+1} - \mathbf{x}_k, \tag{10}$$

$$\mathbf{y}_k = \mathbf{g}_{k+1} - \mathbf{g}_k, \tag{11}$$

$$\rho_k = \frac{1}{\mathbf{y}_k^T \mathbf{s}_k}, \tag{12}$$

where $\mathbf{g}_k = \nabla f(\mathbf{x}_k)$ denotes the gradient of f at $\mathbf{x}_k$. The algorithm evaluates the approximate Newton's direction $\mathbf{d}_k$ as follows:

$$\mathbf{d}_k = -\mathbf{H}_k\mathbf{g}_k, \tag{13}$$

where $\mathbf{H}_k$ is the inverse of the Hessian matrix. At the k-th iteration, the algorithm finds the step length which minimises the function f in the direction of $\mathbf{d}_k$. For the next iteration, the update of the inverse of the Hessian matrix is conducted as follows:

$$\mathbf{H}_{k+1} = \left(\mathbf{I} - \rho_k\mathbf{s}_k\mathbf{y}_k^T\right)\mathbf{H}_k\left(\mathbf{I} - \rho_k\mathbf{y}_k\mathbf{s}_k^T\right) + \rho_k\mathbf{s}_k\mathbf{s}_k^T, \tag{14}$$

where $\mathbf{I}$ denotes the identity matrix. Naive L-BFGS-B means conducting the L-BFGS-B algorithm to find the optimum pressing distances $\mathbf{c}_{\text{opt}}$ by changing the pressing distances in a finite element model, which can be a computationally costly and ineffective method.

Ten randomly distorted geometries, which are not included in the straightening database, were generated and utilised for the performance evaluation of the proposed straightening stroke decision algorithm. The geometries for the performance evaluation also have maximum absolute distortion measurement ranges from 6 mm to 9 mm, which is the same as the database.

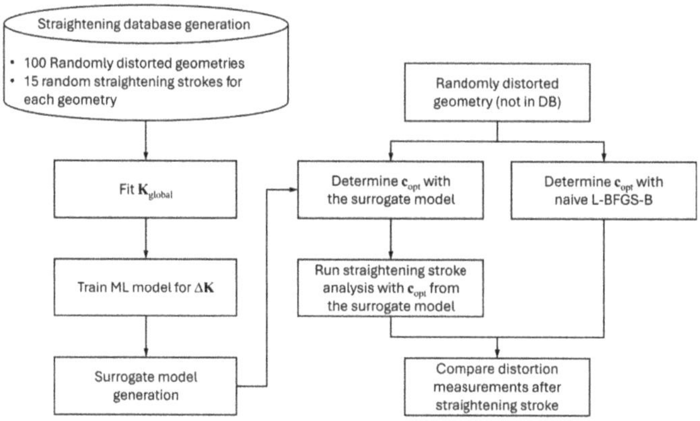

Figure 7. Flow chart of the numerical experiment procedure.

The finite element models were established to conduct the simulations to make the database, conduct the straightening stroke with $\mathbf{c}_{\text{opt}}$ calculated by the surrogate model, and conduct the naive L-BFGS-B. The finite element models of the simple box, the centre spine, and the side member are depicted in Figure 8. As shown in the figures and mentioned in Section 3.1, there are pairs of straightening pins at each straightening pin location. The finite element models were developed using the commercial finite element analysis software Abaqus 2020 HF6. The simple box model (Figure 8a) has 4680 nodes and 4655 linear quadrilateral shell with reduced integration (S4R) elements. The centre spine model (Figure 8b) has 235,138 nodes and 767,196 linear tetrahedral (C3D4) elements. The side member model (Figure 8c) has 48,182 nodes and 159,466 C3D4 elements. The straightening pins are set as a rigid body.

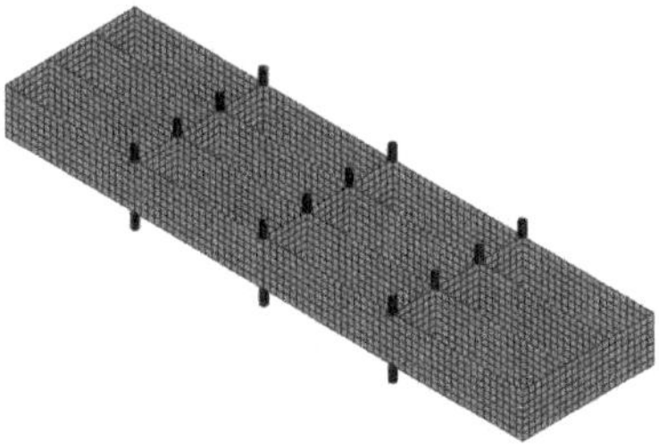

(a)

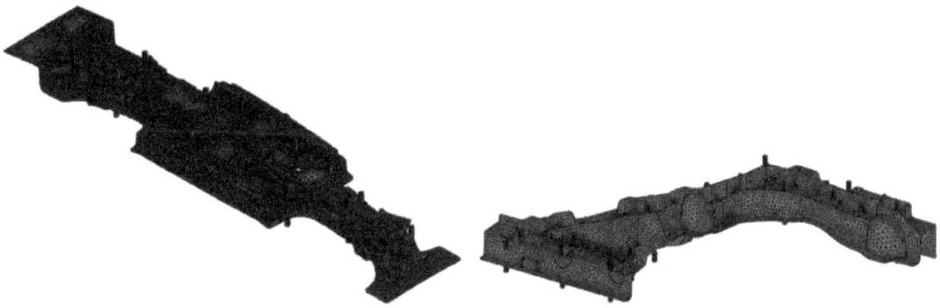

(b) (c)

Figure 8. Finite element models of the target components. Each straightening pin location has a pair of straightening pins. (**a**) Simple box. (**b**) Centre spine. (**c**) Side member.

3.3. Machine Learning Models for $\Delta\mathbf{K}$

As mentioned in Section 2.3, $\Delta\mathbf{K}$, the change in the stiffness matrix, was modelled using machine learning models. Two machine learning models, a deep neural network (DNN) model and XGBoost [23] model, were adopted to model $\Delta\mathbf{K}$. The DNN and XGBoost were chosen since these are known to have satisfactory performance on regression problems [22,25].

The architecture of the developed DNN model is described in Figure 9. The DNN model was established with Keras [26] on Python 3.9. The input of the DNN model is the current distortion measurements **d**, and the output is a row of $\Delta\mathbf{K}$. There are three hidden layers with sizes of $2 \times \dim(\mathbf{C})$, $3 \times \dim(\mathbf{C})$, and $2 \times \dim(\mathbf{C})$. The number of hidden layers was chosen as three based on the work of Asghari et al. [27] since the DNN model with three hidden layers showed fine performance on the regression problem. The rectified linear unit (ReLU) activation function [28] was adopted for the hidden layers. The ReLU function is defined as follows:

$$f(x) = \max(0, x). \tag{15}$$

The training was conducted using Adam optimiser [29] with a learning rate of 0.0001 and a batch size of 100. The number of epochs was set as 1000. The loss function for the training is Equation (8).

XGBoost is well suited for regression problems due to its inherent design and capabilities. Its core functionality revolves around building an ensemble of decision trees, each focusing on correcting the errors of its predecessors. This sequential learning approach allows XGBoost to effectively capture complex relationships within data and accurately predict continuous numerical values, making it a powerful tool for regression problems. The XGBoost model for $\Delta\mathbf{K}$ has the same input and output as the DNN model. The XGBoost model utilised in this study adopted L1 and L2 regularisation terms, which were added to the loss function. The L1 and L2 regularisation terms are defined as follows:

$$L1 = \alpha \sum_{j=1}^{T} |w_j|, \tag{16}$$

$$L2 = \frac{1}{2}\lambda \sum_{j=1}^{T} w_j^2, \tag{17}$$

where T is the number of leaves in each tree, w_j is the weight at the leaves, and α and λ are the controlling parameters for the L1 and L2 regularisation terms, respectively. The model's parameters are as follows: the tree's maximum depth is 8, the L1 regularisation term on weights is 0.5, the L2 regularisation term on weights is 0.3, the number of estimators is 3000, and the learning rate is 0.01. The hyperparameters were obtained using the package Hyperopt [30].

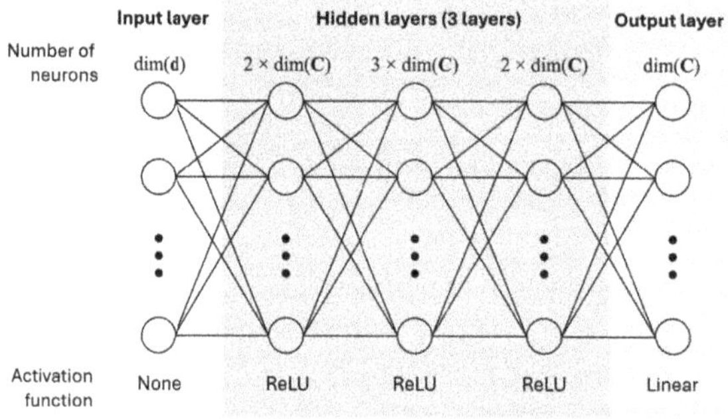

Figure 9. The deep neural network (DNN) architecture for modelling $\Delta\mathbf{K}$.

The XGBoost and DNN models were trained using the same straightening database, and the surrogate models from each machine learning model were generated. The database was partitioned for training, validation, and testing datasets. The ratios of the datasets were 64%, 16%, and 20% for training, validation, and testing, respectively. Table 1 shows the average loss of training, validation, and test datasets of the trained DNN and XGBoost models for each target geometry, where the loss is defined in Equation (8). The average losses show that the trained models can predict the change in the stiffness matrix without loss of generality.

Table 1. Average loss of training, validation, and test datasets of the trained DNN and XGBoost models for each target geometry.

						Unit: 10^{-3} mm
	Simple Box		Centre Spine		Side Member	
Dataset	DNN	XGBoost	DNN	XGBoost	DNN	XGBoost
Training	0.4715	0.4648	7.0851	7.1964	8.4046	8.4928
Validation	0.4918	0.4937	7.0366	7.2356	8.5231	8.7319
Test	0.4779	0.4783	7.2750	7.2438	8.5846	8.5119

4. Results and Discussion

The numerical experiments were conducted for the three target components in Figure 6 as mentioned in the previous section. For each target component, 1500 straightening strokes were simulated, and $\mathbf{K}_{\text{global}}$ and the machine learning models for $\Delta\mathbf{K}$ were trained as shown in the left side of Figure 7. Ten randomly distorted geometries, which are not included in the training database, were generated for each component to evaluate the performance of the proposed straightening stroke decision algorithm.

The performances of the proposed algorithm with the machine learning model for $\Delta\mathbf{K}$ and without it (which means only $\mathbf{K}_{\text{global}}$ was considered) were compared. The DNN model was selected for the performance comparison with the algorithm with $\mathbf{K}_{\text{global}}$ only.

The maximum distortion measurements of the simple box, after the straightening stroke decided using only $\mathbf{K}_{\text{global}}$ and the surrogate model with the DNN model, are depicted in Figure 10. For all ten randomly distorted shapes, the surrogate model with the DNN model showed better performance than the results that considered $\mathbf{K}_{\text{global}}$ only as the maximum distortion measurements of the straightened geometries were low with the surrogate model for all shapes. The average of the maximum distortion measurements after the straightening stroke was 0.0935 mm for the model with $\mathbf{K}_{\text{global}}$ only and 0.0751 mm for the surrogate model with the DNN model. The distortion measurement distributions of the simple box Shape no. 10 after the straightening strokes are presented on the right side of Figure 10. The distortion measurement distributions were considerably different since the model that considered $\mathbf{K}_{\text{global}}$ only could not take account of the change in the stiffness due to the geometry. In addition, this made the straightening stroke determined with only $\mathbf{K}_{\text{global}}$ less effective than the one determined by the surrogate model with the DNN model.

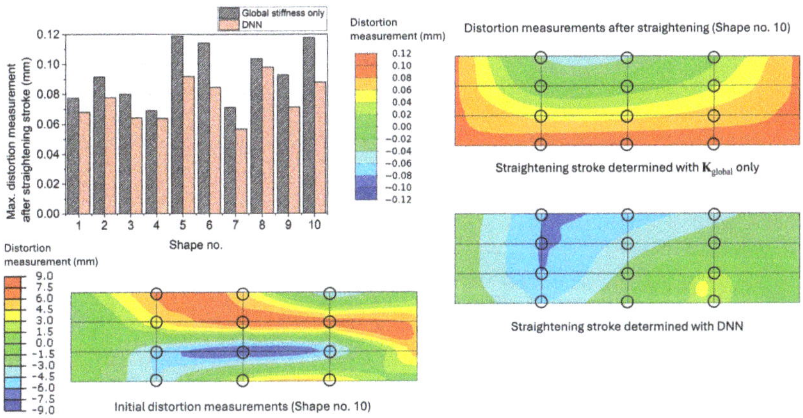

Figure 10. Maximum distortion measurements of the simple box after the straightening stroke decided using the global stiffness only and the surrogate model with the DNN model. The distortion measurement distributions of the initial and straightened geometries of Shape no. 10 are presented as an example. The black circles denote the locations of the straightening pins and distortion measurement.

Figure 11 shows the maximum distortion measurements of the centre spine after the straightening stroke determined with only $\mathbf{K}_{\text{global}}$ and the surrogate model with the DNN model. The average of the maximum distortion measurements after the straightening stroke was 0.1156 mm for the model with $\mathbf{K}_{\text{global}}$ only and 0.0909 mm for the surrogate model with the DNN model. Similar to the simple box cases, the straightening strokes determined with the DNN model were more effective for all ten shapes. The magnitude of the distortion measurements for the straightened geometries was higher than the simple box since the centre spine has a complex geometry, and the number of straightening pins to determine is greater than for the simple box. On the right side of Figure 11, the distortion measurement distributions of the centre spine Shape no. 8 after the straightening are presented. Similar to the simple box, the overall distributions were different from each other, and the DNN model showed better performance than the model with only $\mathbf{K}_{\text{global}}$.

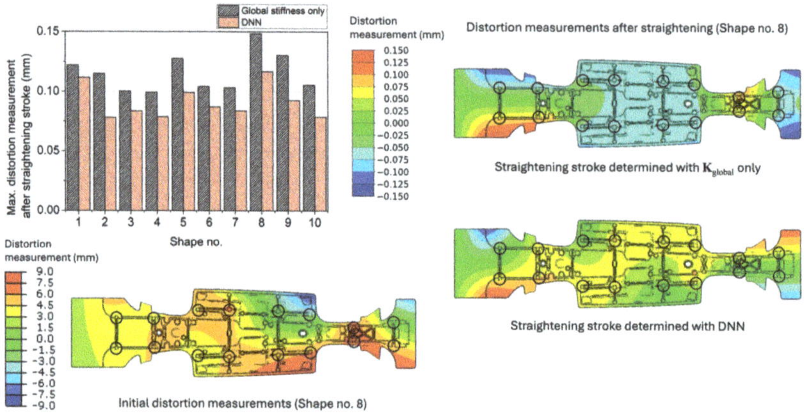

Figure 11. Maximum distortion measurements of the centre spine after the straightening stroke decided using the global stiffness only and the surrogate model with the DNN model. The distortion measurement distributions of the initial and straightened geometries of Shape no. 8 are presented as an example. The black circles denote the locations of the straightening pins and distortion measurement.

Similar results were derived for the side member as shown in Figure 12. The average of the maximum distortion measurements after the straightening stroke was 0.1550 mm for the model with $\mathbf{K}_{global}$ only and 0.1192 mm for the surrogate model with the DNN model. The side member showed the highest magnitude of the distortion measurements for the straightened geometries since the geometry's size and complexity were the most significant among the target geometries. The distortion measurement distributions of the side member Shape no. 1 are presented on the right side of Figure 12. As shown in the simple box and the centre spine cases, the distortion measurement distributions were significantly different from each other since there is a difference in whether the current geometry was considered or not.

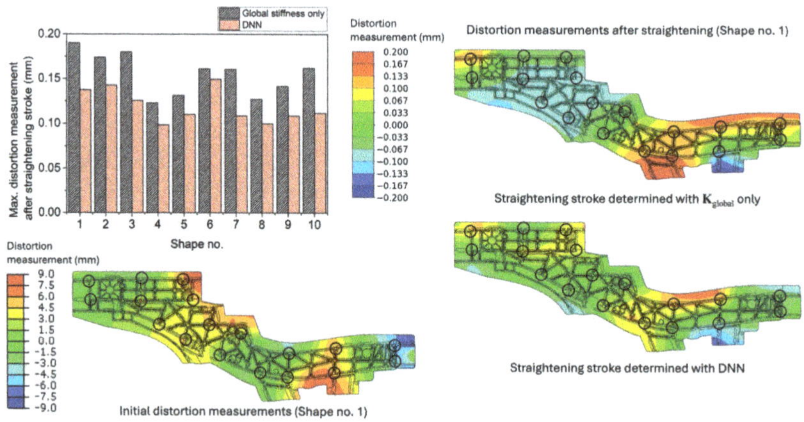

Figure 12. Maximum distortion measurements of the side member after the straightening stroke decided using the global stiffness only and the surrogate model with the DNN model. The distortion measurement distributions of the initial and straightened geometries of Shape no. 1 are presented as an example. The black circles denote the locations of the straightening pins and distortion measurement.

The straightening results derived from the model with only $\mathbf{K}_{global}$ and from the surrogate model with the DNN model showed that the surrogate model was superior to the model with only $\mathbf{K}_{global}$. Considering the current geometry's effect on the component's stiffness is crucial since the surrogate model provided more satisfactory straightening results from all target geometries. The surrogate model showed that the average of the maximum distortion measurements after the straightening stroke improved by 21.4% for all target geometries compared with the model with only $\mathbf{K}_{global}$. However, it should be noted that establishing the global stiffness matrix $\mathbf{K}_{global}$ is also an essential part of the surrogate model since the model is also based on global stiffness. An improper global stiffness matrix will make the algorithm decide the straightening stroke inappropriately, whether the effect of the current geometry is considered or not.

The straightening stroke determined from the naive L-BFGS-B mentioned in Section 3.2 could be treated as a local optimum solution since the naive L-BFGS-B does not utilise any surrogate model and attempts to find the optimal point from the finite element analysis results directly. In this study, the straightening results obtained from the naive L-BFGS-B were considered global optimum solutions because the search space was too broad to conduct a global optimisation process without a surrogate model. The performances of the surrogate models with the DNN model and with the XGBoost model were evaluated by comparing their results with those of the naive L-BFGS-B.

Figure 13 shows the maximum distortion measurements of the simple box after the straightening stroke determined using the naive L-BFGS-B and the surrogate models with the DNN model and with the XGBoost model. The average of the maximum distortion measurements after the straightening stroke was 0.0609 mm for the naive L-BFGS-B and 0.0763 mm for the surrogate model with the XGBoost model. The DNN model showed slightly better performance than the XGBoost model based on the straightening results of the simple box. The distortion measurement distributions of the simple box Shape no. 8 are presented on the right side of Figure 13. The surrogate models showed similar distortion measurement distributions, whilst the distribution of the naive L-BFGS-B was dissimilar from that of the surrogate models.

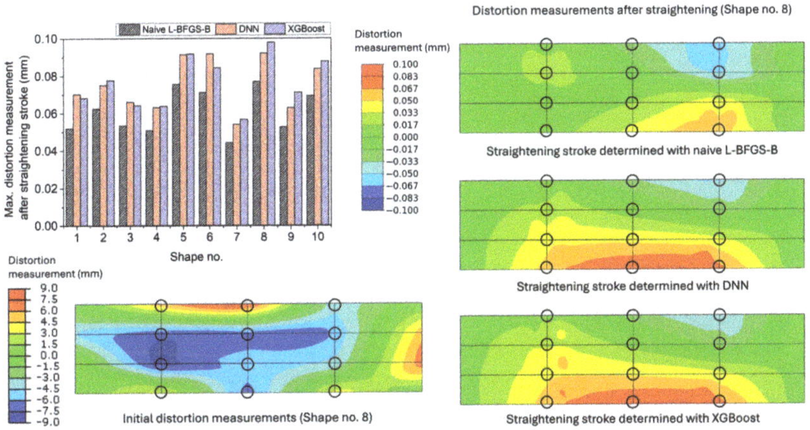

Figure 13. Maximum distortion measurements of the simple box after the straightening stroke decided using the naive L-BFGS-B and the surrogate models with the DNN model and with the XGBoost model. The distortion measurement distributions of the initial and straightened geometries of Shape no. 8 are presented as an example. The black circles denote the locations of the straightening pins and distortion measurement.

The maximum distortion measurements of the centre spine, after the straightening stroke decided using the naive L-BFGS-B and the surrogate models with the DNN model and with the XGBoost model, are shown in Figure 14. The average of the maximum distortion measurements after the straightening stroke was 0.0720 mm for the naive L-BFGS-B and 0.0911 mm for the surrogate model with the XGBoost model. The DNN model performance was slightly better than that of the XGBoost model. However, the difference in the performance between the surrogate models was negligible. The distortion measurement distributions of the centre spine Shape no. 1 are shown on the right side of Figure 14. Similar to the simple box results, the distortion measurement distributions of the DNN and the XGBoost model were similar to each other. In contrast, the naive L-BFGS-B showed a different pattern of distribution.

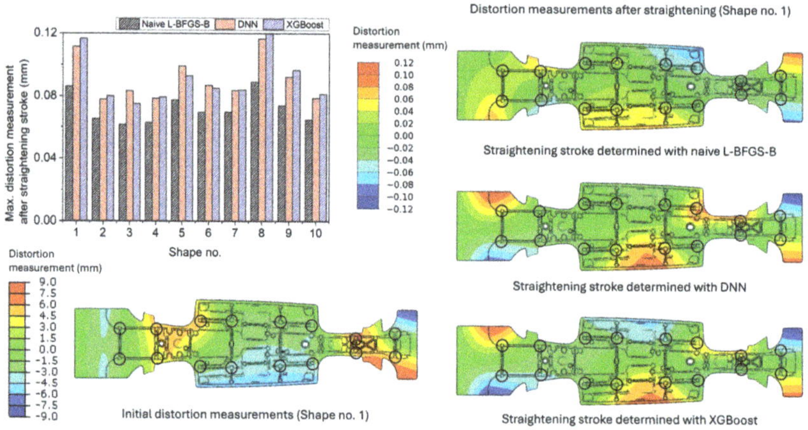

Figure 14. Maximum distortion measurements of the centre spine after the straightening stroke decided using the naive L-BFGS-B and the surrogate models with the DNN model and with the XGBoost model. The distortion measurement distributions of the initial and straightened geometries of Shape no. 1 are presented as an example. The black circles denote the locations of the straightening pins and distortion measurement.

The side member's maximum distortion measurements after the straightening stroke decided using the naive L-BFGS-B and the surrogate models with the DNN model and with the XGBoost model are shown in Figure 15. The average of the maximum distortion measurements after the straightening stroke was 0.0886 mm for the naive L-BFGS-B and 0.1205 mm for the surrogate model with the XGBoost model. The magnitude of the distortion measurements was the largest among the target geometries for the same reason discussed above. Like the centre spine, the DNN model showed slightly better performance than the XGBoost model. The distortion measurement distributions of the side member Shape no. 2 are depicted on the right side of Figure 15. The straightening behaviour was similar between the DNN and XGBoost models, whereas the naive L-BFGS-B showed different behaviour.

The maximum distortion measurements after the straightening stroke for each method and model are organised in Appendix A. Student's t-tests were conducted to evaluate the efficiency of each proposed model for the straightening process. The naive L-BFGS-B was excluded from the tests since it was used to make reference results for each target geometry because the naive L-BFGS-B can be considered to determine the optimum straightening stroke. In the statistical analysis, the significance level was set to be $\alpha = 0.05$ and the standard deviations were assumed to be different.

Table 2 shows the p-values of the two-tailed tests with the null hypothesis of the same mean for the surrogate models with the DNN and XGBoost. In Table 2, H_0 is the null hypothesis, and μ_1 and μ_2 are the means of the maximum distortion measurements after the straightening stroke for the surrogate models with the DNN and XGBoost, respectively.

For all geometries, the p-values are greater than the significance level α, which means that the null hypothesis should not be rejected and the means of the maximum distortion measurements were statistically identical.

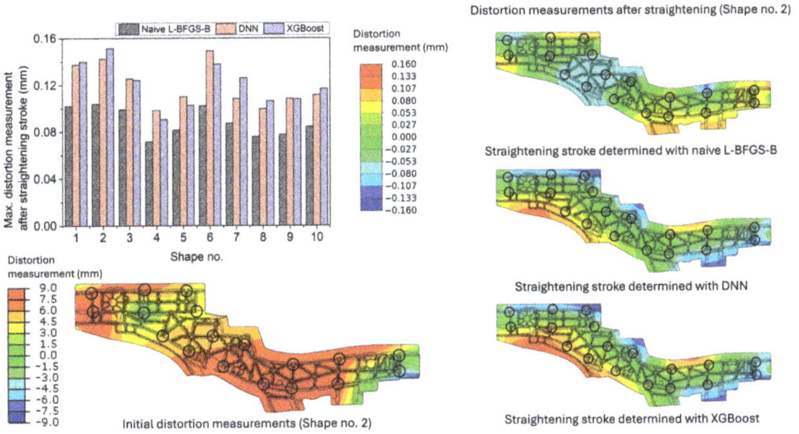

Figure 15. Maximum distortion measurements of the side member after the straightening stroke decided using the naive L-BFGS-B and the surrogate models with the DNN model and with the XGBoost model. The distortion measurement distributions of the initial and straightened geometries of Shape no. 2 are presented as an example. The black circles denote the locations of the straightening pins and distortion measurement.

Table 2. p-values of the two-tailed tests with the null hypothesis of the same mean for the surrogate models with the DNN and XGBoost.

H_0^a: $\mu_1 = \mu_2$			
	Simple Box	**Centre Spine**	**Side Member**
p-value	0.8372	0.9718	0.8752

Table 3 shows the p-values of the left-tailed tests with the null hypothesis of H_0^a: $\mu_1 \geq \mu_2$, where μ_1 and mu_2 are the means of the maximum distortion measurements after the strengthening stroke for the surrogate model with the DNN and the model with the global stiffness only, respectively. The p-values are less than the significance level for all geometries, which means the null hypothesis should be rejected. It shows that the straightening performances were statistically better with the DNN model since the mean is smaller for the surrogate model with the DNN model.

Table 3. p-values of the left-tailed tests with the null hypothesis of H_0^b: $\mu_1 \geq \mu_2$, where μ_1 and mu_2 are the means of the maximum distortion measurements after the strengthening stroke for the surrogate model with the DNN and the model with the global stiffness only, respectively.

H_0^b: $\mu_1 \geq \mu_2$			
	Simple Box	**Centre Spine**	**Side Member**
p-value	0.0123	0.0009	0.0007

The surrogate model with global stiffness could only be used for straightening. However, the straightening performance can be improved when the DNN and XGBoost models are utilised in the surrogate model with the global stiffness matrix, and the statistical analysis results support this. In addition, the maximum distortion measurements after the straightening stroke were reduced by 26.6% on average when the DNN or XGBoost model was adopted in the surrogate model.

Compared to the straightening results from naive L-BFGS-B, the maximum distortion measurements after the straightening stroke from DNN and XGBoost models differed by 28.0% and 29.3%, respectively, on average. The surrogate models with the DNN model and with the XGBoost model showed similar performances for the target geometries, while the DNN model showed slightly better performance on average. On the other hand, it cannot be said that the DNN model is always preferable to the XGBoost model since the DNN model did not show better straightening performance for all cases. In addition, the statistical analysis results indicate that the performances of the DNN and XGBoost models were statistically identical. The straightening results showed that the DNN and XGBoost models are adequate to model $\Delta\mathbf{K}$, the change in the component's stiffness. In addition, the surrogate models could decide the straightening strokes so that the maximum distortion after the straightening becomes 0.02% of the largest dimension of each target geometry.

The distortion measurement distributions from the surrogate models were similar for all target components, while the distribution from the naive L-BFGS-B was different. This shows that the DNN and XGBoost models estimated $\Delta\mathbf{K}$ similarly, whilst the naive L-BFGS-B estimated it directly from the finite element analyses. The estimated stiffness matrix can differ from the real stiffness matrix of the current geometry. However, the naive L-BFGS-B is computationally expensive, which means that it cannot be utilised to operate the straightening process. On the contrary, the surrogate model can provide admissible solutions for the straightening process with an adequately fitted $\mathbf{K}_{\text{global}}$ and a trained machine learning model.

5. Conclusions

This paper proposes a method for straightening distorted giga-cast thin-walled components. The straightening process, which corrects the distorted part by pressing it with numerous straightening pins, is described, and the concept of a straightening machine to carry out the process is suggested. An algorithm which decides the straightening stroke is suggested. The straightening stroke decision algorithm has two primary models: the polynomial model representing the component's global stiffness and the machine learning model for the change in the stiffness due to the distorted geometry. Two machine learning models, DNN and XGBoost, were selected to model $\Delta\mathbf{K}$. The numerical experiments, which make distortion and straighten the component with the finite element analyses, were conducted to evaluate the performance of the proposed algorithm. Three components were adopted for the numerical experiments: the simple box, centre spine, and side member. From the results of the numerical experiment, the conclusions can be summarised as follows:

- The effect of the current geometry should be considered when deciding the straightening stroke to make the straightening strokes effective. The global stiffness matrix can be used alone for straightening, but the determined straightening strokes might be less effective.
- The proposed DNN and XGBoost models adequately modelled the stiffness change due to the current geometry. Their performances on the straightening stroke decision were similar; meanwhile, the DNN model showed slightly better performance.
- Regarding the straightening results from the naive L-BFGS-B as the optimum, the maximum distortion measurements after the straightening strokes decided from the surrogate models differed by 28.7% on average. The surrogate models could provide admissible straightening stroke results, considering that the naive L-BFGS-B process is computationally expensive, which means that it is not appropriate to operate the straightening process.
- The surrogate models decided the straightening strokes so that the maximum remaining distortion became 0.02% of the largest dimension of each target geometry for all

target components. The suggested straightening method and algorithm were suitable for straightening distorted large thin-walled components.

However, the proposed straightening method was only examined using numerical experiments. As future work, we suggest developing a real straightening machine and applying the straightening process to real large-sized thin-walled die-cast components to examine the proposed method for actual application. In addition, we only considered cases in which straightening pin locations and distortion measurement locations were identical. In future work, cases with different straightening pin locations and distortion measurements need to be considered and analysed.

Author Contributions: Conceptualisation, N.K.; methodology, D.P. and J.P.; software, D.P. and J.P.; validation, D.P. and J.P.; formal analysis, D.P. and J.P.; investigation, D.P. and J.P.; resources, N.K.; data curation, D.P. and J.P.; writing—original draft preparation, D.P.; writing—review and editing, J.P. and N.K.; visualisation, D.P.; supervision, N.K.; project administration, N.K.; funding acquisition, N.K. All authors have read and agreed to the published version of the manuscript.

Funding: This research received no external funding.

Institutional Review Board Statement: Not applicable.

Informed Consent Statement: Not applicable.

Data Availability Statement: Data are contained within the article.

Conflicts of Interest: The authors declare no conflicts of interest.

Appendix A. Tables for Maximum Distortion Measurements after the Straightening Stroke

The following are the tables of maximum distortion measurements after the straightening stroke decided using the naive L-BFGS-B, the global stiffness only, and the surrogate models with the DNN and XGBoost models.

Table A1. Maximum distortion measurements after the straightening stroke for the simple box.

				Unit: mm
Shape No.	**Naive L-BFGS-B**	**K_{global} Only**	**DNN**	**XGBoost**
1	0.0520	0.0774	0.0702	0.0681
2	0.0625	0.0917	0.0752	0.0778
3	0.0534	0.0800	0.0660	0.0642
4	0.0507	0.0690	0.0631	0.0637
5	0.0756	0.1185	0.0914	0.0918
6	0.0712	0.1140	0.0918	0.0844
7	0.0442	0.0710	0.0541	0.0567
8	0.0770	0.1035	0.0921	0.0979
9	0.0526	0.0927	0.0628	0.0712
10	0.0693	0.1177	0.0837	0.0877

Table A2. Maximum distortion measurements after the straightening stroke for the centre spine.

				Unit: mm
Shape No.	**Naive L-BFGS-B**	**K_{global} Only**	**DNN**	**XGBoost**
1	0.0862	0.1220	0.1116	0.1168
2	0.0652	0.1149	0.0780	0.0802
3	0.0617	0.1003	0.0833	0.0750
4	0.0628	0.0993	0.0785	0.0794
5	0.0776	0.1276	0.0993	0.0930
6	0.0694	0.1042	0.0869	0.0850
7	0.0696	0.1032	0.0837	0.0840
8	0.0891	0.1483	0.1166	0.1199
9	0.0738	0.1303	0.0923	0.0966
10	0.0647	0.1055	0.0785	0.0813

Table A3. Maximum distortion measurements after the straightening stroke for the side member.

				Unit: mm
Shape No.	**Naive L-BFGS-B**	**K_{global} Only**	**DNN**	**XGBoost**
1	0.1021	0.1902	0.1375	0.1401
2	0.1040	0.1742	0.1425	0.1514
3	0.0992	0.1801	0.1255	0.1242
4	0.0715	0.1230	0.0984	0.0905
5	0.0815	0.1313	0.1102	0.1028
6	0.1024	0.1609	0.1493	0.1380
7	0.0872	0.1603	0.1086	0.1263
8	0.0758	0.1268	0.0998	0.1067
9	0.0776	0.1415	0.1085	0.1082
10	0.0844	0.1619	0.1117	0.1170

References

1. Baser, T.A.; Umay, E.; Akıncı, V. New Trends in Aluminum Die Casting Alloys for Automotive Applications. *Eurasia Proc. Sci. Technol. Eng. Math.* **2022**, *21*, 79–87. [CrossRef]
2. Lehmhus, D. Advances in Metal Casting Technology: A Review of State of the Art, Challenges and Trends—Part I: Changing Markets, Changing Products. *Metals* **2022**, *12*, 1959. [CrossRef]
3. Triller, J.; Lopez, M.L.; Nossek, M.; Frenzel, M.A. Multidisciplinary optimization of automotive mega-castings merging classical structural optimization with response-surface-based optimization enhanced by machine learning. *Sci. Rep.* **2023**, *13*, 21678. [CrossRef] [PubMed]
4. Xu, Y.; Zhang, Y.; Li, J.; Wang, P.; Wang, J.; Chen, H.; Sun, H.; Zhan, H. Effects of Cr-addition on ageing response of an Al–Si–Mg die cast alloy. *Mater. Sci. Eng. A* **2024**, *892*, 146058. [CrossRef]
5. Li, T.; Song, J.; Zhang, A.; You, G.; Yang, Y.; Jiang, B.; Qin, X.; Xu, C.; Pan, F. Progress and prospects in Mg-alloy super-sized high pressure die casting for automotive structural components. *J. Magnes. Alloy.* **2023**, *11*, 4166–4180. [CrossRef]
6. Li, Z.; Tan, H.; Jarfors, A.E.; Jansson, P.; Lattanzi, L. Smart-Cast: An AI-Based System for Semisolid Casting Process Control. *Procedia Comput. Sci.* **2024**, *232*, 2440–2447. [CrossRef]
7. Timelli, G.; Lohne, O.; Arnberg, L.; Laukli, H.I. Effect of solution heat treatments on the microstructure and mechanical properties of a die-cast AlSi7MgMn alloy. *Metall. Mater. Trans. A* **2008**, *39*, 1747–1758. [CrossRef]
8. Anglada, E.; Meléndez, A.; Vicario, I.; Idoiaga, J.K.; Mugarza, A.; Arratibel, E. Prediction and validation of shape distortions in the simulation of high pressure die casting. *J. Manuf. Process.* **2018**, *33*, 228–237. [CrossRef]
9. Hofer, P.; Kaschnitz, E.; Schumacher, P. Distortion and residual stress in high-pressure die castings: Simulation and measurements. *Jom* **2014**, *66*, 1638–1646. [CrossRef]
10. Couto, C.P.; Revilla, R.I.; Politano, R.; Costa, I.; Panossian, Z.; De Graeve, I.; Rossi, J.L.; Terryn, H. Influence of austenitisation temperatures during hot stamping on the local electrochemical behaviour of 22MnB5 steel coated with hot-dip Al-Si. *Corros. Sci.* **2021**, *190*, 109673. [CrossRef]
11. Khan, H.A.; Butt, S.U.; Baqai, A.A.; Saeed, H.A. A new approach to predict the flexibility and precision of manufacturing systems using geometric constraints and small displacement torsors. *Procedia Manuf.* **2018**, *17*, 294–301. [CrossRef]
12. Guan, Q. 9—Control of buckling distortions in plates and shells. In *Processes and Mechanisms of Welding Residual Stress and Distortion*; Feng, Z., Ed.; Woodhead Publishing Series in Welding and Other Joining Technologies; Woodhead Publishing: Sawston, UK, 2005; pp. 295–343. [CrossRef]

13. Hartlieb, A.; Hartlieb, M. The Impact of Giga-Castings on Car Manufacturing and Aluminum Content. *Light Met. Age* **2023**, *3*, 18–23.
14. Cross, E.J.; Gibson, S.J.; Jones, M.R.; Pitchforth, D.J.; Zhang, S.; Rogers, T.J. Physics-Informed Machine Learning for Structural Health Monitoring. In *Structural Health Monitoring Based on Data Science Techniques*; Cury, A., Ribeiro, D., Ubertini, F., Todd, M.D., Eds.; Springer International Publishing: Cham, Switzerland, 2022; pp. 347–367. [CrossRef]
15. Stoll, A.; Benner, P. Machine learning for material characterization with an application for predicting mechanical properties. *GAMM-Mitteilungen* **2021**, *44*, e202100003. [CrossRef]
16. Liu, J.; Zhang, S.; Fan, H. A two-stage hybrid credit risk prediction model based on XGBoost and graph-based deep neural network. *Expert Syst. Appl.* **2022**, *195*, 116624. [CrossRef]
17. Dai, W.; Jiang, Y.; Mou, C.; Zhang, C. An Integrative Paradigm for Enhanced Stroke Prediction: Synergizing XGBoost and xDeepFM Algorithms. In Proceedings of the 2023 6th International Conference on Big Data Technologies, ICBDT '23, New York, NY, USA, 22–24 September 2023; pp. 28–32. [CrossRef]
18. Salb, M.; Jovanovic, L.; Bacanin, N.; Antonijevic, M.; Zivkovic, M.; Budimirovic, N.; Abualigah, L. Enhancing Internet of Things Network Security Using Hybrid CNN and XGBoost Model Tuned via Modified Reptile Search Algorithm. *Appl. Sci.* **2023**, *13*, 2687. [CrossRef]
19. Park, S.; Kim, J.; Park, J.; Burgner-Kahrs, J.; Noh, G. Design of patterns in tubular robots using DNN-metaheuristics optimization. *Int. J. Mech. Sci.* **2023**, *251*, 108352. [CrossRef]
20. Kolar, D.; Lisjak, D.; Pająk, M.; Gudlin, M. Intelligent Fault Diagnosis of Rotary Machinery by Convolutional Neural Network with Automatic Hyper-Parameters Tuning Using Bayesian Optimization. *Sensors* **2021**, *21*, 2411. [CrossRef] [PubMed]
21. Lin, J.; Qi, C.; Wan, H.; Min, J.; Chen, J.; Zhang, K.; Zhang, L. Prediction of cross-tension strength of self-piercing riveted joints using finite element simulation and XGBoost algorithm. *Chin. J. Mech. Eng.* **2021**, *34*, 36. [CrossRef]
22. Hashemi, A.; Jang, J.; Beheshti, J. A Machine Learning-Based Surrogate Finite Element Model for Estimating Dynamic Response of Mechanical Systems. *IEEE Access* **2023**, *11*, 54509–54525. [CrossRef]
23. Chen, T.; Guestrin, C. XGBoost: A Scalable Tree Boosting System. In Proceedings of the 22nd ACM SIGKDD International Conference on Knowledge Discovery and Data Mining, KDD '16, New York, NY, USA, 13–17 August 2016; pp. 785–794. [CrossRef]
24. Zhu, C.; Byrd, R.H.; Lu, P.; Nocedal, J. Algorithm 778: L-BFGS-B: Fortran subroutines for large-scale bound-constrained optimization. *ACM Trans. Math. Softw.* **1997**, *23*, 550–560. [CrossRef]
25. Shehadeh, A.; Alshboul, O.; Al Mamlook, R.E.; Hamedat, O. Machine learning models for predicting the residual value of heavy construction equipment: An evaluation of modified decision tree, LightGBM, and XGBoost regression. *Autom. Constr.* **2021**, *129*, 103827. [CrossRef]
26. Chollet, F. Keras. 2015. Available online: https://keras.io (accessed on 11 January 2024).
27. Asghari, V.; Leung, Y.F.; Hsu, S.C. Deep neural network based framework for complex correlations in engineering metrics. *Adv. Eng. Inform.* **2020**, *44*, 101058. [CrossRef]
28. Fukushima, K. Visual Feature Extraction by a Multilayered Network of Analog Threshold Elements. *IEEE Trans. Syst. Sci. Cybern.* **1969**, *5*, 322–333. [CrossRef]
29. Kingma, D.P.; Ba, J. Adam: A Method for Stochastic Optimization. *arXiv* **2017**, arXiv:1412.6980.
30. Bergstra, J.; Yamins, D.; Cox, D. Making a Science of Model Search: Hyperparameter Optimization in Hundreds of Dimensions for Vision Architectures. In Proceedings of the 30th International Conference on Machine Learning, Atlanta, GA, USA, 17–19 June 2013; Volume 28, pp. 115–123.

materials

MDPI

Article

Analysis of Mechanical Characteristics of the Swing Angle Milling Head of a Heavy Computer Numerical Control Milling Machine and Research on the Light Weight of a Gimbal

Youzheng Cui [1,2,3,*], Chengxin Liu [1], Haijing Mu [1], Hui Jiang [4], Fengxia Xu [1,2,3], Yinfeng Liu [1,4] and Qingming Hu [1,5,*]

1 School of Mechanical and Electronic Engineering, Qiqihar University, Qiqihar 161006, China
2 The Engineering Technology Research Center for Precision Manufacturing Equipment and Industrial Perception of Heilongjiang Province, Qiqihar 161006, China
3 The Collaborative Innovation Center for Intelligent Manufacturing Equipment Industrialization, Qiqihar 161006, China
4 Qiqihar Heavy CNC Equipment Co., Ltd., Qiqihar 161005, China
5 School of Mechatronics Engineering, Harbin Institute of Technology, Harbin 150001, China
* Correspondence: 02734@qqhru.edu.cn (Y.C.); 02510@qqhru.edu.cn (Q.H.)

Abstract: As the key component of a five-axis CNC planer-type milling machine, the integral mechanical property of the A/C swing angle milling head directly affects the machining accuracy and stability of the milling machine. Taking the mechanical A/C swing-angle milling head of a five-axis numerical-control gantry milling machine as the research object, the stress deformation characteristics and natural frequency of the swing-angle milling head under actual working conditions were studied using finite-element analysis. Based on the analytical results, it was determined that the cardan frame, with its large mass proportion and strong rigidity of the whole milling head, is the object to be optimized. The topological optimization of the cardan frame, in which achieving the minimum flexibility was the optimization objective, was carried out to determine the quality reduction area. By comparing the simulation results of the cardan frames of three different rib plate structures, it was shown that the cardan frame performance of the ten-type rib plate structure was optimal. The analytical results showed that, when the cardan frame met the design requirements for stiffness and strength, the mass after optimization was reduced by 13.67% compared with the mass before optimization, the first-order natural frequency was increased by 7.9%, and the maximum response amplitude was reduced in all directions to avoid resonance, which was beneficial to the improvement of the dynamic characteristics of the whole machine. At the same time, the rationality and effectiveness of the lightweight design method of the cardan frame were verified, which has strong engineering practicality. The research results provide an important theoretical basis for the optimization of other machine tool gimbals and have important practical significance and application value.

Keywords: heavy milling machines; milling head; finite-element analysis; universal frame; topological optimization

1. Introduction

Topological optimization is widely used in machinery, aerospace and other fields due to its high design freedom and wider design space [1]. The double-swing angle milling head is one of the core components of the five-axis CNC planer milling machine, and its dynamic and static characteristics directly affect the overall working performance of the machine tool. Research on key parts of the machine tool via topological optimization is an important means for lightweight research [2,3]. As the cardan frame is the heaviest part of the swing angle milling head, the realization of its lightweight design has an important impact on improving the machining accuracy of the whole swing angle milling head, thus having important practical significance.

Scholars at home and abroad have carried out extensive research on many key parts of machine tools. Some scholars [4] have carried out optimization research on the structure of the swing angle milling head's cardan frame. The results before and after optimization have shown that the new cardan frame has a good optimization effect while the rigidity and strength meet the design requirements. Francesco A [5] et al. realized the lightweight design of machine tool components using mixed lightweight materials to improve the machining performance. Raz K [6] et al. optimized the milling head's topology, and the results showed that the milling head mass was reduced and the rigidity was improved. TriM J et al. [7] studied the sliding table of a three-axis vertical milling machine to reduce the energy consumption of the machine tool and successfully realized a lightweight design. Shindo R et al. achieved a lightweight machine tool structure by combining topological optimization and shape optimization methods on the basis of not following the basic design principles for the non-optimal design problem of the machine tool (i.e., a bending structure that cannot be avoided). Compared with the traditional design, the optimization method proposed in the study greatly improved the machine tool's natural frequency and efficiency [8]. Li Congbo [9] and others comprehensively considered the energy consumption and dynamic and static performance of the machine tool. On the premise of ensuring both the static and dynamic performance, they not only reduced the energy consumption of the optimized structure, but also achieved a lightweight structural design. Petra L [10] et al. designed and studied the moving parts of the gantry-type CNC machine tool and performed a topological optimization of these moving elements. Under the premise of maintaining the same rigidity, compared with the original structural design, its stability was guaranteed, although the quality was reduced by about 25%. TRIEN et al. [11] proposed a lightweight design of the machine tool using a genetic algorithm, and the results showed that the energy consumption of the machine tool was reduced and that the rigidity met the requirements. Ko and Lee et al. [12] focused on studying the constrained stiffness optimization problem, considering the maximum allowable displacement. Based on genetic algorithms, they conducted optimization research on the size and weight of five-axis machine tools. Abdelwahab et al. [13] used topology optimization methods and AM manufacturing processes to reduce the weight of the machine tool box structure, while meeting the design requirements in terms of stiffness and strength. Kim, Seung-Gi et al. [14] realized the optimal design of important structures such as the machine tool's ram and column parts via the topological optimization method, and they achieved good rigidity and weight. In order to improve the performance of the planer milling machine, BES et al. [15] studied the milling machine through the analytic hierarchy process. The research showed that the method proposed in the article can better realize the selection and structural optimization of the gantry. The optimized designs presented by the above scholars based on finite-element simulation analyses demonstrate the advantages of lightweight machine tool parts and also provide a reference for follow-up research.

The combination of finite-element analysis and topological optimization methods can effectively achieve the goal of a lightweight design for machine tool components. At the same time, based on the topology optimization results and the production process, different rib plate structures can be introduced into the reduced-quality area of machine tool components to improve their static and dynamic performance. For example, Dimitrov et al. [16] improved the stiffness of a structure by welding or casting reinforcement ribs, and they achieved their goal of removing unnecessary materials on the fuselage by using finite-element analysis and topological optimization methods. Tzu et al. [17] used a combination of finite-element methods and topological optimization methods to analyze and optimize the operating structure of CNC machine tools in order to propose better optimization design solutions for key components of the machine tool. The work of Ullah et al. [18] is crucial for the design and improvement of precision machinery based on static and dynamic analyses. By comparing the data with the results of the finite-element analysis (FEA), the static and dynamic characteristics, accuracy and efficiency of the finite-element model were verified. On the basis of improving the dynamic characteristics of machine tools, scholars such as Guido Stoppler have optimized the topology of the crossbeam structure to achieve a lightweight performance [19]. Hu Shijun et al. [20] proposed an "X" rib plate as a new beam with a good performance, verifying its performance by comparing the static and dynamic performance of the commonly used five-rib plate structures. The results showed that the performance of the optimized machine was improved to different degrees and that the beam mass was reduced by 2.1%. Bakker C et al. [21] proposed a method for optimizing the stiffened plate and its structural layout to improve the performance of the stiffened plate shell. Jiang Shufeng et al. [22] designed and researched the DVT250 machine tool crossbeam, and their results show that the diamond rib plate structure, as the optimized crossbeam structure, has the best lightweight effect. Liu Chengying et al. [23] optimized the standing column of the machine tool by means of multiple optimization methods and selected the design scheme of a W-shaped plate rib with optimal parameters to improve the natural frequency of the machine tool and realize its lightweight design. Based on the above literature analysis, it can be concluded that the combination of finite-element analysis and topology optimization, combined with the introduction of a rib plate structure, can provide theoretical reference and methodological support for the optimization and lightweight design of the swing angle milling head structure studied in this paper.

Based on the research and analysis of the above literature, it can be seen that more research has been carried out on the spindle boxes, rib plates and beams of machine tools, while less research has been carried out on the gimbals of swing angle milling heads. Furthermore, topology optimization has been mainly carried out for the gimbals of direct drive swing angle milling heads. At the same time, the gimbal accounts for a large proportion of the quality of the whole swing angle milling head, and it is also the basis for the installation of other parts. Therefore, it is very necessary to perform lightweight research on the gimbal of the swing angle milling head. In this paper, the mechanical A/C swing angle milling head of a five-axis CNC gantry milling machine is taken as the research object. Based on static and modal analyses, the gimbal structure with the strongest rigidity and the largest mass proportion of the milling head is identified as the object to be optimized. The topology optimization research is carried out for the case of less deformation (i.e.,higher strength) of the upper end of the gimbal, and a comparative analysis of the gimbal with three kinds of rib plate structures is carried out. It is found that the performance of the ten-type rib plate structure is better. After optimization, under the premise of ensuring that the stiffness and strength of the optimized universal joint meet the design requirements, the purpose of achieving a lightweight design is fulfilled. This research combines the topology optimization method with the rib plate structure, which can not only reduce the mass of the swing angle milling head, but also improve the dynamic characteristics of the whole machine. It has an important theoretical basis and practical significance for the optimization design of subsequent structures and for improving machine performance.

2. A/C Swing Angle Milling Head Structural Analysis and Finite-Element Analysis Pretreatment

2.1. Structural Analysis, Simplification and Mesh Generation of the Milling Head Model

This article takes the swing angle milling head of a five-axis CNC gantry milling machine as the research object. Its structure is composed of a transmission mechanism, a universal frame, an A-axis (main shaft), the A-axis housing (main shaft seat), front-end connectors and other parts. The overall structure is shown in Figure 1. The overall size of the A/C double swing angle milling head is 1550 mm × 830 mm × 930 mm.

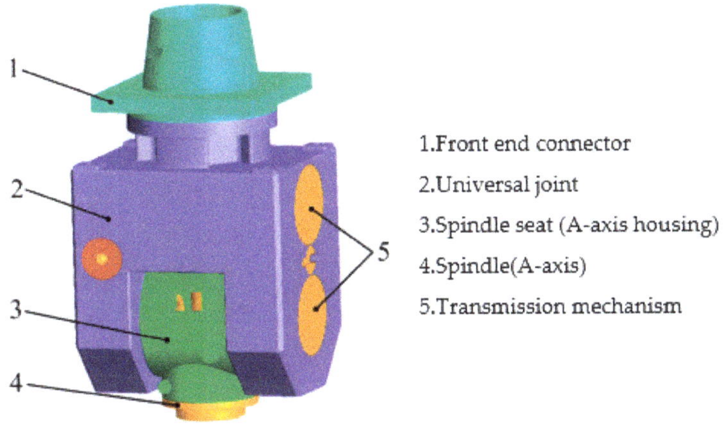

Figure 1. Model diagram of a mechanical A/C swing angle milling head.

In order to study the static characteristics of the swing angle milling head under different working conditions, three typical milling head angles (i.e., 0°, 45° and 90°) are selected for comparative analysis, as shown in Figure 2.

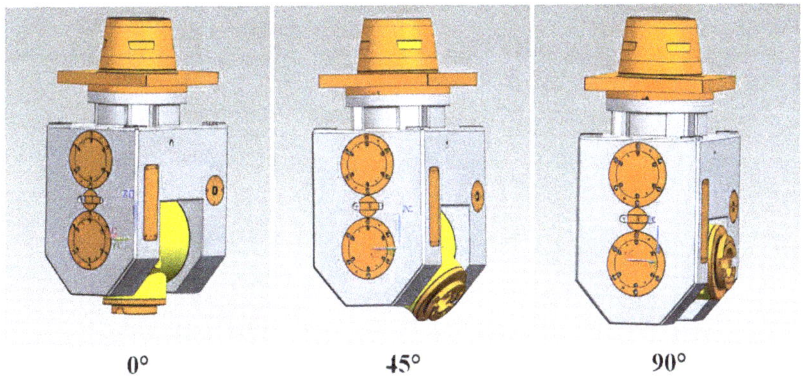

Figure 2. Three angles of the mechanical A/C swing angle milling head.

Due to the complexity of the internal structure of the A/C swing angle milling head, this paper simplifies the relatively minor process features in the digital model of the swing angle milling head, such as the chamfer, fine hole, undercut, oil channel, etc., in order to reduce the calculation time of the finite-element analysis. Then, the simplified swing angle milling head model is imported into ANSYS Workbench (2021 R1, ANSYS company, Pittsburgh, PA, USA) for mesh generation. In order to obtain ideal calculation results, the A/C swing angle milling head model is discretized by using the automatic mesh generation method, and the number of nodes and elements in the model is 976,796 and 572,014, respectively. The finite-element mesh generation model of the swing angle milling head is shown in Figure 3.

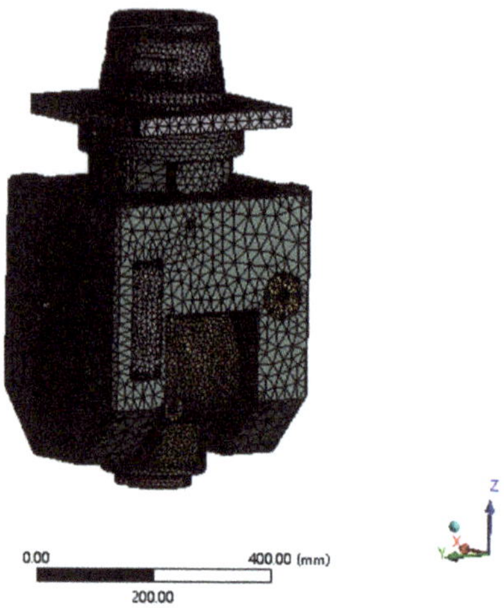

Figure 3. Finite-element mesh generation of a mechanical A/C swing angle milling head.

2.2. Material Properties of the Milling Head

In this paper, the material used for the box and rotary mechanism of the A/C swing angle milling head is ductile cast QT600-3, and the material of all shafting components is 38CrMoAl. Among the components, the gear and bearing inside the milling head are made of alloy structural steel. See Table 1 for the specific parameters of the materials used in the milling head simulation.

Table 1. Specific parameters of materials used for milling head simulation.

Material Properties	38CrMoAl	QT600-3	Alloy Structural Steel
Elastic modulus E (GPa)	209	158	206
Poisson's ratio (μ)	0.295	0.285	0.3
Density (kg/m^3)	7266	7120	7850

2.3. Load and Boundary Conditions of Swing Angle Milling Head

The external loads to be borne by the A/C swing angle milling head during the machining process mainly include the milling force and gravity. Since the gravitational load can be applied as set parameters in the simulation software and directly added for use, it is only necessary to calculate the cutting force of the milling head according to the empirical formula. Taking the A/C swing angle milling head assembled on a heavy CNC planer-type milling machine as an example, tungsten steel (a hard alloy) is selected as the cutter material and bearing steel is selected as the machined material. The symmetric end milling method is adopted, and the total milling force of the mechanical A/C swing angle milling head of the five-axis CNC planer-type milling machine can be calculated according to the following empirical formula [24]:

$$\begin{cases} F_C = C_{FC} \times Z \times D^{-1.1} \times {a_e}^{1.1} \times a_p^{0.95} \times f_z^{0.8} \times K_{FC} \\ K_{FC} = \left(\frac{\sigma_b}{637}\right)^{0.3} \end{cases} \quad (1)$$

In the formula, F_C is the total milling force, C_{FC} is the milling force coefficient, K_{FC} is the correction coefficient, D is the milling cutter diameter, Z is the number of milling cutter teeth, f_z is the feed rate per tooth, a_e is the cutting width, a_p is the number of back tooth cutters and σ_b is the tensile strength. By substituting relevant milling parameters in Table 2, the total milling force is 1184.01. The components of the milling force acting on a mechanical A/C swing angle milling head in all directions are as follows: transverse feed force F_e, longitudinal feed force F_h and vertical feed force F_v. The geometric relationship with the total milling force is set as follows:

$$\begin{cases} F_e = 0.9F_c \\ F_h = 0.3F_c \\ F_v = 0.5F_c \end{cases} \quad (2)$$

Table 2. Table of milling parameters for the swing angle milling head.

Name	f_z (mm)	a_e (mm)	a_p (mm)	C_{FC}	D (mm)	Z
Parameter value	0.2	35	2	773	60	4

After calculation, the total milling force is decomposed into three milling forces along the moving direction of the workpiece, namely, the transverse feed force $F_e = 1065.6$ N, the longitudinal feed force $F_h = 355.2$ N and the vertical feed force $F_v = 592$ N.

Before conducting static analysis, it is necessary to add loads and constraints. A fixed constraint is applied to the upper-end face of the swing angle milling head, and then the load step method in ANSYS is used to add a standard gravitational load to the connection surface between the swing angle milling head and the slider. The milling force is applied in the form of a remote force in the model, with the point of action set at a distance of 100 mm from the tip of the swing angle milling head [25]. The deformation caused by the longitudinal feed force F_h and vertical feed force F_v is relatively small and has little influence on the machining accuracy of the milling head, and thus can be ignored. Therefore, in this paper, the effect of the cross-feed force Fe on milling head deformation is selected. An infeed force Fe is added to the model in the form of a remote force. Figure 4 shows the constraints and loadings of the milling head under gravitational and milling forces in the Y direction when the milling head's A-axis is at 0°. Note that gravitational and milling forces need to be loaded separately when the load is applied.

Figure 4. Restraint and load model at 0° for the A-axis of the milling head.

3. Analysis of Static and Dynamic Characteristics

3.1. Static Analysis of Swing Angle Milling Head

Since the deformation of the milling head structure caused by the gravitational load can be eliminated, only the static simulation analysis results of the milling force deformation are given here. The results under the combined action of the output gravity and milling force obtained through finite-element analysis are shown in Figure 5. Through the simulation analysis of milling head angles of 0°, 45° and 90°, it can be seen that the maximum deformation areas of the three angles are distributed in the same area in the milling head, that is, the connecting tool shank. The distribution of the maximum stress is also the same, except it is distributed in the cutter head area of the milling head's A-axis, which is consistent with the actual working conditions, indicating that the analysis's conclusion is reasonable and effective. Therefore, in this paper, the limit angle under working conditions (i.e., the A-axis at 90°) is selected based on three typical angles as examples, and the analytical results are shown in Figure 5a,b.

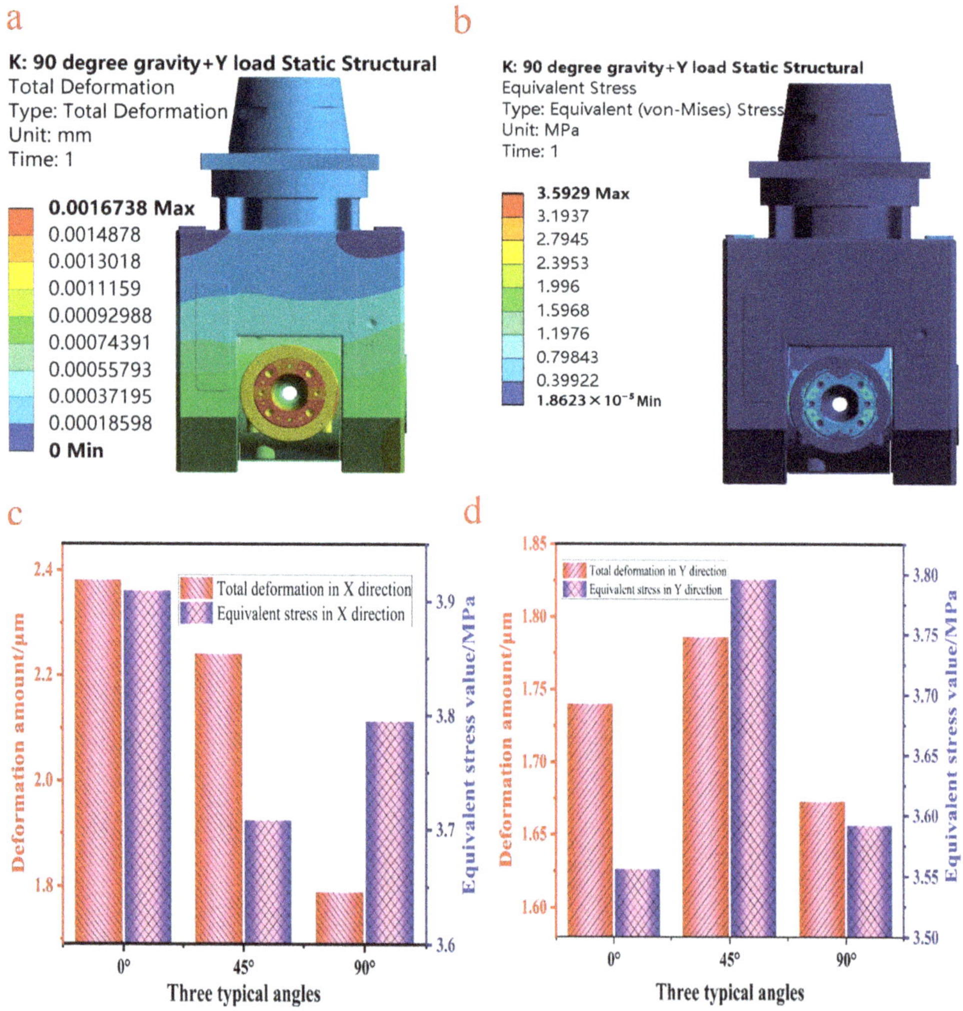

Figure 5. Results diagram of different angles of the swing angle milling head under the action of gravity and milling force in the (X, Y) direction. (**a**) Deformation cloud map at an angle of 90° under gravity and Y load. (**b**) Stress cloud map at an angle of 90° under gravity and Y load. (**c**) Stress and deformation values at different angles under gravity and X load. (**d**) Stress and deformation values at different angles under gravity and Y load.

According to the deformation cloud chart in Figure 5a, the maximum deformation area of the A-axis of the swing angle milling head is mainly concentrated at the connecting tool handle under the action of the milling force in the Y direction, while the area next to the maximum deformation is located at the bottom of both sides of the milling head, which has room for degradation. At the same time, the upper part of the milling head has less deformation and a better stiffness, and is thus the key optimization area. According to the equivalent stress nephogram in Figure 5b, the maximum stress of the milling head is 3.59 Mpa, which is far lower than the strength limit of the milling head material. Considering the influence of the A-axis swing angle further, the deformation at different angles is analyzed. According to the stress and deformation values of the X-load and Y-load in Figure 5c,d, the total deformation and equivalent stress of the A-axis of the swing angle milling head fluctuate at different angles. The total deformation fluctuates most obviously under the action of the X-axis milling force: its fluctuation value reaches 0.592 μm. The equivalent stress value under the action of the Y-axis milling force fluctuates greatly, and its fluctuation value reaches 0.24 Mpa. When the milling head's A-axis is at 0° under the action of the X-direction milling force, the maximum deformation is 2.381 μm. In addition, when the milling head's A-axis is set at different angles, the maximum deformation area is also concentrated on the connecting tool handle. The maximum stress of the milling head under different angles occurs in the A-axis cutter head area, which is consistent with the actual working conditions. This shows that the area with better stiffness and less stress (the upper part of the milling head) can be optimized to prepare for the subsequent lightweight design.

3.2. Modal Analysis of Swing Angle Milling Head

ANSYS Workbench was used to conduct modal analysis on the double swing angle milling head of the machine tool. The dynamic performance of the swing angle milling head is basically determined by the low-order natural frequency [26], so only the first two modes of the swing angle milling head under typical working conditions are extracted in this study. The modal analysis results at an angle of 0° are shown in Figure 6 and Table 3.

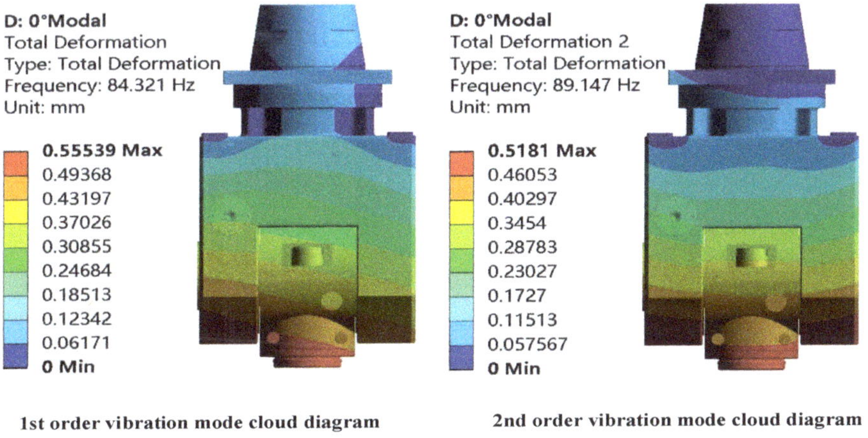

Figure 6. The first two vibration modes of the mechanical A/C double swing angle milling head at an angle of 0°.

Table 3. The first two modal characteristics of the milling head at an angle of 0°.

Order	Natural Frequency/Hz	Description of Milling Head Vibration Mode
1	84.32	The milling head mainly composed of a universal gimbal rotates along the Z-axis and swings along the X-axis
2	89.14	The middle and lower ends of the milling head based on a universal gimbal swing left and right along the Y-axis

From the analysis results in Figure 6 and Table 3, it can be seen that the first-order modal vibration mode of the swing angle milling head is mainly represented by the overall rotation and swing of the cardan frame. The maximum deformation is concentrated on the single side of the lower end of the cardan frame and the A-axis area. The natural frequency is 84.32 Hz, and the deformation is relatively large. The maximum deformation value is 555.39 μm. The second-order vibration mode is mainly manifested as the integral swing dominated by the cardan frame. The maximum deformation is concentrated on both sides of the lower end of the cardan frame and the A-axis area. The natural frequency is 89.14 Hz. The analysis shows that the initial design of the cardan frame structure is reasonable, and the rigidity of the upper part of the cardan frame is good, so the topological optimization can be further studied.

The static and modal analyses show that the structure of the swing angle milling head gimbal will affect the machining accuracy of the five-axis CNC planer milling machine, Therefore, it is necessary to optimize and improve its structure. On the premise that the stiffness and strength of the gimbal meet the design requirements, reduce the quality of the gimbal, which is beneficial to the improvement of the dynamic characteristics of the whole machine.

4. Optimization Design of Universal Carriage

Topology optimization is a powerful tool for searching for lightweight structures. This process can transform the optimal topology problem into an exploration and optimization of the material distribution problem in a designated design space [27,28] and can also maximize economic benefits. This article selects the maximum rigidity of the universal joint as the optimization objective, taking volume as the constraint condition, and integrates the theoretical foundation of the SIMP variable density method. The mathematical model for topology optimization is expressed as follows:

$$\begin{cases} \min C = U^T K U = \sum\limits_{e=1}^{N} u^e k^e u^e \\ s.t\ V = fV_0 = \sum\limits_{e=1}^{N} X_e V_e \leq V^* \\ F = KU\ (0 < x_{\min} \leq x^e \leq x_{\max}) \end{cases} \quad (3)$$

where C is the objective function (overall flexibility of the universal joint); F is the force vector; U is the displacement matrix of the universal joint; K is the total stiffness matrix of the universal joint structure; V_0 is the initial volume function; f is the optimized volume ratio; V is the optimized volume function; V_e is the optimized volume of the gimbal unit; X_e is the design variable (unit relative density); $x_{\min}$ and $x_{\max}$ are the minimum and maximum limit values of unit relative density; and u^e is the unit displacement column vector.

4.1. Topology Optimization and Model Reconstruction of Universal Frame

After applying constraints and loads to the gimbal model for initial static analysis, the topology optimization module is used to optimize the gimbal's topology. During the optimization process, the material of the universal joint is defined as ductile iron QT600-3, with a density of 7.12 g/cm^3, a Poisson's ratio of 0.285 and an elastic modulus of 158 GPa.

The universal joint model is divided using the automatic grid division method, with a unit size of 10 mm. The grid model is shown in Figure 7. In order to avoid interference with the assembly of other structural components due to changes in the optimized structure of the universal joint, relevant areas where other components are installed on the universal joint and the areas on both sides of the universal joint are designated as non-optimized areas, while the remaining areas are designated as optimization objects. Figure 8 shows the division of the universal joint optimization area, with red indicating the exclusion area and blue indicating the optimization area.

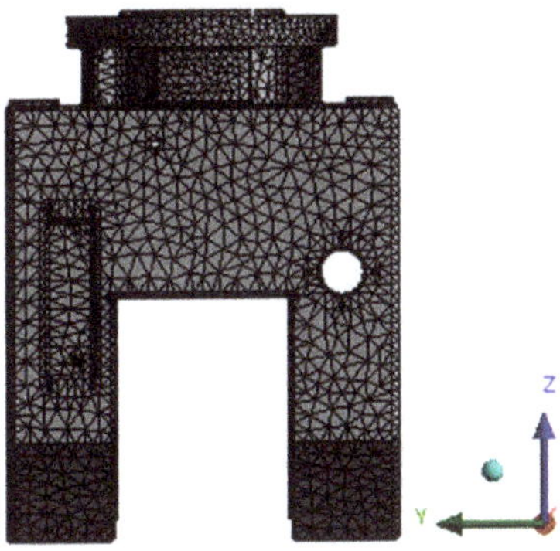

Figure 7. Mesh model of the universal joint.

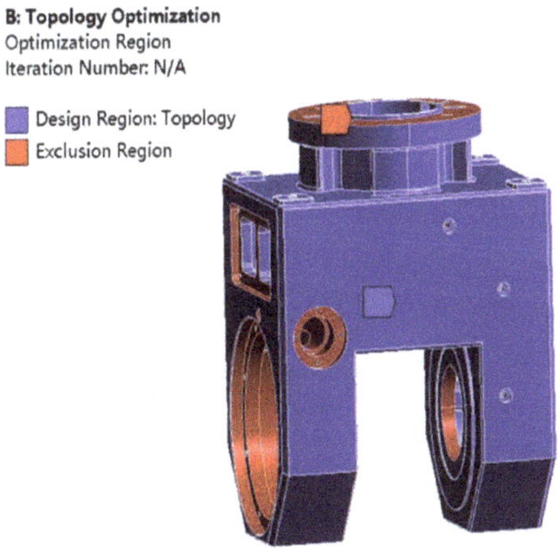

Figure 8. Schematic diagram of optimized and non-optimized areas for the gimbal.

The constraints and loads imposed by the universal joint are shown in Figure 9. Constraints are added at the upper end of the universal joint, and the two toroidal surfaces in the middle of the upper end are subjected to a positive pressure of 700 N and 1000 N, respectively. A total of 9850 N of vertical downward pressure is applied to the four semicircular surfaces at the lower end. The material removal rate was set at 20% as the optimization objective for calculation, and after 11 iterations, the density cloud map of the optimized universal joint was obtained, as shown in Figure 10.

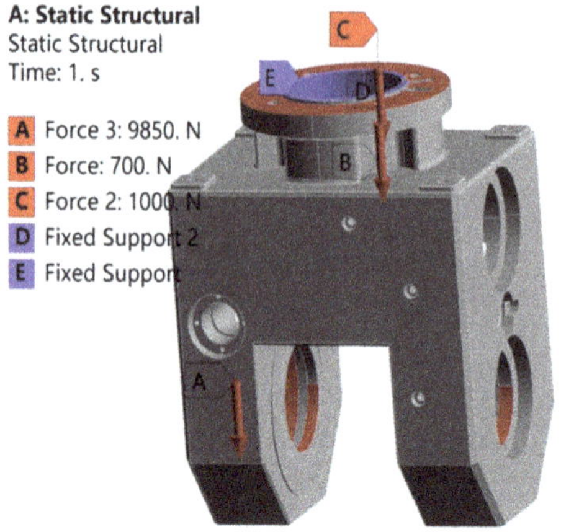

Figure 9. Load and constraint diagram of universal joint.

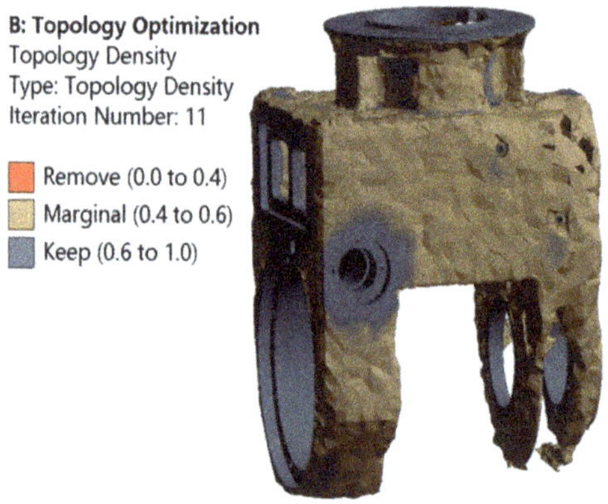

Figure 10. Density cloud map after optimizing the universal joint frame.

From Figure 10, it can be seen that the removal of material from the universal joint is mainly distributed at the upper and lower ends of both sides of the universal joint. Due to its irregular shape at the bottom and because the most important structural materials were retained, it can be seen from the static characteristic cloud diagram in the previous section that the maximum deformation is seen at the connection of the milling head and the tool handle, while the deformation at the bottom ends of both sides of the universal joint is relatively small. Therefore, the trapezoidal transition method at the bottom ends of both sides of the universal joint can be changed to arc transitions. At the same time, some material can also be removed from the area at the upper end of the universal joint. According to the modal analysis results in the previous section, different types of rib plates are considered in the design to achieve the design purpose due to the good rigidity of this area. Based on the above analysis, the optimization results of the swing angle milling head directly confirm the accuracy of the previous static mechanics and modal analysis results and can provide a reference for subsequent reconstruction design.

In order to reduce the quality of the universal joint and better align with the design requirements in terms of stiffness and strength, the materials on the upper sides of both ends of the universal joint are modified and redesigned based on the topology optimization results, resulting in a manufactured and assembled universal joint model. When designing, it is usually necessary to pay attention to the following two points. First, due to the different layout forms of the reinforcement plates, they will play different roles. For example, transverse reinforcement plates will have a bending resistance effect, and longitudinal reinforcement plates will have a torsional resistance effect. Therefore, selecting the appropriate reinforcement plate is particularly crucial. Second, the complexity of the rib plate structure and whether it is conducive to production and other factors must be fully considered. In response to these concerns, this article selects three commonly used rib plate layouts as the rib plate layouts on both sides of the universal joint. Figure 11 shows the universal joint with a "well"-shaped rib plate, "ten"-shaped rib plate and "X"-shaped rib plate configuration, respectively.

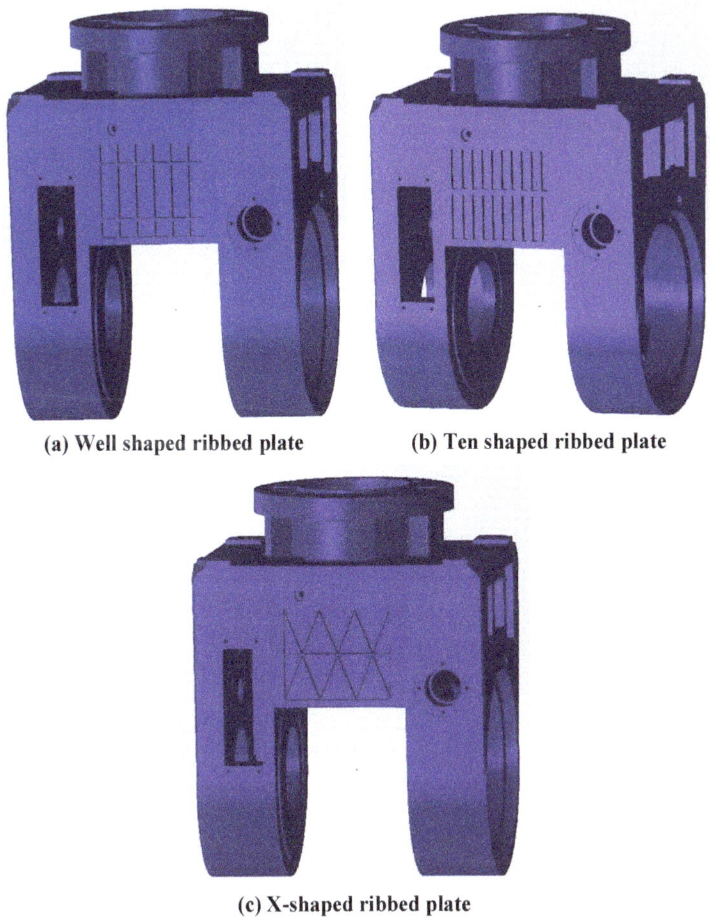

Figure 11. Universal frames with different rib plate configurations.

The static and modal analysis results of the three types of rib plate configurations of the universal joint were obtained through finite-element analysis, as shown in Figure 12. From Figure 12, it can be seen that the minimum deformation and equivalent stress among the three rib plate configurations of the universal joint is the "ten"-shaped rib plate. Compared with the other two rib plate structures, the "ten"-shaped rib plate configuration has the best static stiffness in the universal joint. Moreover, compared to the other two rib plate configurations, the ten-shaped rib plate configuration of the universal joint has the highest first-order natural frequency and the best dynamic stiffness. Analysis shows that the ten-shaped rib plate configuration of the universal joint has a better static and dynamic performance than the other two rib plate configurations. Therefore, this article uses the ten-shaped rib plate structure to replace the materials on both sides of the original universal joint.

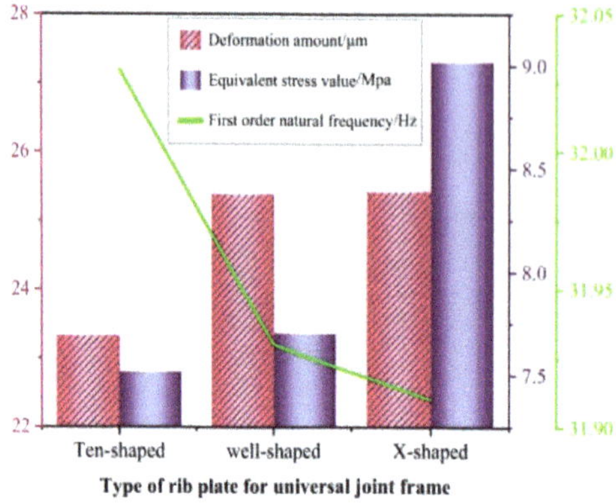

Figure 12. Static and modal analysis results of three types of rib plate configurations of universal frames.

4.2. *Performance Analysis of Optimized Universal Joint Model*

The reconstructed universal joint model was analyzed using Workbench finite-element simulation software, and its stress–strain, mass and other parameters are shown in Figure 13. From Figure 13, it can be seen that the total deformation of the optimized universal joint increased by 14.21%, which meets the precision requirements of the swing angle milling head and does not have a significant impact on the performance of the entire machine. The equivalent stress decreased by 0.39%, and the mass decreased from 22.45 kg to 19.38 kg, with a weight reduction ratio of 13.67%. The optimization of the design was achieved, and the effect was significant.

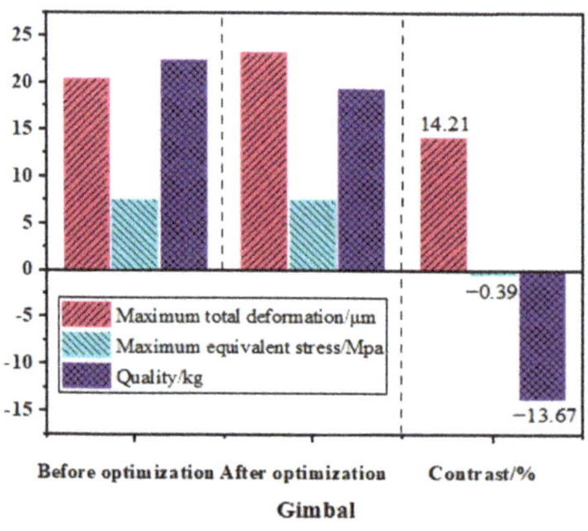

Figure 13. Comparison of static characteristics and mass before and after optimization of universal joint frame.

The same material properties and constraints as the original model were loaded for modal analysis, and the first six natural frequencies of the universal joint model with a "ten" rib plate configuration were recorded. The natural frequencies of the universal joint model before and after optimization are shown in Figure 14.

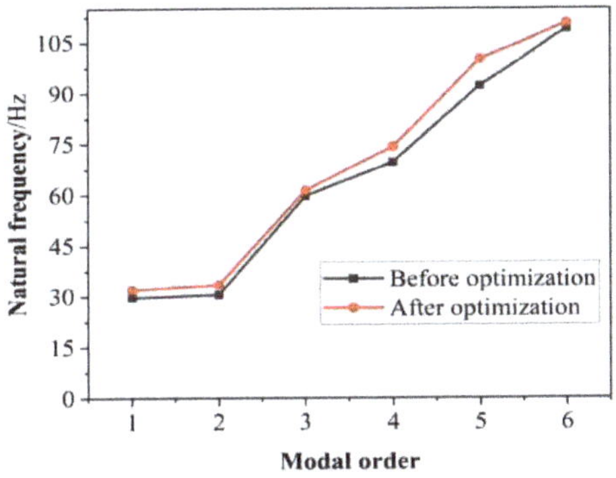

Figure 14. Comparison diagram of the first 6 natural frequencies of the optimized front and rear universal joint.

From Figure 14, it can be seen that the natural frequency of the optimized swing angle milling head universal joint increases to varying degrees. The natural frequencies of the milling head universal joint from the first to sixth orders have been increased by 7.9%, 9.92%, 2.81%, 6.18%, 8.53% and 1.54%, respectively. Among them, the fundamental frequency of the universal joint is significantly increased, reducing the probability of resonance. The first natural frequency of the optimized universal joint is 32.03 Hz, which is higher than the 25 Hz corresponding to the highest speed of the mechanical swing angle milling head under normal working conditions. This further indicates that the improved universal joint's natural frequency will not affect the normal operation of the swing angle milling head.

Figure 15 shows the harmonic response analysis results of the universal joint in all directions before and after optimization. By comparing the response spectral lines before and after optimization, it can be seen that the frequencies at the peak points of the universal joint in the X, Y and Z directions are basically consistent with the first six modal results, which indirectly verifies the accuracy of the modal analysis. The peak values corresponding to the fundamental frequency in the X and Y directions are significantly reduced. The Y direction is relatively stable in the frequency range of 70–100 Hz, while the Z direction is relatively stable in the frequency range of 0–50 Hz. At the same time, a backward shift in the peak positions occurs to varying degrees in the three directions, which indicates a significant improvement in the dynamic performance of the universal joint and verifies the rationality of the optimization results of the universal joint.

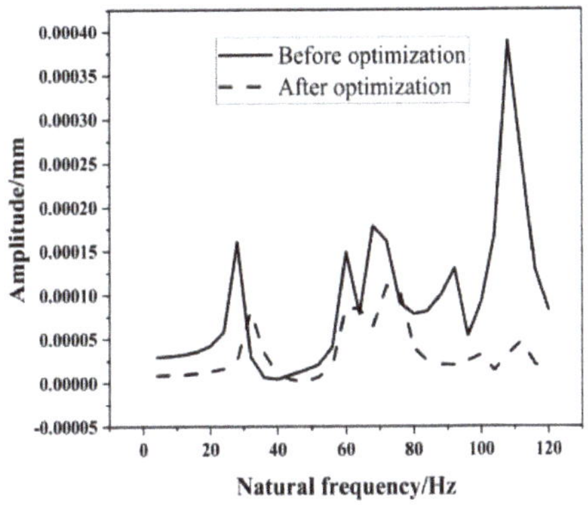

(a)

Figure 15. *Cont.*

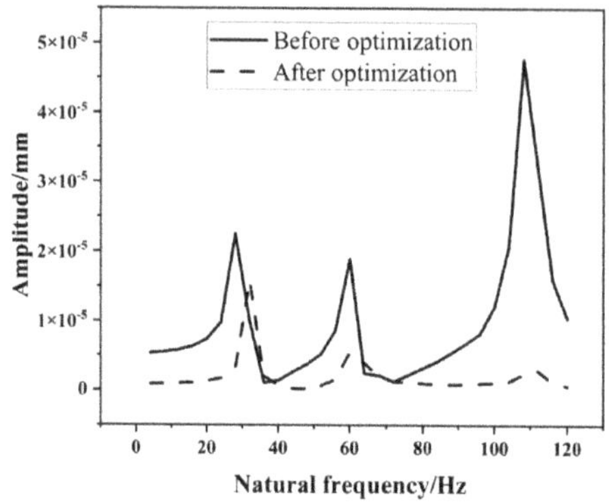

(**b**)

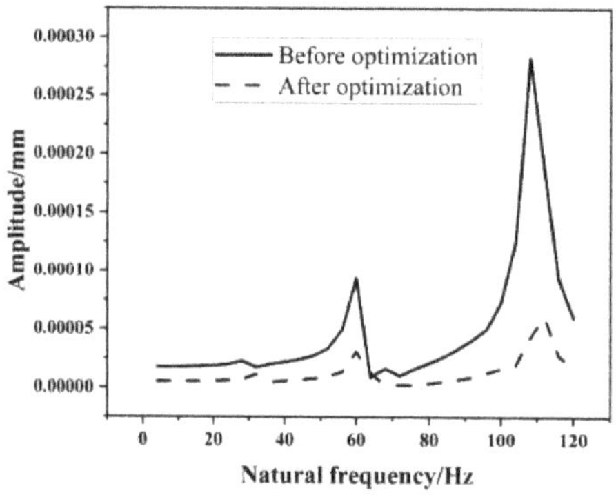

(**c**)

Figure 15. Comparison of response harmonic lines of the universal carriage: (**a**) X direction, (**b**) Y direction, (**c**) Z direction.

5. Conclusions and Prospect

In order to reduce the mass of the swing angle milling head and improve the dynamic characteristics of the whole machine, the mechanical characteristics of the swing angle milling head of a heavy-duty CNC milling machine are studied in this paper. Based on the cloud chart of the static and dynamic characteristics, it can be seen that the gimbal structure with the largest proportion of the mass of the milling head and the strongest rigidity is the research object to be optimized. The gimbal is then optimized and improved using the topology optimization method. The following conclusions can be drawn:

(1) By analyzing the static and dynamic performance of the gimbal with different stiffened plate structures, combined with the optimization results, the "ten"-type stiffened plate structure with the best performance is finally selected as the optimized structure of the upper end of the gimbal.

(2) Through the analysis of the results of the gimbal before and after optimization, it is shown that, under the premise that the stiffness and strength meet the design requirements, the weight reduction ratio of the optimized gimbal structure is 13.67%, and the weight reduction effect is obvious.

(3) After optimization, the first six natural frequencies of the gimbal are increased, among which the first natural frequency is increased by 7.9%, which can effectively avoid resonance. After optimization, the maximum response amplitudes of the gimbal in

the X, Y and Z directions are reduced, which is conducive to the improvement of the dynamic performance of the whole machine.

Since this paper only studies the gimbal with the greatest mass and the highest stiffness in the swing angle milling head, in-depth topology optimization research will be carried out for other key components of the swing angle milling head to achieve the optimal mechanical properties of the entire milling head. In addition, the optimization method proposed in this paper provides a theoretical basis and reference for the lightweight research of other key components (such as the beam, spindle box, column, etc.) in similar machine tools. This research provides an important theoretical reference and methodological support for improving the static and dynamic characteristics of machine tools.

Author Contributions: Conceptualization, Y.C. and H.M.; methodology, Y.C., C.L. and Q.H.; software, Y.C., Y.L., C.L., Q.H. and F.X.; investigation, Y.C.,Y.L., Q.H. and H.M.; resources, Y.C., H.J. and F.X.; data curation, Y.C., H.M., H.J. and F.X.; writing—original draft preparation, Y.C.; writing—review and editing, Y.C.; supervision, H.J. and Y.L.; project administration, Y.C. and Q.H.; funding acquisition, Y.C. All authors have read and agreed to the published version of the manuscript.

Funding: This research was funded by the Research Projects of Basic Scientific Research Business Expenses of Provincial Colleges and Universities in Heilongjiang Province (145109410).

Institutional Review Board Statement: Not applicable.

Informed Consent Statement: Not applicable.

Data Availability Statement: All data included in this study are available upon request by contact with the corresponding author.

Conflicts of Interest: Hui Jiang was employed by the company Qiqihar Heavy CNC Equipment Co., Ltd. The remaining authors declare that the research was conducted in the absence of any commercial or financial relationships that could be construed as a potential conflict of interest.

References

1. Chen, Q.; Zhang, S.; Li, H.; Zhou, J.; Zhang, J. Discrete variable topology optimization based on cellular mosaic grid. *Modul. Mach. Tools Autom. Process. Technol.* **2023**, *6*, 123–127.
2. Weule, H.; Fleischer, J.; Neithardt, W.; Emmrich, D.; Just, D. Structural Optimization of Machine Tools including the static and dynamic Workspace Behavior. In Proceedings of the 36th CIRP-International Seminar on Manufacturing Systems, Saarbrucken, Germany, 3–5 June 2003; Volume 12, pp. 56–60.
3. Sigmund, O.; Maute, K. Topology optimization approaches: A comparative review. *Struct. Multidiscip. Optim.* **2013**, *48*, 1031–1055. [CrossRef]
4. Xie, L.; Gao, S. Lightweight Design of Tool Holder Bracket Based on Topology Optimization. *Mech. Des. Manuf. Eng.* **2019**, *48*, 15–18.
5. Aggogeri, F.; Borboni, A.; Merlo, A.; Pellegrini, N.; Ricatto, R. Vibration Damping Analysis of Lightweight Structures in Machine Tools. *Materials* **2016**, *10*, 297. [CrossRef] [PubMed]
6. Raz, K.; Chval, Z.; Stepanek, M. Topographic Optimization of the Milling Head. In Proceedings of the International Conference on Mechanical, System and Control Engineering, Kazan, Russia, 28–30 May 2021; Spring: Singapore, 2021; pp. 171–177.
7. Trebe, M.J.; Zhao, F.; Sutherland, J.W. Modelling the effect of slide table mass on machine tool energy conversation: The role of light weighting. *J. Manuf. Syst.* **2022**, *62*, 668–680. [CrossRef]
8. Shindo, R.; Nishiwaki, S. Latest machine tool structure design technology for ultrarefining machine. *Int. J. Autom. Technol.* **2020**, *14*, 304–310. [CrossRef]
9. Ji, Q.; Li, C.; Zhu, D.; Jin, Y.; Lv, Y.; He, J. Structural design optimization of moving component in CNCMachine tool for energy saving. *J. Clean. Prod.* **2020**, *246*, 118976. [CrossRef]
10. Petra, L.; Stan, G. Topographic optimization of the mobile element structure of Gantry type CNC machine using FEA method. In *IOP Conference Series: Materials Science and Engineering*; IOP Publishing: Bristol, UK, 2021; Volume 1182, p. 012061.
11. Triene Mathhow, J.; Zhao, F.; Suthejohn, W. Genetic Optimization for the Design of a Machine Tool Slide Table for Reduced Energy Consumption. *J. Manuf. Sci. Eng.* **2021**, *3*, 1–31.
12. Ko, S.; Lee, D. Stiffness optimization of 5-axis machine tool for improving surface roughness of 3D printed products. *J. Mech. Sci. Technol.* **2017**, *31*, 3355–3369. [CrossRef]
13. Abdelwahab, M.M.; Tsavdaridis, K.D. Application of Topology Optimisation to Steel Node-Connections and Additive Manufacturing. In *Industrializing Additive Manufacturing: Proceedings of AMPA2020*; Springer International Publishing: New York City, NY, USA, 2021.

14. Kim, S.G.; Kim, J.H.; Kim, S.H.; Youn, J.W. Design optimization of the rib structure of a 5-Axis multi-functional machine tool considering static stiffness. *J. Korean Soc. Manuf. Technol. Eng.* **2016**, *25*, 313–320.
15. Beamati, S.R.; Dabbagh, V.; Aminl, H.; Sarhan, A.A.D.; Akbari, J.; Shukor, M.H.A.; Ong, Z. Multi-objective selection and structural optimization of the gantry in a gantry machine tool for IMPROVING static, dynamic, and weight and cost performance. *Concurr. Eng.* **2016**, *24*, 83–93.
16. Dimitrov, D.M.; Slavov, S.D.; Yordanov, K.K. Stiffness design of machine tools structures by topology management optimization approach. In *IOP Conference Series: Materials Science and Engineering*; IOP Publishing: Bristol, UK, 2019; Volume 564, p. 012071.
17. Chan, T.C.; Reddy SV, V.S.; Ullah, A.; Roy, B. Effect of spatial moving structure and topology optimization of the CNC turning machine tools. *Int. J. Adv. Manuf. Technol.* **2023**, *129*, 2969–2987. [CrossRef]
18. Chan, T.C.; Ullah, A.; Roy, B.; Chang, S.-L. Finite element analysis and structure optimization of a gantry-type high-precision machine tool. *Sci. Rep.* **2023**, *13*, 13006. [CrossRef] [PubMed]
19. Stopper, G.; Douglas, S. *Numerical Methods in Fine Element Analysis*; Pretentice-Hall, Inch.: London, UK, 2010.
20. Hu, S.; Gu, H. Lightweight design of beam based on finite element analysis of machine tool. *Comb. Mach. Tool Autom. Process. Technol.* **2020**, 139–143.
21. Bakker, C.; Zhang, L.; Higginson, K.; van Keulen, F. Simultaneous optimization of topology and layout of modifier stingers on shells and plates. *Struct. Multi-Discip. Optim.* **2021**, *64*, 3147–3161. [CrossRef]
22. Jiang, S.; Jia, R.; Wang, J.; Gao, F. Research on Lightweight Design of Beam Structure of Double-column Vertical Lathe. *Mod. Manuf. Eng.* **2022**, *4*, 143–148.
23. Liu, C.; Tan, F.; Wang, L.; Cai, Z. Research on Optimization Design of Column Structure for Dynamic Performance of Machine Tool. *J. Mech. Eng.* **2016**, *52*, 161–168. [CrossRef]
24. Lu, J.; Sun, J. *Principles of Metal Cutting and Tools*; China Machinery Industry Press: Beijing, China, 2011.
25. Wang, Z.; Li, T.; Chen, S. Mechanical analysis of direct drive a/c double swing angle heavy milling head. *Mech. Des.* **2012**, *29*, 83–86.
26. Tang, W.; Yi, H.; Xing, Y. Structural analysis of machining center bed. *Mech. Strength* **1998**, *20*, 11–18.
27. Dbouk, T. A review about the engineering design of optimal heat transfer systems using topology optimization. *Appl. Therm. Eng.* **2017**, *112*, 841–854. [CrossRef]
28. Mirzendehdel, A.M.; Behandish, M.; Nelaturi, S. Topology optimization with accessibility constraint for multi-axis machining. *Comput. Aided Des.* **2020**, *122*, 102825. [CrossRef]

materials

MDPI

Article

Modification of Epoxides with Metallic Fillers—Mechanical Properties after Ageing in Aqueous Environments

Anna Rudawska [1,2,†], Jakub Szabelski [1,2,*,†], Mariaenrica Frigione [2] and Valentina Brunella [3]

1 Faculty of Mechanical Engineering, Lublin University of Technology, Nadbystrzycka 36, 20-618 Lublin, Poland; a.rudawska@pollub.pl
2 Department of Innovation Engineering, University of Salento, Via Arnesano, 73100 Lecce, Italy; mariaenrica.frigione@unisalento.it
3 Department of Chemistry, University of Torino, Via P. Giuria 7, 10125 Torino, Italy; valentina.brunella@unito.it
* Correspondence: j.szabelski@pollub.pl
† Serve as the visiting professor during the research.

Abstract: The aim of this research was a comparative analysis of selected mechanical properties of epoxy compounds that were modified with metallic fillers and aged in aqueous environments. The tested epoxy compounds consisted of three components: styrene modified epoxy resin based on Bisphenol A, triethylenetetramine curing agent (resin/curing agent ratio of 100:10) and two types of metallic fillers in the form of particles: aluminum alloy (EN AW-2024–AlCu4Mg1) and tin-phosphor bronze (CuSn10P). Samples were subjected to ageing in 4 water environments: low-, medium- and high-mineralized natural water and in a sugar-containing solution for 1, 2 and 3 months. The epoxy samples were subjected to compressive strength tests in accordance with the ISO 604:2002 standard. It was observed that, among others, the compositions seasoned in low-mineralized water usually achieved the highest average compressive strength. As for filler type, using the bronze filler (CuSn10P) usually achieved the highest average compressive strength results.

Keywords: epoxy; adhesives; modification; metallic fillers; ageing; aqueous environments; mechanical properties; compressive strength

1. Introduction

Modifications of polymeric materials are aimed at improving selected properties of the cured material [1]. They are most commonly carried out to modify the crosslinking reaction and influence the obtained network structure [2] in order to extend the life of the composition, reduce the exothermic effect of the crosslinking reaction [3] or accelerate the crosslinking process [4]. In addition, they can provide, for example, a reduction in manufacturing costs [5]. Their tasks are also to reduce the brittleness of the cured resin [6,7]; increase mechanical strength [8,9]; increase resistance to dynamic loads [10]; improve chemical [11], dielectric [12,13] and thermal resistance [14,15]; minimize flammability [16] and smokiness [17] and in general, expand the range of applications.

For polymer materials, there are several types of modifications: chemical modifications, physical modifications and physical–chemical modifications. Chemical modifications involve, for example, changing the type and amount of hardener to adjust the curing time, strength, chemical and low/high temperature resistance of the material [18–21]. In physical modifications of epoxy resins, the most commonly used agents include fillers, stabilizers, plasticizers and thinners. To expand the range of applications, special modifiers or other additives are used to change the properties of the substance. Typically, about 20% of the weight of artificial materials is simply modifying agents. With the help of some fillers, the fluidity of adhesives can be adjusted [22]. It is possible to increase the viscosity of the composition to the point that a paste is formed. When fillers are used in epoxy compounds, they reduce shrinkage and at the same time increase the strength of the joint (even doubling it). A distinction is made between fillers of the often-used type: (quartz sand or aluminum oxide, or slightly more expensive—silver flakes) and very expensive ones (such as boron fibers).

Adding reduced graphene oxide (RGO) up to 0.5 wt% increased the glass transition temperature, the modulus of elasticity and the strength of the epoxy adhesive and reduced the adhesive's strain at the break at the ambient temperature, but the ultimate strain of RGO-reinforced adhesives decreased slightly [23]. Various types of biofillers [24–27], such as waste linseed cake, can also be used in small amounts to improve the value of the material's storage modulus [28]. Biochar derived from wheat straw can be used to double the Young's modulus and tensile strength, and biochar from miscanthus can be used to increase the elongation by 40% [6]. Other biochars are also used [29,30]: bamboo fibers [31], palm fibers [32], microcork particles [33], etc.

Aluminum oxide, silica and asbestos (inorganic fillers) increase strength and also contribute to heat resistance [14,34,35]. Fillers are also used to reduce or increase density. The introduction of a filler into the composition (by choosing the right type, as well as its granulation and pre-treatment) can also change chemical resistance and water absorption. By using silica, the rate of water vapor diffusion is reduced, while by using titanium white, the adsorption of solvent vapors is reduced.

However, the use of modifiers can lead to undesirable effects. The most important of these is increased viscosity of the composition, which makes it more difficult to process. In addition, problems such as poor adhesion, blistering, unevenness or discoloration on the surface and other potentially unfavorable hazards can occur.

The operating environment of polymeric materials has a significant impact on their strength and durability. Aging is associated with changes in the characteristics of the polymer material due to factors such as UV radiation [36–38], temperature and its variation [39–43], humidity/water environment [44–51], chemical agents [52–55] or simply long storage time [56]. Long-term exposure to atmospheric factors can lead to a reduction in their tensile, compressive and bending strength, etc. Some materials may become more brittle or react with a reduction in elasticity. In order to minimize the effects of the seasoning environment on the mechanical properties of polymeric materials, additives such as UV stabilizers and antioxidants, protective coatings and structural changes are used to minimize the material's exposure to adverse conditions. It is then important to regularly test materials to monitor changes in mechanical properties. Thus, it seems important to study the behavior of epoxy materials with different types of fillers, seasoned in varying environments, as this comparison can provide important insights into the effect of seasoning time and type of operating environment on the strength of the target epoxy material, such as in [40].

Many construction solutions require materials with significant compression strength and adhesive properties, as they are intended as joining or sealing materials [57]. In many cases, the function of bonding and sealing is also combined with resistance to external factors, e.g., moisture or water (including rain). This situation may apply to the construction of building elements. Therefore, the aim of this work was to propose the preparation of modified epoxy materials, which are assumed to have greater strength compared to unmodified materials. Additionally, the compressive strength of such elements in various water environments and also in a sugar-containing environment was assessed. The aim was to obtain information that could be used when selecting materials for construction solutions.

In this paper, a comparative analysis of selected mechanical properties of epoxy compounds modified with metallic fillers (aluminum alloy and tin-phosphorus bronze) that were seasoned in aqueous environments was carried out. Five experimental environments were analyzed as aqueous environments: three aqueous environments with different degrees of mineralization (low-mineralized water, medium-mineralized water and high-mineralized water), a carbonated beverage with high sugar content and an ambient environment in which the seasoning process was not implemented—for comparison purposes. The experimental study provided information on the properties of modified epoxy compounds exposed to aqueous environments, which can be used in the design of adhesive joints and the development of adhesive technology.

2. Materials and Methods

2.1. Epoxy Compound

The following components were used in epoxy compounds: a styrene-modified epoxy resin based on a bisphenol A (Epidian 53) with an epoxide number of min. 0.41 mol/100 g (DGEBA and styrene) and a triethylenetetramine curing agent (Z-1 trade name) with an amine number of min. 1100 mg KOH/g. Both main components of the epoxy composition were produced by the "Organika-Sarzyna" Chemical Plant in Nowa Sarzyna, Poland.

The selected properties of the epoxy resin and curing agent are shown in Table 1.

Table 1. Properties of epoxy resin.

Properties	Value
Flash point	75 [°C]
Boiling point	1410 [°C]
Density at 20 °C	1.11–1.15 g/cm^3
Viscosity at 20 °C	900–1500 mPa·s
Epoxy number	0.41 mol/100 g
Gelation time with Z-1 curing agent	60 min

A triethylenetetramine curing agent (amine type) used most often for compounds with low molecular weight epoxy resins. Table 2 shows the properties of amine curing agent.

Table 2. Properties of triethylenetetramine curing agent.

Properties	Value
Amine number	min. 1100 mgKOH/g
Boiling point at 13 hPa	143 °C
Boiling point at 66.5 hPa	183 °C
Viscosity at 25 °C	20–30 mPa·s
Density at 25 °C	0.98 g/cm^3

Two types of particulate materials were used as metallic fillers: aluminum alloy (EN AW-2024 AlCu4Mg1) and tin-phosphorus bronze (CuSn10P).

Aluminum alloy (EN AW-2024) is a lightweight metal with low oxidation resistance and is difficult to weld. It is characterized by a large amount of copper and high fatigue strength. These materials find important applications in the automotive and aerospace industries, as well as in the construction industry—if there is no risk of corrosion. The physical properties of the aluminum alloy used are shown in Table 3. The particle size of the aluminum alloy was 0.08 ÷ 0.10 mm.

Table 3. Physical properties of aluminum alloy (EN AW-2024 AlCu4Mg1).

Properties	Value
Density	2.78 g/cm^3
Tensile strength Rm	360–435 MPa
Hardness	104–123 HB

Tin-phosphorus bronze (CuSn10P) is an alloy of copper with tin and an admixture of phosphorus. The copper content is approximately 90%, the tin content is 1–9% and the phosphorus content is 0.1–0.3%. It is typically used in the manufacture of components for heavy-duty and high-speed operation, e.g., corrosion-prone bearings; machine parts including shafts, bushings and other components; and chemical fittings. Since tin-phosphorus bronze is characterized by very good self-lubricating properties, it is used for the manufacture of poorly lubricated components, such as those that are difficult to access and exposed to long intervals of operation without maintenance. The particle size of the tin-phosphorus bronze was 0.05 ÷ 0.10 mm. The basic physical properties of bronze are shown in Table 4.

Table 4. Physical properties of tin-phosphorus bronze (CuSn10P).

Properties	Value
Density	7.5–9.3 g/cm^3
Casting shrinkage	1.5–2.5%
Melting point	940–1084 °C

The main components of the epoxy composition, the resin and curing agent were mixed, taking into account the appropriate stoichiometric ratio recommended for these specific components. For this resin and curing agent, the stoichiometric resin:curing agent ratio is 100:10. The compounds adopted in the research plan (Table 5) were prepared by mixing the ingredients for 2 min using a table-top drill at 730 rpm and then placed in molds with walls coated with release agent. The epoxy compound samples used in the study were produced in a cylindrical mold of diameter ∅12 ± 1 mm and length 37 ± 1 mm (Figure 1).

Table 5. Epoxy compounds.

Resin and Curing Agent	Filler	Denotation
DGEBA + styrene (Epidian 53) (100 g) and TETA (Z-1) (10 g) (10:1)	-	E53/Z-1
	1% aluminum alloy AW-2024 AlCu4Mg1	E53/Z-1/Al
	1% tin CuSn10P	E53/Z-1/Br

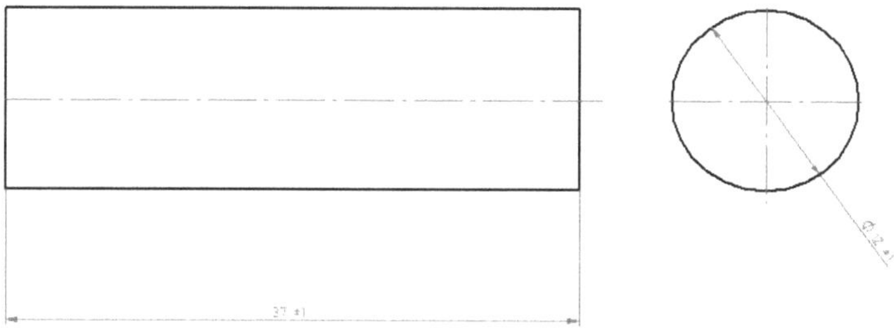

Figure 1. The cylindrical epoxy compound sample.

The following conditions were used during curing: temperature: 20 ± 2 °C, humidity: 26 ± 1% and curing time: 7 days. A total of 78 samples of each epoxy compound were made.

2.2. Test Environment

Five test environments were used for the study. The aqueous environments encompassed water classified into three different groups according to the current Regulation of the Polish Minister of Health of 31 March 2011 on natural mineral waters, spring waters and table waters, Annex 5—"Criteria for chemical classification used in the labelling of natural mineral waters". These were: low-mineralized water, medium-mineralized water and high-mineralized water (Table 6). A carbonated beverage with a high sugar content was used as a fourth environment. In addition, some samples were analyzed in ambient environments (without a seasoning process). Table 7 shows the labelled operating environments.

Table 6. Aquatic environment parameters.

Aqueous Environments	Low-Mineralized Water (LMW)	Medium-Mineralized Water (MMW)	High-Mineralized Water (HMW)
Trade name	Spring water Aleksandria	Natural mineral water from the Sudety	CECHINI MUSZYNA
Manufacturer	Polska Woda, Ozorków, Poland	Mineral, Gorzanów, Poland	Cechini, Muszyna, Poland
Mineral quantities	50–500 mg/L	500–1500 mg/L	1500–4000 mg/L
CATIONS			
Calcium	52.1 mg/L	198.0 mg/L	325.8 mg/L
Magnesium	7.3 mg/L	18.4 mg/L	61.6 mg/L
Sodium	2.5 mg/L	94.3 mg/L	33.2 mg/L
Potassium	0.86 mg/L	7.3 mg/L	3.7 mg/L
ANIONS			
Bicarbonate	143.4 mg/L	812.0 mg/L	1314.0 mg/L
Sulphate	27.8 mg/L	54.0 mg/L	16.5 mg/L
Chlorine	10.3 mg/L	40.0 mg/L	8.9 mg/L
Fluorine	0.13 mg/L	0.49 mg/L	0.11 mg/L

Table 7. Labeling of ageing environments.

Type of Environment	Label
Ambient conditions	O
Low-mineralized water	LMW
Medium-mineralized water	MMW
High-mineralized water	HMW
Carbonated high sugar beverage	C

The water types used for testing are extracted from underground and show purity and the absence of chemical as well as bacterial contamination. They all have a one-year expiry date set by the manufacturers (LMW: Polska Woda, Ozorków, Poland, MMW: Mineral, Gorzanów, Poland, HMW: Cechini, Muszyna, Poland) and specific storage conditions (dark and cool places). The ingredients of the sweet drink used in the study are water, glucose-fructose syrup, sugar, color, ammonia-sulphite caramel, phosphoric acid, acidity regulator and flavors. It is also unlike the waters presented above in being carbonated. This drink, according to the manufacturer's recommendation, should be stored in shaded and cool areas.

2.3. Experimental Research Methodology

The modified epoxy samples were cured for 7 days at 20 $\pm$ 2 °C and humidity: 26 $\pm$ 1%. Series of 6 cured samples were placed in varying operating environments including one series in ambient conditions. The seasoning times for each series were: 1, 2 and 3 months. The seasoning conditions (shaded, cool location) were identical for all samples.

After fabrication, the specimens of the cured epoxy compounds were milled to remove surface irregularities, which was necessary for proper strength testing. Compressive strength tests were carried out on a ZWICK/ROELL Z150 machine (Zwick Roel, Ulm, Germany) according to the ISO 604:2002 Plastics—Determination of compressive properties [58], with constant parameters for all the specimens. Due to the fact that for many construction materials, e.g., those used in the building industry, the basic mechanical property is compressive strength, an attempt was made to determine this type of strength to compressive load.

3. Results

3.1. Variability of Obtained Results

The coefficient of variation (COV), as a measure of the variability of the data in a series of samples, describes the degree of dispersion or variability of the data (standard deviation SD) in relation to its mean value (MV). It is expressed by the Formula (1):

$$COV = (SD/MV) \times 100\% \quad (1)$$

A coefficient of variation of less than 15% is estimated to define a series with low variability, i.e., those in which the individual values are relatively close to the mean value and there are no large discrepancies between them.

Table 8 shows the summary results of the distribution of results in the individual series, depending on the material tested, the environment and the seasoning time.

Table 8. Coefficient of variation of sample series.

Environment	Seasoning Time	Epoxy Compound E53/Z-1	 E53/Z-1/Al	 E53/Z-1/Br
C	1 month	2.16%	3.52%	6.34%
	2 months	2.13%	3.33%	5.08%
	3 months	1.69%	8.06%	7.66%
LMW	1 month	3.47%	4.10%	3.65%
	2 months	3.36%	4.57%	5.86%
	3 months	2.32%	7.32%	3.23%
MMW	1 month	3.61%	7.85%	4.62%
	2 months	3.11%	4.48%	3.85%
	3 months	2.25%	5.99%	8.90%
HMW	1 month	2.21%	1.95%	5.64%
	2 months	1.01%	3.09%	3.23%
	3 months	2.60%	1.47%	6.51%
	Mean COV	2.49%	4.65%	5.38%

It can be observed that all the series studied had a very low variability, somewhere between 1 and 9%. At the same time, there is a clear difference between the variability of the results of the individual series depending on the tested epoxy material. The E53/Z-1/Al showed almost twice the average variability of the unmodified epoxy compound (E53/Z-1), while E53/Z-1/Br showed more than twice as much variability. Still, the results of the dispersions make it possible to describe all series as having low variability. This indicates, on the one hand, a high consistency of results and, indirectly, also a correctly selected minimum number of samples in each series.

A more detailed analysis of the dependence of the degree of dispersion of the results in relation to the seasoning time and the seasoning environment did not show similar relationships.

3.2. Comparative Analysis of Test Results within the Same Seasoning Time

3.2.1. Test Results for Unseasoned Epoxy Compound (O)

The results of the unseasoned epoxy compound samples that were not seasoned in aqueous environments were analyzed. Figure 2 shows the strength results of the samples of the epoxy compound tested: E53/Z-1, E53/Z-1/Al, E53/Z-1/Br.

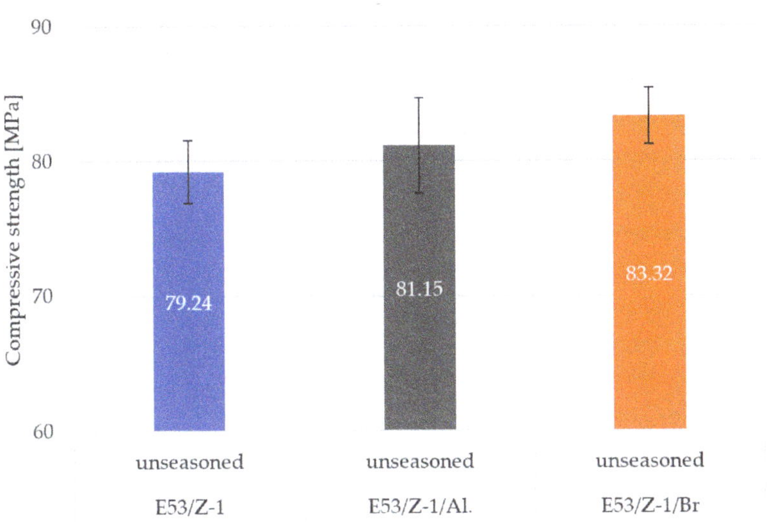

Figure 2. Comparative analysis of strength results of unseasoned epoxy compound in an aqueous environment.

The samples with the epoxy compound Epidian 53 with Z-1 curing agent and bronze filler (E53/Z-1/Br) clearly showed the highest average compressive strength of 83.32 MPa. The difference between the average strength of the strongest and weakest samples was only 5%. The scatter in the results obtained with the E53/Z-1/Al epoxy compound was noticeably greater than with the other two epoxy compounds.

3.2.2. Test Results for Epoxy Compounds Seasoned in Different Environments

Table 9 shows a summary of the compressive strength of epoxy compounds results obtained, depending on the environment and seasoning time.

Table 9. Compressive strength (mean values) of epoxy compounds.

Environment	Seasoning Time	Epoxy Compound		
		E53/Z-1	**E53/Z-1/Al**	**E53/Z-1/Br**
C	1 month	79.90	80.88	81.72
	2 months	77.60	81.03	79.73
	3 months	75.63	80.98	80.30
LMW	1 month	84.30	82.37	84.28
	2 months	82.67	85.42	84.50
	3 months	84.22	83.06	90.58
MMW	1 month	83.00	79.30	87.65
	2 months	82.00	78.72	85.84
	3 months	83.38	79.90	83.44
HMW	1 month	84.68	80.67	82.53
	2 months	81.90	80.27	86.08
	3 months	80.87	81.52	83.02

When analyzing the average results obtained after 1 month of seasoning (Table 9), the following patterns were noted. Samples E53/Z-1/Al had the lowest average strength (79.30 MPa), while samples E53/Z-1/Br had the highest (87.65 MPa). The difference in average strength between the two is 8.35 MPa, representing 9.53%. Both values were obtained in a medium mineralized water environment. In high-mineralized water, the difference between the smallest (E53/Z-1/Al (80.67 MPa) and the largest mean (E53/Z-1 84.68 MPa) is only 4 MPa (4.74%), while in low-mineralized water and the CO_2-saturated high-sugar beverage, the differences were only about half the size.

The individual compositions responded differently to the seasoning environment. E53/Z-1 between the highest (in high-mineralized water 84.68 MPa) and lowest average strength (for the high-sugar beverage 79.90 MPa) achieved a difference of 4.78 MPa (5.65%). For the E53/Z-1/Al composition, the difference was 3.73% (max. in low-mineralized water 82.37 MPa, min. in medium-mineralized water 79.30 MPa) and E53/Z-1/Br 5.93 MPa, i.e., 6.75% (min. in high-sugar beverage 81.72 MPa, max. in medium-mineralized water 87.65 MPa).

Looking in detail at the results obtained from the samples after 2 months of seasoning, it can be seen that the highest average compressive strength among the compositions tested was obtained for samples seasoned in highly mineralized water in the case of the E53/Z-1/Br composition (86.08 MPa), while the lowest was obtained for the E53/Z-1 composition seasoned in a high-sugar beverage (77.6 MPa). The difference between these extreme averages was 8.48 MPa, which represents 9.8%. The behavior of the tested compositions in low-mineralized water and the high-sugar beverage was quite similar, as can be seen by only a 3.2% and 4.23% difference between the largest and smallest results obtained. However, in the case of the high-mineralized water, it was 6.75%, and in the medium-mineralized water, it was 8.3% (78.72 MPa–E53/Z-1/Al vs. 85.84 MPa E53/Z-1/Br).

Analyzing the results obtained after 3 months of seasoning, the following observations can be noted. There is a clear difference of up to 16.5% between the most and least durable composition (E53/Z-1 in high sugar beverage 75.63 MPa vs. E53/Z-1/Br in low mineralized water 90.58 MPa). Within the individual seasoning environments, the differences were significantly smaller: 8.3% in the low-mineralized water, 6.6% in the high-sugar beverage, 5.4% in the medium-mineralized water and 2.6% in the high-mineralized water.

3.3. Comparative Analysis of Research Results within a Common Seasoning Environment

An attempt to summarize all the results obtained for the compressive strength of the epoxy material samples grouped according to the type of environment and seasoning time is shown in Figure 3.

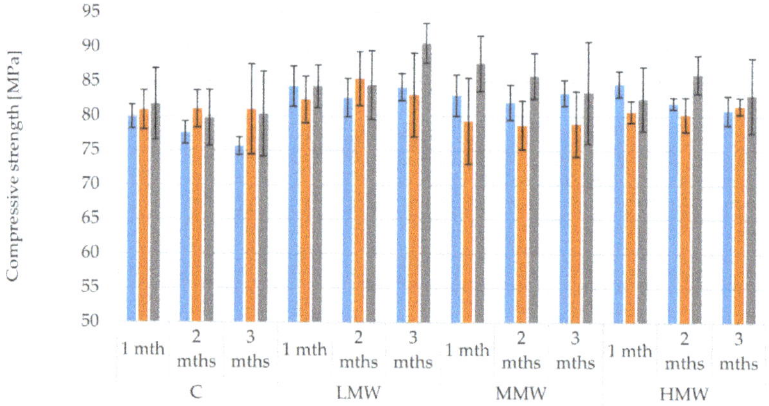

	C 1 mth	C 2 mths	C 3 mths	LMW 1 mth	LMW 2 mths	LMW 3 mths	MMW 1 mth	MMW 2 mths	MMW 3 mths	HMW 1 mth	HMW 2 mths	HMW 3 mths
E53/Z-1	79.90	77.60	75.63	84.30	82.67	84.22	83.00	82.00	83.38	84.68	81.90	80.87
E53/Z-1/Al	80.88	81.03	80.98	82.37	85.42	83.06	79.30	78.72	78.90	80.67	80.27	81.52
E53/Z-1/Br	81.72	79.73	80.30	84.28	84.50	90.58	87.65	85.84	83.44	82.53	86.08	83.02

Figure 3. Comparative analysis of compressive strength results of epoxy compounds specimens grouped by seasoning medium.

Analyzing the results obtained after seasoning in low-mineralized water, it was noted that, irrespective of the seasoning time, the difference between the highest (E53/Z-1/Br for 3 months: 90.58 MPa) and lowest (E53/Z-1/Al for 1 month: 82.37 MPa) average compressive strength obtained was 8.21 MPa (9.1%). Within the individual epoxy compounds, in fact, a notable difference was only observed between the samples made from E53/Z-1/Br seasoned for 1 and 3 months: namely, the 3-month samples proved to be 6.3 MPa stronger (7%). The other two compositions had differences between minimum and maximum average compressive strengths in the range of 2–3%.

In the case of medium mineralized water, the higher average strength of the epoxy samples filled with tin-phosphorus bronze was clearly seen, especially at the beginning of the seasoning. The difference between the least durable material (E53/Z-1/Al seasoned 2 months—78.72 MPa) and the most durable (E53/Z-1/Br seasoned 1 month—87.65 MPa) exceeded 10% (8.93 MPa). Comparing the average values obtained for the different materials, it is evident that the influence of the seasoning of the undoped material (E53/Z-1) and aluminum-doped material (E53/Z-1/Al) is very small: 0.7–1.2%. Only the samples in which bronze (E53/Z-1/Br) was used achieved a decrease in average compressive strength of 4.2 MPa (approx. 5%).

Samples aged in highly mineralized water showed average strength changes with seasoning time. Globally, a 6.7% difference was registered between the strongest and weakest samples—the average strength of the bronze-doped epoxy was 5.8 MPa higher than that of the aluminum-filled epoxy. It was the E53/Z-1/Al samples that were the most stable during seasoning, showing only a 1.5% difference between the seasoning series. A 4.1–4.5% difference (3.55 MPa–3.81 MPa) was observed between the undoped and bronze-doped samples.

Using aluminum filler again resulted in the greatest stability of the average compressive strength results when the material was seasoned in a CO_2-saturated, high-sugar beverage. These values were essentially unchanged over time (0.1% difference). A slightly larger change was registered for epoxy with bronze (E53/Z-1/Br)—2.4%. In contrast, for the unmodified specimens, degradation progressing with seasoning was observed, changing by 4.3 MPa (5.3%). Comparing all materials with each other, the greatest difference was observed between the unmodified samples (75.63 MPa at month 3 of seasoning) and the modified bronze, which reached an average strength of 91.72 MPa after one month of seasoning. This gives a 7.5% difference between the averages.

3.4. Changes in Average Compressive Strength of Individual Epoxy Compounds

Figure 4 shows the change in compressive strength compared to the unseasoned material specimens (Figure 2). A detailed analysis of the strength changes as a function of seasoning time and environment is presented for the individual compositions later in this paper.

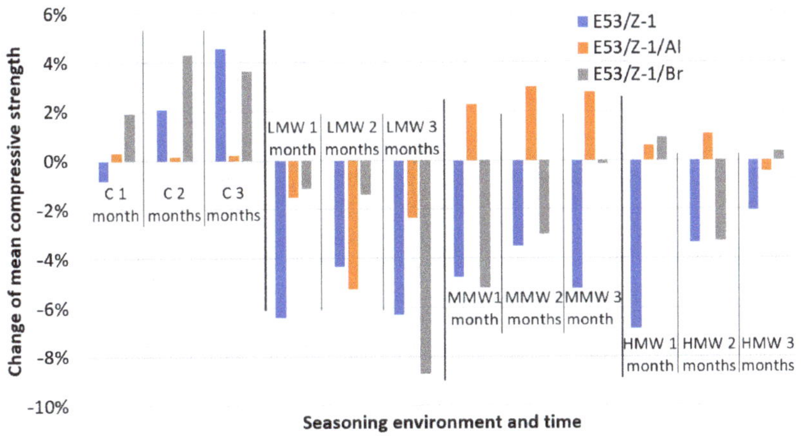

Figure 4. Change in mean compressive strength of tested samples of epoxy compounds.

3.4.1. Unmodified Epoxy Compound E53/Z-1

The results of the compressive strength tests were analyzed according to the type of epoxy compound—E53/Z-1, which was seasoned in four environments for one, two and three months, and epoxy compounds that were not seasoned. The samples that were not seasoned achieved an average compressive strength of 79.24 MPa. The seasoned epoxy compound showed little change in average strength in two cases: when seasoned in low- and medium-mineralized water. In highly mineralized water and the Cola beverage, a progressive decrease in average compressive strength of up to approximately 5% was recorded with seasoning time. When analyzing the differences between the various compositional behaviors in the seasoning environments studied, it is worth noting that the epoxy compound showed a particularly large deterioration in compressive strength during seasoning in the carbonated high-sugar beverage, with the average values in this group being noticeably lower than in the mineralized water groups.

3.4.2. Epoxy Compound Filled with Aluminum E53/Z-1/Al

The aluminum-doped composition, as noted above, proved to be quite stable and resistant to seasoning in the analyzed environments. In case of low-mineralized water—the difference in average compressive strength during seasoning did not exceed 3.6%. Water with a higher mineral content caused even less of a change in average strength: medium—0.5%, high-mineralized—1.5%. Similarly, a carbonated beverage with high sugar content registered a 0.2% change. Regarding the overall effect of the seasoning environment on the average compressive strength value, it is clear that lower values were obtained in the medium mineralized water compared to the other waters and the beverage. Meanwhile, the highest values were obtained in the group of samples seasoned in low mineralized water. However, it is difficult to draw detailed conclusions on this basis, as precisely these two groups (samples seasoned in low- and medium-mineralized water) were characterized by clearly higher standard deviation values.

3.4.3. Epoxy Compound with Tin-Phosphor Bronze E53/Z-1/Br

Analyzing the results obtained, it can be seen that quite large differences were registered within the samples seasoned in different environments. Between the least and most durable variants, the difference reached 12%. Clearly, the lowest values of average compressive strength were obtained with the epoxy compound in the environment of a carbonated beverage with high sugar content. And in this environment, the results obtained were the most consistent, with a min-max difference of 2.4%. The less mineralized the water used for seasoning, the greater the differences within the group. That is, for highly mineralized water, 3.5 MPa (4.1%) was recorded between the lowest and highest mean strength values, for medium mineralized water it was 4.2 MPa (4.8%) and low mineralized water showed a difference of 6.3 MPa (7%).

4. Discussion

The studied epoxy compounds were based on Epidian 53 epoxy resin, Z-1 curing agent and fillers in the form of particles of aluminum alloy (EN AW-2024) and tin-phosphorus bronze (CuSn10P). The cured compounds were seasoned in four experimental environments—low-, medium-, and high-mineralized water and also in a carbonated high-sugar beverage for periods of one, two and three months.

The application of fillers to polymeric materials, in this case epoxy, was carried out in order to obtain better properties of the cross-linked material. It may happen that for the price of improving selected characteristics, others may deteriorate. The selection of the right filler and its amount for use in the epoxy is therefore crucial to achieving the desired goals. Hence, experimental studies on the effect of selected fillers on the various characteristics of specific epoxy compounds are needed.

Epoxy materials can be degraded by water sorption. The mechanism of this interaction has been analyzed in many works. When testing the effect of water on epoxy coatings, using infrared spectroscopy and dielectric measurements, the appearance of hydrogen bonds was observed during hydrothermal aging [59]. The analysis of water sorption of various epoxy compounds by gravimetric measurements and positron annihilation lifetime spectroscopy (PALS), on the other hand, showed that, regardless of the test temperature, the free volume fraction is not a decisive factor regulating the equilibrium water content, while the polarity of the system is an essential factor [60].

The degree of saline water absorption by epoxy coatings and the swelling of these coatings, regardless of the solution temperature, were lower for coatings with physical aging than for coatings without aging. Thermodynamic analysis of the diffusion parameters showed that thickening of the polymer matrix induced by physical aging and, consequently, a higher number of interactions between the polar groups of polymer chains, which reduce the number of polar groups for water diffusion, are the main factors regulating the process of water absorption [61]. Similar results were obtained for adhesive joints made with aged epoxy adhesives. Under cyclic aging conditions, the moisture diffusion constant and maximum moisture content increase with aging cycles and mechanical properties such as elastic modulus and tensile strength decrease as the number of aging cycles increases [62].

The progress of the crosslinking process is also important. It has been shown that depending on the curing temperature of DGEBA/TETA compounds, despite the achievement of complete crosslinking, different curing conditions affect the free volume and activation volume associated with viscoelastic properties. Compounds cured at temperatures below Tg exhibit higher mechanical properties and are less susceptible to water ingress than systems cured at temperatures above Tg [63].

In the present research, degradation of the average compressive strength of epoxy compound samples with metallic fillers was recorded in almost all of the analyzed cases of aging the epoxy material in a mineral water environment. Only the epoxy compound filled with aluminum resisted deterioration of properties when it was seasoned in medium and highly mineralized water. On the other hand, ageing in the environment of a carbonated beverage with high sugar content did not lead to such a clear decrease in the compressive strength properties.

To conclude, when ageing the epoxy compounds in low-mineralized water, the results show a slight improvement in strength properties. This may be due to the presence of fewer mineral components. These components generally have a negative effect on the strength of epoxy compounds.

The right choice of filler type affects the mechanical strength of the epoxy material. The study showed that bronze was the most favorable filler. The results may have indicated a higher strength due to a better distribution of bronze particles in the cylindrical molds used to shape the samples from the tested composition.

The topic of modification of epoxy compounds with metallic fillers is extensive and needs to be further explored. Conducting further research will allow us to continuously improve and gain more knowledge about the modification of epoxy compounds with metallic fillers.

5. Summary

This study was conducted to test the compressive strength of epoxy compounds with metallic fillers. Analysis of the results established that the environment and seasoning time affect the compressive strength results. The following conclusions can be drawn:

- The epoxy compound that has been seasoned in low-mineralized water usually obtains the highest strength results: the highest average compressive strength was achieved by E53/Z-1/Br samples, which were seasoned in low-mineralized water for a period of three months; the strength of this epoxy compound is 90.58 MPa,
- The seasoning environment—in the form of a carbonated beverage—is not particularly beneficial to obtaining high-strength results; it was during the seasoning of the epoxy compound in this environment that the lowest strength results usually occurred,

- The epoxy compound with bronze filler (CuSn10P) tended to obtain the highest compressive strength results,
- Samples of epoxy compounds without filler usually achieved the lowest strength results: the lowest average strength was achieved by the E53/Z-1 epoxy compound, which was seasoned for a period of three months in a carbonated high-sugar beverage and reached an average of only 75.63 MPa.

The practical recommendations resulting from the research include:

- In general, it can be seen that the content of minerals in the aqueous environment influences the mechanical properties of epoxy materials, and a smaller amount of minerals is recommended,
- A sugar-containing environment negatively affects the properties of epoxy materials; therefore, exposing such materials to this type of environment should be avoided,
- During modification, the type of modifying agent is important, which is why it is necessary to analyze the use of epoxy materials in specific construction solutions along with the required strength properties.

Author Contributions: Conceptualization, A.R. and J.S.; methodology, A.R.; software, J.S.; validation, A.R., J.S., M.F. and V.B.; formal analysis, A.R. and J.S.; investigation, A.R.; resources, A.R.; data curation, J.S.; writing—original draft preparation, A.R.; writing—review and editing, J.S.; visualization, J.S.; supervision, M.F. and V.B., project administration, A.R.; funding acquisition, A.R. All authors have read and agreed to the published version of the manuscript.

Funding: This research received no external funding.

Institutional Review Board Statement: Not applicable.

Informed Consent Statement: Not applicable.

Data Availability Statement: Data are contained within the article.

Conflicts of Interest: The authors declare no conflict of interest.

References

1. Mohan, P. A Critical Review: The Modification, Properties, and Applications of Epoxy Resins. *Polym. Plast. Technol. Eng.* **2013**, *52*, 107–125. [CrossRef]
2. Vidil, T.; Tournilhac, F.; Musso, S.; Robisson, A.; Leibler, L. Control of Reactions and Network Structures of Epoxy Thermosets. *Prog. Polym. Sci.* **2016**, *62*, 126–179. [CrossRef]
3. Prolongo, M.G.; Salom, C.; Arribas, C.; Sánchez-Cabezudo, M.; Masegosa, R.M.; Prolongo, S.G. Influence of Graphene Nanoplatelets on Curing and Mechanical Properties of Graphene/Epoxy Nanocomposites. *J. Therm. Anal. Calorim.* **2016**, *125*, 629–636. [CrossRef]
4. Romano, V.; Naddeo, C.; Vertuccio, L.; Lafdi, K.; Guadagno, L. Experimental Evaluation and Modeling of Thermal Conductivity of Tetrafunctional Epoxy Resin Containing Different Carbon Nanostructures. *Polym. Eng. Sci.* **2017**, *57*, 779–786. [CrossRef]
5. Tichý, M.; Müller, M.; Valášek, P.; Miturska, I. Effect of Rubber Powder from Waste Tyre Rubbers on Mechanical Properties of One-Component Polyurethane Putty. *Agron. Res.* **2018**, *16*, 1257–1265. [CrossRef]
6. Bartoli, M.; Giorcelli, M.; Rosso, C.; Rovere, M.; Jagdale, P.; Tagliaferro, A. Influence of Commercial Biochar Fillers on Brittleness/Ductility of Epoxy Resin Composites. *Appl. Sci.* **2019**, *9*, 3109. [CrossRef]
7. Ridzuan, M.J.M.; Abdul Majid, M.S.; Khasri, A.; Gan, E.H.D.; Razlan, Z.M.; Syahrullail, S. Effect of Pineapple Leaf (PALF), Napier, and Hemp Fibres as Filler on the Scratch Resistance of Epoxy Composites. *J. Mater. Res. Technol.* **2019**, *8*, 5384–5395. [CrossRef]
8. Giorcelli, M.; Khan, A.; Pugno, N.M.; Rosso, C.; Tagliaferro, A. Biochar as a Cheap and Environmental Friendly Filler Able to Improve Polymer Mechanical Properties. *Biomass Bioenergy* **2019**, *120*, 219–223. [CrossRef]
9. Iyer, T.; Nayak, S.Y.; Hiremath, A.; Heckadka, S.S.; Jaideep, J.P. Influence of TiO_2 Nanoparticle Modification on the Mechanical Properties of Basalt-Reinforced Epoxy Composites. *Cogent Eng.* **2023**, *10*, 2227397. [CrossRef]
10. Müller, M.; Valášek, P. Impact Strength of Filled Polymer Materials. *Mater. Sci. Forum* **2017**, *883*, 46–50. [CrossRef]
11. Petrović, Z.; Stojaković, N. Study of Epoxy Resin—Filler Interaction. *Polym. Compos.* **1988**, *9*, 42–50. [CrossRef]
12. Giorcelli, M.; Savi, P.; Khan, A.; Tagliaferro, A. Analysis of Biochar with Different Pyrolysis Temperatures Used as Filler in Epoxy Resin Composites. *Biomass Bioenergy* **2019**, *122*, 466–471. [CrossRef]
13. Li, Y.; Wong, C.P. Recent Advances of Conductive Adhesives as a Lead-Free Alternative in Electronic Packaging: Materials, Processing, Reliability and Applications. *Mater. Sci. Eng. R Rep.* **2006**, *51*, 1–35. [CrossRef]

14. Zotti, A.; Zuppolini, S.; Borriello, A.; Zarrelli, M. Effect of SiO2@polydopamine Core/Shell Nanoparticles as Multifunctional Filler for an Aeronautical Epoxy Resin. *Mater. Today Proc.* **2019**, *34*, 117–120. [CrossRef]
15. Jia, Z.-R.; Gao, Z.-G.; Lan, D.; Cheng, Y.-H.; Wu, G.-L.; Wu, H.-J. Effects of Filler Loading and Surface Modification on Electrical and Thermal Properties of Epoxy/Montmorillonite Composite. *Chin. Phys. B* **2018**, *27*, 117806. [CrossRef]
16. Bao, Q.; Wang, B.; Liu, Y.; Wang, Q.; Yang, Z. Epoxy Resin Flame Retarded and Toughed via Flexible Siloxane Chain Containing Phosphaphenanthrene. *Polym. Degrad. Stab.* **2020**, *172*, 109055. [CrossRef]
17. Sałasińska, K.; Kirpluks, M.; Cabulis, P.; Kovalovs, A.; Skukis, E.; Kozikowski, P.; Celiński, M.; Mizera, K.; Gałecka, M.; Kalnins, K.; et al. Experimental Investigation of the Mechanical Properties and Fire Behavior of Epoxy Composites Reinforced by Fabrics and Powder Fillers. *Processes* **2021**, *9*, 738. [CrossRef]
18. Szabelski, J.; Karpiński, R.; Jonak, J.; Frigione, M. Adhesive Joint Degradation Due to Hardener-to-Epoxy Ratio Inaccuracy under Varying Curing and Thermal Operating Conditions. *Materials* **2022**, *15*, 7765. [CrossRef]
19. Szabelski, J.; Karpiński, R.; Krakowski, P.; Jojczuk, M.; Jonak, J.; Nogalski, A. Analysis of the Effect of Component Ratio Imbalances on Selected Mechanical Properties of Seasoned, Medium Viscosity Bone Cements. *Materials* **2022**, *15*, 5577. [CrossRef]
20. Szabelski, J. Effect of Incorrect Mix Ratio on Strength of Two Component Adhesive Butt-Joints Tested at Elevated Temperature. *MATEC Web Conf.* **2018**, *244*, 01019. [CrossRef]
21. Sharma, R.; Mavani, A.M.; Tanna, V.L. Indigenous Development of Epoxy Resin System for Cryogenic Services and Fusion Application. *Fusion Sci. Technol.* **2023**, 1–14. [CrossRef]
22. Chen, Y.; Zhang, C.; Yang, W.; Shi, X.; He, Y.; Chen, X. Study on the Relationship Between Thermal Conductivity and Viscosity of Al2O3/BN Epoxy Composite Insulating Materials. *Gaodianya Jishu High Volt. Eng.* **2019**, *45*, 2715–2721. [CrossRef]
23. Khabaz-Aghdam, A.; Behjat, B.; Marques, E.A.S.; Carbas, R.J.C.; Silva, L.F.M.D.; Roghani-Mamaqani, H. Effect of Reduced Graphene Oxide on Mechanical Behavior of an Epoxy Adhesive in Glassy and Rubbery States. *J. Compos. Mater.* **2021**, *55*, 3839–3848. [CrossRef]
24. Rudawska, A.; Szabelski, J.; Miturska-Barańska, I.; Doluk, E. Biological Reinforcement of Epoxies as Structural Adhesives. In *Structural Adhesives: Properties, Characterization and Applications*; John Wiley: Hoboken, NJ, USA, 2023; pp. 31–104. ISBN 978-1-394-17560-4.
25. Baroncini, E.A.; Kumar Yadav, S.; Palmese, G.R.; Stanzione, J.F. Recent Advances in Bio-based Epoxy Resins and Bio-based Epoxy Curing Agents. *J Appl. Polym. Sci* **2016**, *133*, app.44103. [CrossRef]
26. Fernando, P.H.L.; Antonino, L.D.; Garcia, G.E.S.; De Sousa Júnior, R.R.; Neto, A.V.; Nakamoto, F.Y.; Dos Santos, D.J. Effects of the Incorporation of Modified Kraft Lignin on the Mechanical Properties of Epoxy Adhesive: Experimental and Theoretical Approaches. *J. Adhes.* **2023**, 1–13. [CrossRef]
27. Sienkiewicz, A.; Czub, P. Modification of Epoxy Compositions by the Application of Various Fillers of Natural Origin. *Materials* **2023**, *16*, 3149. [CrossRef]
28. Matykiewicz, D.; Mysiukiewicz, O. Epoxy Composites Reinforced with Natural Fillers Such as Flax Fiber and Linseed Cakes (Rapid Communication). *Polim. Polym.* **2020**, *65*, 828–832. [CrossRef]
29. Khan, A.; Savi, P.; Quaranta, S.; Rovere, M.; Giorcelli, M.; Tagliaferro, A.; Rosso, C.; Jia, C.Q. Low-Cost Carbon Fillers to Improve Mechanical Properties and Conductivity of Epoxy Composites. *Polymers* **2017**, *9*, 642. [CrossRef]
30. Saba, N.; Jawaid, M.; Alothman, O.Y.; Paridah, M.; Hassan, A. Recent Advances in Epoxy Resin, Natural Fiber-Reinforced Epoxy Composites and Their Applications. *J. Reinf. Plast. Compos.* **2016**, *35*, 447–470. [CrossRef]
31. Hu, H.; Cheng, X.; Liu, Z.; Chen, X.; Piao, R.; Xiang, C. Influence of Bamboo Fibers Weight Fraction on the Quasi-Static and Dynamic Compressive Responses of Epoxy Matrix Composites. *J. Nat. Fibers* **2023**, *20*, 2246646. [CrossRef]
32. Akhavan-Safar, A.; Barbosa, A.Q.; Ayatollahi, M.R.; da Silva, L.F.M. Influence of Microcork Particles on the Lap Shear Strength of an Epoxy Adhesive Subjected to Fatigue Loading and Different Environmental Conditions. *Proc. Inst. Mech. Eng. Part L J. Mater. Des. Appl.* **2020**, *234*, 851–858. [CrossRef]
33. Akhavan-Safar, A.; Delzendehrooy, F.; Ayatollahi, M.; da Silva, L.F.M. Influence of Date Palm Tree Fibers on the Tensile Fracture Energy of an Epoxy-Based Adhesive. *J. Nat. Fibers* **2022**, *19*, 14379–14395. [CrossRef]
34. Szewczak, A.; Szeląg, M. Physico-Mechanical and Rheological Properties of Epoxy Adhesives Modified by Microsilica and Sonication Process. *Materials* **2020**, *13*, 5310. [CrossRef] [PubMed]
35. Khanum, K.K.; Kostalas, A.; Jayaram, S.; Ulcar, J. Synergistic Effect of Nano-Silica and Micro-Silica Fillers on the Rheological Properties of Epoxy Resin. In Proceedings of the 2020 IEEE Industry Applications Society Annual Meeting, Detroit, MA, USA, 10–16 October 2020.
36. Francis, U.U.; Philip, A.N. Degradation of Epoxy Reinforced Banana Fibers and Eggshell Particles Hybrid Composite High-Voltage Insulators via Accelerated UV Aging Processes. *Chem. Data Collect.* **2022**, *38*, 100842. [CrossRef]
37. Korkmaz, Y.; Gültekin, K. Effect of UV Irradiation on Epoxy Adhesives and Adhesively Bonded Joints Reinforced with BN and B4C Nanoparticles. *Polym. Degrad. Stab.* **2022**, *202*, 110004. [CrossRef]
38. Torres, M.; Burdin, L.; Rentería-Rodríguez, A.V.; Franco-Urquiza, E.A. Degradation of Epoxy–Particles Composites Exposed to UV and Gamma Radiation. *Chemistry* **2023**, *5*, 559–570. [CrossRef]
39. de Castro Lopes, F.V.B.; Akhavan-Safar, A.; Carbas, R.J.C.; Marques, E.A.S.; Goyal, R.; Jennings, J.; da Silva, L.F.M. The Interaction of Loading Mode and Humidity on the Properties Degradation of an Epoxy Adhesive Subjected to Strength, Fracture, and Fatigue Tests. *J. Appl. Polym. Sci.* **2023**, *140*, e53490. [CrossRef]

40. da Costa, J.A.; Akhavan-Safar, A.; Marques, E.A.S.; Carbas, R.J.C.; da Silva, L.F.M. Effects of Cyclic Ageing on the Tensile Properties and Diffusion Coefficients of an Epoxy-Based Adhesive. *Proc. Inst. Mech. Eng. Part L J. Mater. Des. Appl.* **2021**, *235*, 1451–1460. [CrossRef]
41. Rudawska, A.; Abdel Wahab, M.; Szabelski, J.; Miturska, I.; Doluk, E. The Strength of Rigid and Flexible Adhesive Joints at Room Temperature and after Thermal Shocks. *Lect. Notes Civ. Eng.* **2021**, *110*, 229–241. [CrossRef]
42. Szabelski, J.; Karpiński, R.; Machrowska, A. Application of an Artificial Neural Network in the Modelling of Heat Curing Effects on the Strength of Adhesive Joints at Elevated Temperature with Imprecise Adhesive Mix Ratios. *Materials* **2022**, *15*, 721. [CrossRef]
43. Xiao, C.; Guo, Y.; Tang, Y.; Ding, J.; Zhang, X.; Zheng, K.; Tian, X. Epoxy Composite with Significantly Improved Thermal Conductivity by Constructing a Vertically Aligned Three-Dimensional Network of Silicon Carbide Nanowires/ Boron Nitride Nanosheets. *Compos. Part B Eng.* **2020**, *187*, 107855. [CrossRef]
44. Morgan, R.J.; O'neal, J.E.; Fanter, D.L. The Effect of Moisture on the Physical and Mechanical Integrity of Epoxies. *J. Mater. Sci.* **1980**, *15*, 751–764. [CrossRef]
45. De'Nève, B.; Shanahan, M.E.R. Water Absorption by an Epoxy Resin and Its Effect on the Mechanical Properties and Infra-Red Spectra. *Polymer* **1993**, *34*, 5099–5105. [CrossRef]
46. Reifsnider, K.; Sendeckyj, G.; Wang, S.; Johnson, W.; Stinchcomb, W.; Pagano, N.; Woo, M.; Piggott, M. Water Absorption of Resins and Composites: II. Diffusion in Carbon and Glass Reinforced Epoxies. *J. Compos. Technol. Res.* **1987**, *9*, 162. [CrossRef]
47. Maggana, C.; Pissis, P. Water Sorption and Diffusion Studies in an Epoxy Resin System. *J. Polym. Sci. B Polym. Phys.* **1999**, *37*, 1165–1182. [CrossRef]
48. Rudawska, A.; Frigione, M. Aging Effects of Aqueous Environment on Mechanical Properties of Calcium Carbonate-Modified Epoxy Resin. *Polymers* **2020**, *12*, 2541. [CrossRef]
49. Rudawska, A. Mechanical Properties of Epoxy Compounds Based on Bisphenol a Aged in Aqueous Environments. *Polymers* **2021**, *13*, 952. [CrossRef] [PubMed]
50. Zhang, J.; Lu, W.-G.; Yan, H.; Zhao, Z.-B.; Xu, L.; Ye, J.-H.; Li, W. Improvement of Wear-Resistance and Anti-Corrosion of Waterborne Epoxy Coating by Synergistic Modification of Glass Flake with Phytic Acid and Zn2+. *Ceram. Int.* **2023**, *49*, 17910–17920. [CrossRef]
51. Tian, Y.; Xie, Y.; Dai, F.; Huang, H.; Zhong, L.; Zhang, X. Ammonium-Grafted Graphene Oxide for Enhanced Corrosion Resistance of Waterborne Epoxy Coatings. *Surf. Coat. Technol.* **2020**, *383*, 125227. [CrossRef]
52. Kumar, B.G.; Singh, R.P.; Nakamura, T. Degradation of Carbon Fiber-Reinforced Epoxy Composites by Ultraviolet Radiation and Condensation. *J. Compos. Mater.* **2002**, *36*, 2713–2733. [CrossRef]
53. Kojnoková, T.; Nový, F.; Markovičová, L. The Study of Chemical and Thermal Influences of the Environment on the Degradation of Mechanical Properties of Carbon Composite with Epoxy Resin. *Polymers* **2022**, *14*, 3245. [CrossRef]
54. Amaro, A.M.; Reis, P.N.B.; Neto, M.A.; Louro, C. Effects of Alkaline and Acid Solutions on Glass/Epoxy Composites. *Polym. Degrad. Stab.* **2013**, *98*, 853–862. [CrossRef]
55. Amaro, A.M.; Reis, P.N.B.; Neto, M.A.; Louro, C. Effect of Different Acid Solutions on Glass/Epoxy Composites. *J. Reinf. Plast. Compos.* **2013**, *32*, 1018–1029. [CrossRef]
56. Rudawska, A.; Miturska, I.; Szabelski, J.; Abdel Wahab, M.; Stančeková, D.; Čuboňová, N.; Madleňák, R. The Impact of the Selected Exploitation Factors on the Adhesive Joints Strength. *Lect. Notes Mech. Eng.* **2020**, 899–913. [CrossRef]
57. Jin, F.-L.; Li, X.; Park, S.-J. Synthesis and Application of Epoxy Resins: A Review. *J. Ind. Eng. Chem.* **2015**, *29*, 1–11. [CrossRef]
58. *ISO 604*; Plastics—Determination of compressive properties. ISO (International Organization for Standardization): Geneva, Switzerland, 2002.
59. Coniglio, N.; Nguyen, K.; Kurji, R.; Gamboa, E. Characterizing Water Sorption in 100% Solids Epoxy Coatings. *Prog. Org. Coat.* **2013**, *76*, 1168–1177. [CrossRef]
60. Li, L.; Yu, Y.; Wu, Q.; Zhan, G.; Li, S. Effect of Chemical Structure on the Water Sorption of Amine-Cured Epoxy Resins. *Corros. Sci.* **2009**, *51*, 3000–3006. [CrossRef]
61. Elkebir, Y.; Mallarino, S.; Trinh, D.; Touzain, S. Effect of Physical Ageing onto the Water Uptake in Epoxy Coatings. *Electrochim. Acta* **2020**, *337*, 135766. [CrossRef]
62. Moazzami, M.; Ayatollahi, M.R.; Akhavan-Safar, A.; da Silva, L.F.M. Experimental and Numerical Analysis of Cyclic Aging in an Epoxy-Based Adhesive. *Polym. Test.* **2020**, *91*, 106789. [CrossRef]
63. Dang, D.N.; Cohendoz, S.; Mallarino, S.; Feaugas, X.; Touzain, S. Effects of Curing Program on Mechanical Behavior and Water Absorption of DGEBA/TETa Epoxy Network. *J. Appl. Polym. Sci.* **2013**, *129*, 2451–2463. [CrossRef]

materials MDPI

Article

The Effect of the Forging Process on the Microstructure and Mechanical Properties of a New Low-Cost Ti-5Al-1.5Mo-1.8Fe Alloy

Jinbao Hu [1,2], Yiqiang Mu [1,2], Qinsi Xu [1,2,*], Nan Yao [1], Shujun Li [2] and Xiaofei Lei [2,*]

[1] College of Civil Aviation, Shenyang Aerospace University, Shenyang 110136, China; hujinbao0831@163.com (J.H.); 20180053@sau.edu.cn (Y.M.)
[2] Institute of Metal Research, Chinese Academy of Sciences, Shenyang 110016, China
* Correspondence: 20180062@sau.edu.cn (Q.X.); xflei@imr.ac.cn (X.L.)

Abstract: This paper presents results on the microstructure and mechanical properties of a new low-cost titanium alloy Ti-5Al-1.5Mo-1.8Fe after different forging processes. The β phase transformation temperature of this alloy was 950 °C. In this study, the forging temperatures were designed at 920 °C and 980 °C, and the deformation degree ranged from 20% to 60%, with an interval of 20%. This study investigated the impact of the equiaxed α phase and shape of the lamellar microstructure on the tensile characteristics and fracture toughness of an alloy. The research employed a microstructure analysis and static tensile testing to evaluate the effect of forging temperatures and degree of deformation on the microstructure features. The findings revealed that forging temperatures could modify the microstructure characteristics, and the degree of deformation also affected this microstructure. This study demonstrates that a bimodal structure with an equiaxed α phase can be utilized to balance high strength and high ductility, resulting in better overall mechanical properties.

Keywords: a new low-cost titanium alloy; forging; microstructure; mechanical properties; fracture toughness

1. Introduction

Titanium and its alloys are widely used in aerospace, the chemical industry, and load-bearing medical implants due to their exceptional mechanical, corrosive resistance, and biocompatible properties [1–3]. However, the high production cost of these titanium alloys has hindered their widespread use. Recently, researchers have focused on developing low-cost titanium alloys. Among them, Ti-Al-Mo-Fe systems have shown a high, dependable, and competitive mechanical property balance in the hot deformed, annealed, and thermally strengthened stages [4]. Moreover, Ti-Al-Mo-Fe alloy systems have been proven to be a promising alternative for developing complete β-Ti alloys with a low cost, high strength, and low elastic modulus, because a low-cost Mo-Fe master alloy can be used to stabilize the β phase at room temperature [5]. Compared to the majority of beta stable elements, alpha stabilizers are more plentiful and less expensive. Al can increase strength and thermal stability [6–8]. However, an Al content exceeding 6 wt% will reduce the plasticity due to a precipitate brittle Ti_3Al phase [7,9]. Low-priced Mo-Fe master alloys have been used as alloying elements for their cost. Only a 10% weight of molybdenum (Mo) is required to stabilize the β phase at room temperature, making it a potent β-stabilizing element [10]. The addition of Mo reduces the elastic modulus and increases the strength of titanium alloys [11]. The addition of Mo to Ti-5Mn has successfully promoted the mechanical twinning of the alloys and improved their ductility [12]. Among the eutectoid β-stabilizing elements, Fe is the most commonly used substitute for isomorphous β-stabilizers due to its low cost and strong β-stabilizing capabilities [13–15]. Fe is a commonly used strengthening element. Rongjun Xu et al. reported that an addition of Fe to a Ti-Al-Mo-Fe alloy enhanced the room-temperature strength of the alloy and reduced its cost [7]. The addition of Fe to a Ti-Al system enhances the strength and lowers the liquidus temperature of the alloy, improving its thermo-mechanical properties, and Ti-Al-Fe alloy systems have also been effective at reducing costs [9]. However, an Fe content above 2 wt% will reduce the ductility. The elongation of low-cost Ti-5Al-3Mo-2Fe [7] alloys in the as-rolled and annealed state is 8.5%. The elongation of low-cost Ti-5Al-2.5Fe [9] alloys in the as-cast state is only 6%. Controlling the elemental content is critical to regulating an alloy's properties.

The mechanical properties of alloys can be tailored by changing the morphology and distribution of the alpha phases and intermetallic compounds using thermomechanical treatments such as hot forging [9]. Forging is a common processing method for titanium alloys, and previous studies have demonstrated that deformation is an effective approach to achieving superior mechanical properties [16]. Ref. [17] states that the service life and reliability of Ti-6Al-4V alloy components can be effectively improved by designing suitable forging process parameters, including optimizing the forging temperatures while minimizing the scrap and energy consumption generated during the forging production process. Titanium components are forged below the β phase transition temperature to achieve the desired microstructure [18]. Forging plastic-forming processes, commonly used for producing Ti-6Al-4V alloys, can improve their mechanical properties. In Ti-Mo-Fe alloys, combining different plastic deformation mechanisms has been found to enhance the strength–ductility trade-off [19]. A high deformation rate during forging results in finer grains, and the primary strengthening mechanism employed during forging is grain refinement strengthening [17,20]. Lou et al. investigated the impact of different forging parameters on the phase transformation of a Ti-6Al-4V alloy during hot hammer forging [21]. Therefore, controlling the forging temperature is crucial when performing hot hammer forging plastic-forming processes, as it significantly influences the resulting forging properties [17]. Y. Chong et al. revealed that a bimodal microstructure with an equiaxed primary α had an excellent balance between its strength and ductility [22,23].

The material studied in this paper is Ti-5Al-1.5Mo-1.8Fe with an Al content of no more than 6% and an Fe content of no more than 2%. This composition was chosen to ensure that the material exhibited both a sufficient strength and a high ductility. The mechanical properties were investigated by examining the effects of different forging temperatures and deformation degrees. The purpose of this study was to research the influence that different forging temperatures and deformations have on the microstructural and mechanical properties of a new low-cost Ti-5Al-1.5Mo-1.8Fe alloy.

2. Material and Experimental Procedures

This study involved the use of commercially available grade 1 sponge titanium, as well as intermediate Al-Mo and Fe-Mo alloys, and Fe and TiO_2 powder in the preparation of the alloy ingots. These materials were combined according to specific ratios and enclosed in a compacted alloy package, before being utilized as electrodes in a double vacuum consumable arc melting process. This approach was employed to achieve the efficient and cost-effective melting of the ingots. The chemical composition of the alloy used in this study was Ti-5Al-1.5Mo-1.8Fe (in wt.%) and the oxygen content was 0.070 (wt.%).

The Mo equivalent ($[Mo]_{eq}$) [24], according to the following equation:

$$[Mo]_{eq} = 1.0[Mo]+2.9[Fe] - 1.0[Al](wt\%) \quad (1)$$

Equation (1) states that the $[Mo]_{eq}$ of Ti-5Al-1.5Mo-1.8Fe can be 1.76 wt%, which is an α + β two-phase titanium alloy [25]. As is shown in Figure 1, the Ti-5Al-1.5Mo-1.8Fe alloy had an initial microstructure with a widmanstätten morphology, where elongated α grains with a large aspect ratio were tightly packed within primary β grains.

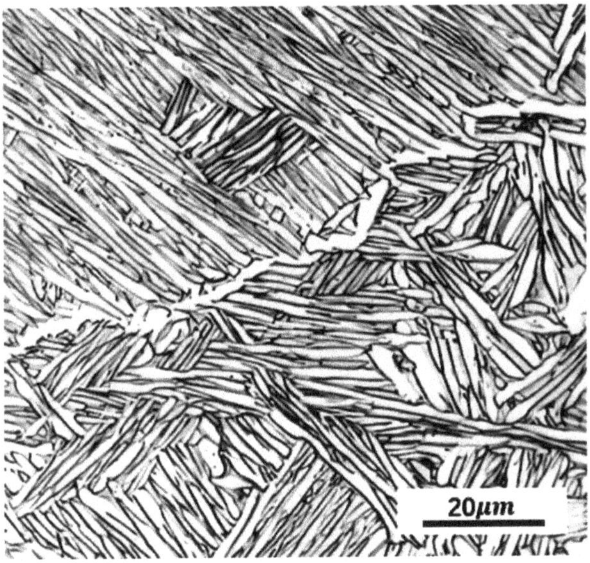

Figure 1. Optical micrograph of as-received Ti-5Al-1.5Mo-1.8Fe alloy.

Temperature is widely acknowledged as a critical factor that significantly impacts the performance in the fabrication of components [26–28]. The β phase transformation temperature was about 950 °C, which was measured using the metallographic method. This translated into two distinct forging techniques for the Ti-5Al-1.5Mo-1.8Fe alloy, namely α + β forging ($T = T_\beta - 30$ °C) and β forging ($T > T_\beta$). Below the β transition temperature was the two-phase region, and above it was the single-phase region. The forging experiment plan is shown in Table 1, and the cooling method was air cooling.

Table 1. Forging experiment scheme of Ti-5Al-1.5Mo-1.8Fe alloy at different forging process.

No.	Forging Types	Forging Temperature/°C	Forging Deformation/%
1	α + β	920	20
2			40
3			60
4	β	980	20
5			40
6			60

Figure 2 shows the shape of the Ti-5Al-1.5Mo-1.8Fe alloy after forging, and the surface had not developed any external cracks.

The phase compositions of the studied alloys were identified using Cu-Kα X-ray diffractometry (XRD) with a diffractometer (Rigaku Smartlab equipment) operating at 45 kV and 200 mA in a 2θ range of 25–75°. The software MDI Jade 6 and OriginPro 2021 were used to process the XRD data. Samples for the microstructural evaluations were obtained from a central plane section and subjected to optical microscopy (OM) and scanning electron microscopy (SEM) analyses. Prior to these analyses, the specimen surfaces were wet ground using SiC papers of 150, 800, 1200, and 3000 grit and polished with SiO_2 emulsion. To facilitate corrosion, the specimens were immersed in a solution containing 5 HF, 10 HNO_3, and 85 H_2O for 20 s. Tensile samples were obtained from each forged sample and subjected to room-temperature tensile testing. Before testing, all the samples were mechanically polished, cleaned, and dried. The tensile tests were conducted at room temperature using a SANS-CMT 5205 testing machine with a crosshead speed of 1mm/min. The shape of the tensile sample is shown in Figure 3. Each condition was evaluated by conducting tensile tests on two parallel specimens, and the ultimate tensile strength (UTS), yield strength (YS), elongation (EI), and reduction in area (RA) were determined from these tests. Subsequently, the plastic deformation and fracture behavior were analyzed using a TESCAN MIRA3 scanning electron microscope.

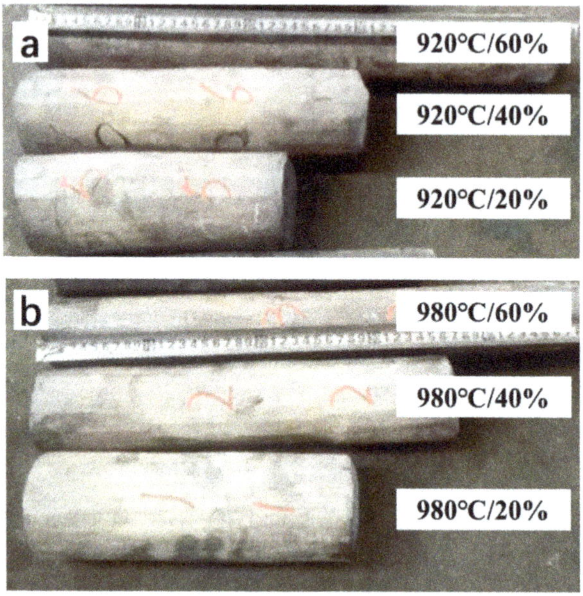

Figure 2. Overall shape of each deformation degree after forging, (**a**) different forging deformation of alloys at 920 °C, (**b**) different forging deformations of alloys at 980 °C.

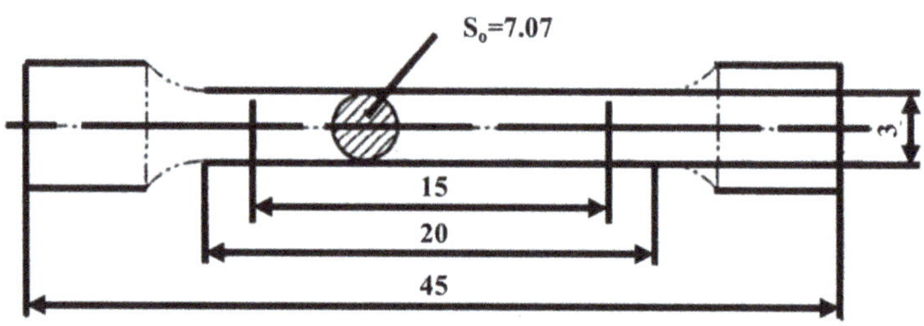

Figure 3. Schematic illustration of specimens used for the tensile tests. All dimensions are in mm.

3. Results and Discussion

3.1. Effect of Processing on Microstructure

Figure 4 shows the XRD patterns of the Ti-5Al-1.5Mo-1.8Fe alloys before and after different degrees of the forging deformation process. The XRD analysis revealed the presence of body-centered cubic (bcc) β and hexagonal closed-packed (hcp) α phases in all of the studied samples. As the forging temperature of the alloy increased, the intensity of the peaks corresponding to the α(002), α(101), α(102), and β(110) phases substantially increased, as shown in Figure 4. As the forging deformation of the alloy increased, the intensity of the peaks corresponding to the α(100), α(002), α(101) and β(110) phases increased substantially, indicating an increase in the lattice density. The lower intensities of the identified α(101), α(110), and β(300) diffraction peaks after the forging deformation processing can be interpreted for the microstructural refinement during this postproduction process [29]. Comparing the different forging deformation of those samples, no distinct change in the peak positions was visible, indicating that the lattice constants were weakly affected by the amount of forging deformation.

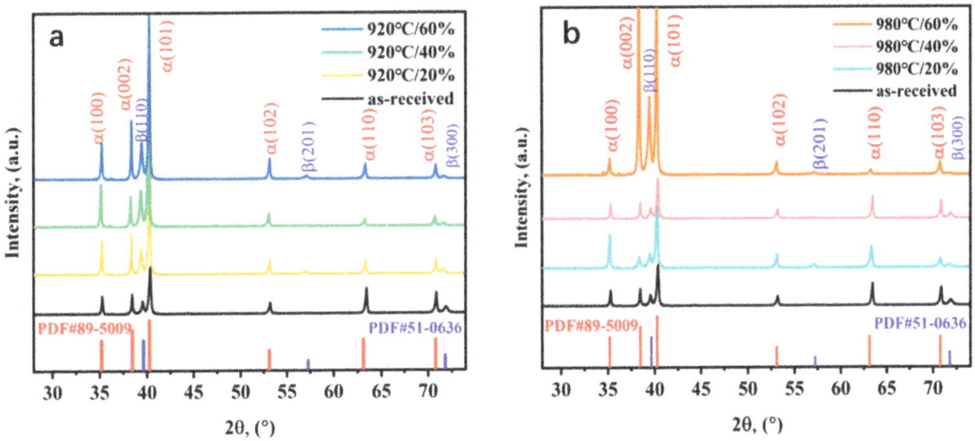

Figure 4. XRD patterns of Ti-5Al-1.5Mo-1.8Fe alloys. (**a**) Forging at 920 °C, (**b**) forging at 980 °C.

The microstructures of the Ti-5Al-1.5Mo-1.8Fe alloys at different deformation temperatures and degrees are shown in Figure 5. The microstructures after forging at 920 °C are shown in Figure 5a–c, and the microstructures after forging at 980 °C are shown in Figure 5d–f. Among them, a and d correspond to the microstructures achieved with a forging deformation degree of 20%, b and e at 40%, and c and f at 60%.

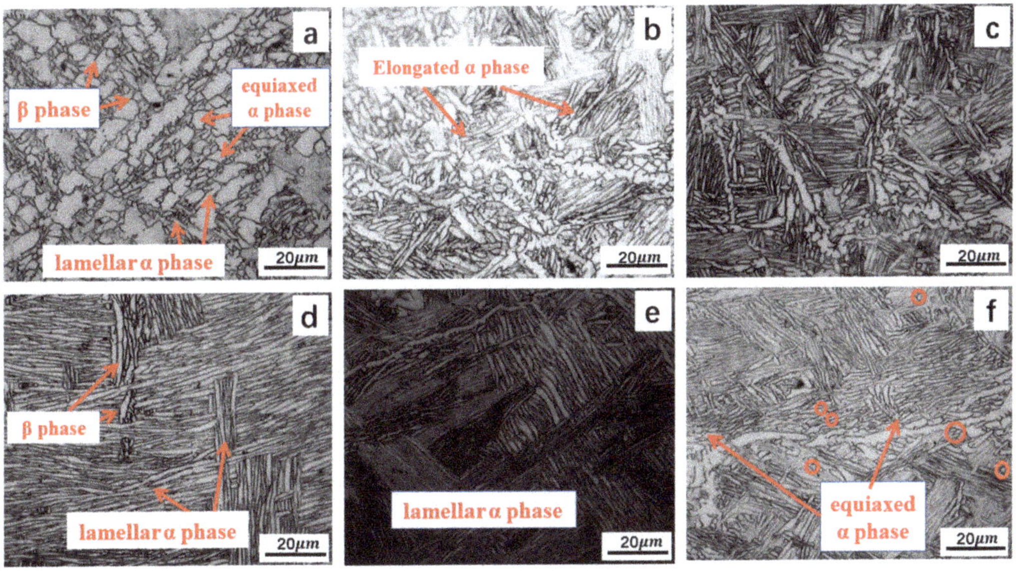

Figure 5. Microstructure under the different forging process: (**a–c**) forged at 920 °C with deformations of 20%, 40%, and 60%, respectively; (**d–f**) forged at 980 °C with deformations of 20%, 40%, and 60%, respectively.

The deformation temperature had a significant impact on the microstructure. The microstructure of the Ti-5Al-1.5Mo-1.8Fe alloy in Figure 5a consisted of equiaxed primary α, lamella α, and inter-crystalline β when forged below the β transition temperature. Increasing the forging temperature over the β transition temperature resulted in a decrease in the primary α content at the same deformation, as seen in Figure 5d. Deformation temperatures up to 980 °C led to a complete transformation of the equiaxed α into a lamellar structure. This lamellar structure could be observed (Figure 5d–f) when the deformation temperature was up to 980 °C. A widmanstätten microstructure formed after air cooling, characterized by coarse β grains with numerous α lamellar structures. During forging in the β single-phase region, the lack of pinning effects from the α grains led to an accelerated diffusion of the alloying elements within the β grain, resulting in rapid grain growth and deformation. The original β grains elongated in the deformation direction and the α phases were arranged in a lamellar structure within the grains [30,31].

The microstructure of a material is significantly impacted by forging deformation [32–34]. As this deformation increases, the degree of the alloy structure fragmentation also increases, leading to a smaller grain size, greater grain density, and increased mutual restraint between grains. The primary α grains were flattened, distorted, and broken-up with an increase in the deformation degree, and the part primary α phase recrystallized dynamically as the degree of deformation increased. As shown in Figure 5a, forging at low deformation rates below the β transition temperature provided enough time for recrystallization, leading to the formation of a larger primary α phase. Based on Figure 5b,c, it was observed that, when the deformation exceeded 40%, the temperature of the alloy surpassed the β transition point. This could result in the transformation of some equiaxed α grains into lamellar α grains and the creation of a basketweave structure in certain areas. It can be inferred that, during the forging process, a portion of the energy absorbed by the material was converted into internal energy of the plastic deformation. The diffusion pathway of this internal energy was dependent on the degree of forging deformation. As the deformation degree increased and the plastic deformation time was shortened, the internal energy could not diffuse to the surrounding medium and remained in the alloy material. Increases in both the distortion energy and dislocation density within the alloy were observed as the degree of deformation increased [35,36].

3.2. *Effect of Parameters on Mechanical Property*

Figure 6 illustrates the room-temperature mechanical properties of the Ti-5Al-1.5Mo-1.8Fe alloy under different forging temperatures and deformation degrees. The yield strength and ultimate tensile strength of the Ti-5Al-1.5Mo-1.8Fe alloy forged at 920 °C were found to be higher than those of the alloy forged at 980 °C at deformation degrees of 20% and 40%. However, forging the Ti-5Al-1.5Mo-1.8Fe alloy at 920 °C did weakly increase its strength, as evidenced by a yield strength increase from 807 MPa to 830 MPa and an ultimate tensile strength increase from 910 MPa to 922 MPa. In contrast, the strength of the alloy with a lamellar structure after forging at 980 °C continued to increase as the forging deformation increased, with a yield strength increase from 792 MPa to 824 MPa and an ultimate tensile strength increase from 892 MPa to 937 MPa. The reduction in area and elongation of the alloy forged at 920 °C were consistently higher than those of the alloy forged at 980 °C, with the maximum values of the elongation and reduction in area reaching 17% and 46%, respectively.

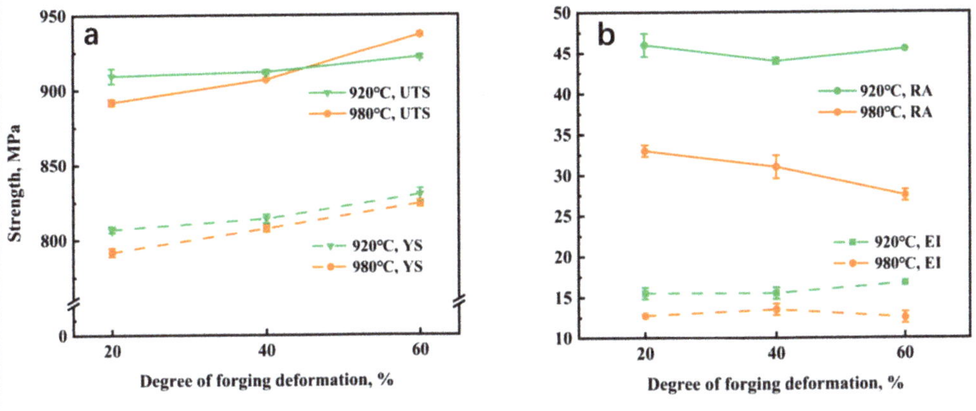

Figure 6. The room-temperature mechanical properties of Ti-5Al-1.5Mo-1.8Fe alloy after different deformation degree: (**a**) ultimate tensile strength (UTS) and the yield strength (YS), (**b**) reduction in area (RA) and elongation (EI).

3.2.1. Forging Temperature

The effect of the forging temperature on the mechanical properties is significant. The Ti-5Al-1.5Mo-1.8Fe alloy showed a better plasticity when forged at 920 °C, which was related to the presence of an equiaxed α phase. In the microstructure containing an equiaxed α phase, the mean-free-path of the α phase played a pivotal role in the tensile ductility of the alloy. The formation and growth of cavities were observed on interphase boundaries between the equiaxed α and transus β phases during the application of small tensile strain. The presence of primary α grains hindered the expansion of these cavities, with an increased primary α content resulting in a decrease in mean-free-path and a subsequent increase in cavity resistance. Consequently, the cavities encountered more barriers during growth, ultimately leading to an enhanced tensile deformation and high tensile ductility. In an alloy microstructure where the β matrix contains an equiaxed α phase, the mean free path of particles can significantly affect the tensile ductility of the alloy [37,38]. When the Ti-5Al-1.5Mo-1.8Fe alloy was forged above the phase transition temperature at 980 °C, the widmanstätten microstructure was formed. This microstructure is characterized by a regular arrangement of lamellar α phases within β grains. The α bundles share the same habit plane, and the slip can easily propagate and form a rapidly developing coarse slip band. Additionally, the presence of dislocation pile-up at the grain boundary α can lead to the premature formation and development of voids. These factors can ultimately weaken the resistance to crack initiation and reduce its plasticity.

Figure 7 shows the tensile fracture morphology of the alloy under different forging processes. The morphologies of the fracture surfaces forged below the β transition temperature at 920 °C are shown in Figure 7a,b. The areas where cracks propagate through the fibrous and shear lip regions are shown in Figure 7a. The macroscopic fracture surface of forging below the β transition temperature appears largely black, with visibility of the fiber region in the center and the cup-shaped shear lip on the outside. Figure 7b displays numerous dimples of varying sizes that correspond to the microscopic characteristics of a ductile fracture. The presence of these dimples indicates that the material possesses a good ductility, which is consistent with high cross-sectional shrinkage and elongation. Figure 7c,d show the microstructure of the tensile fracture of the lamellar structure after forging at a temperature above the β transition temperature (980 °C). When forging in the single-phase region, the microfracture surface exhibits features such as a few dimples, cleavage planes, and tear ridges, indicating induced embrittlement [39,40].

Numerous dimples are observed in the two-phase region of the forged Ti-5Al-1.5Mo-1.8Fe alloy in Figure 8a, as revealed by its microstructure. This observation is consistent with the presence of ductile fracture on the surface. Ref. [4] also showed a similar conclusion, where the micro-level fracture surfaces had dimpled relief, indicating the ductile nature of the fracture. In contrast, cracks are visible in the single-phase region in Figure 8b, propagating along the grain boundaries. These findings confirm that the forging process led to a ductile fracture mechanism in the two-phase region and a quasi-dissociation fracture [40] mechanism in the single-phase region. Ref. [27] also showed a similar conclusion, where fewer and shallower dimples were observed. The fracture mode of the specimens exhibited a quasi-cleavage fracture, characterized by the gradual transformation of the fracture surface morphology into a cleavage plane with limited dimpling. The cleavage planes were linked by tearing edges, resulting in a fracture type that displayed both ductile and brittle fracture traits. The observed fracture morphology is consistent with the mechanical test results presented.

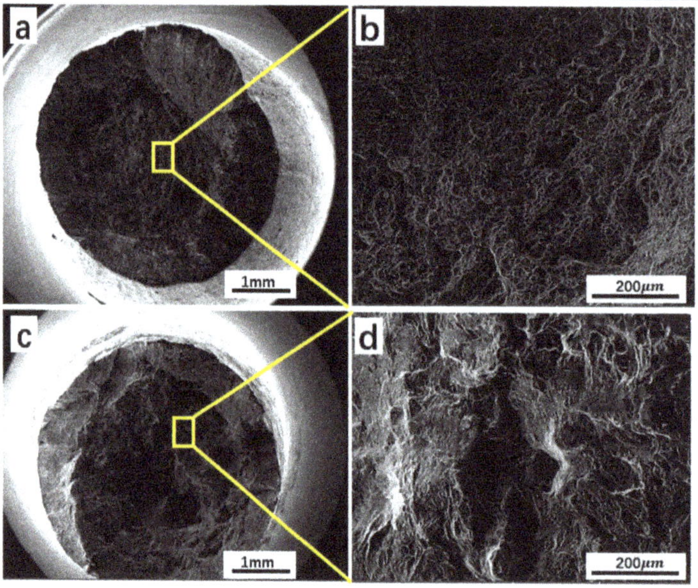

Figure 7. At a deformation of 20%, macroscopic fractography of two-phase region forging (**a**,**b**), macroscopic fractography of single-phase region forging (**c**,**d**).

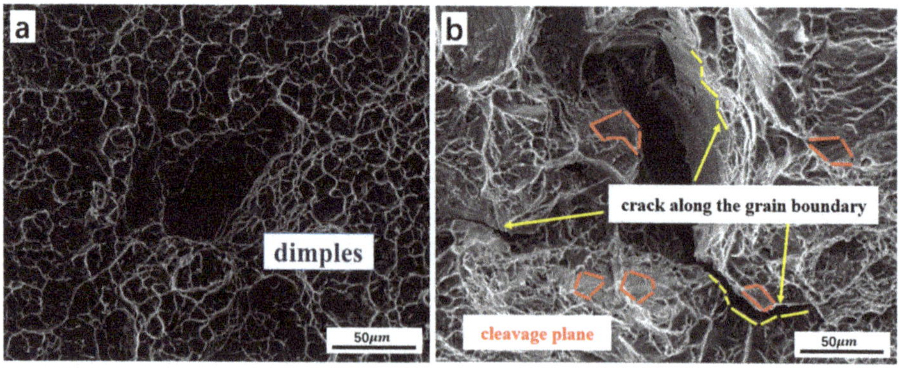

Figure 8. SEM micrographs of fracture surface, (**a**) dimples fracture, and (**b**) cleavage and crack propagating along the grain boundary.

3.2.2. Forging Deformation

At a forging temperature of 920 °C, more primary α grains showed a better plasticity. There was the presence of α colonies and a low degree of interweaving, which facilitates the passage of dislocations and leads to a diminished alloy strength [40]. The forging deformation exceeded 40%, and the plasticity of the alloy decreased because the reduction in the primary α. As the forging deformation increased, grain size refinement and an increased grain boundary area per unit volume prevented dislocation slip and more dislocation buildup occurred in the material, leading to an increased strength [27]. At a forging deformation degree of 60%, the ultimate tensile strength of the alloy forged at 980 °C was higher than that of the alloy forged at 920 °C. This can be explained by the fact that, when forging above the β transition temperature, a lamellar structure characterized by coarse β grains with numerous α lamellars was formed after air cooling, with grains elongated in the deformation direction and the α grains arranged in a lamellar structure within the grains. The resulting lamellar structure exhibited a relatively poor plastic deformation capacity, a phenomenon referred to as "β brittleness" [41].

Increasing the degree of deformation in the Ti-5Al-1.5Mo-1.8Fe alloy could result in the development of a lamellar structure. Refs. [42–44] stated that alloys with a lamellar microstructure exhibit an improved strength due to modifications in the crack direction and the appearance of secondary crack branching. Figure 9a shows that modifying the crack path along the α/β phase boundary in a lamellar structure changes the crack propagation direction. In addition, Figure 9b demonstrates that crossing the boundaries between colonies may trigger the formation of secondary cracks, which require more energy [37]. Ref. [40] also showed a similar conclusion, where secondary cracks and a dissociation surface could be observed simultaneously, showing the characteristics of quasi-dissociation fracture.

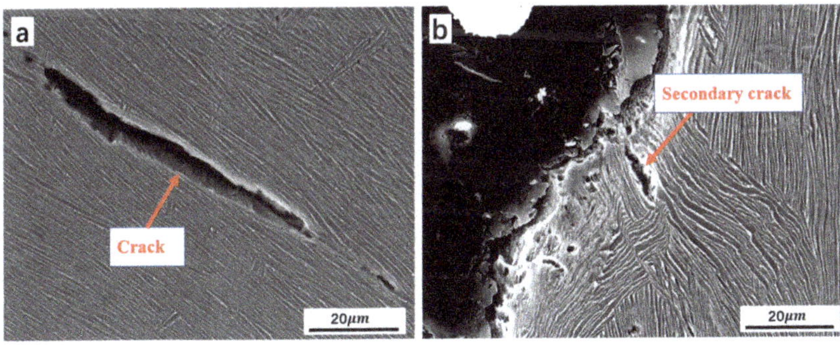

Figure 9. Crack-propagating path in lamellar microstructure (**a**) crack propagating along α/β interface, (**b**) secondary crack during crack propagating.

4. Conclusions

(1) Forging the Ti-5Al-1.5Mo-1.8Fe alloy at 980 °C in the single-phase area resulted in a lamellar structure, while forging at 920 °C in the two-phase region generated a bimodal microstructure with equiaxed grains.

(2) The Ti-5Al-1.5Mo-1.8Fe alloy forged in the two-phase region had a superior plasticity, with a ductile fracture displaying many dimples. The alloy forged in the single-phase zone exhibited an inferior ductility, showing a crack fracture with numerous tear ridges, indicating that cracks propagated along the grain boundaries.

(3) The Ti-5Al-1.5Mo-1.8Fe alloy with the bimodal microstructure indicated a higher plastic deformation capacity compared to the lamellar structure. Although the strength of the lamellar structure increased with deformation, its ductility was compromised due to β-brittleness. The two-phase region forging achieved a balance between high strength and ductility, resulting in better overall mechanical properties.

Author Contributions: Conceptualization, X.L.; Methodology, J.H., Y.M. and S.L.; Software, N.Y.; Validation, Y.M. and Q.X.; Formal analysis, Y.M. and S.L.; Investigation, Q.X.; Data curation, J.H. and N.Y.; Writing—original draft, J.H.; Writing—review & editing, J.H., Y.M. and S.L.; Supervision, Q.X.; Project administration, Q.X. and X.L.; Funding acquisition, X.L. All authors have read and agreed to the published version of the manuscript.

Funding: Opening Project of National Key Laboratory of Shock Wave and Detonation Physics (No. 2022JCJQLB05702).

Data Availability Statement: The raw/processed data required to reproduce these findings cannot be shared at this time because the data also forms part of an ongoing study.

Conflicts of Interest: The authors declare no conflict of interest.

References

1. Geetha, M.; Singh, A.K.; Asokamani, R.; Gogia, A.K. Ti based biomaterials, the ultimate choice for orthopaedic implants—A review. *Prog. Mater. Sci.* **2009**, *54*, 397–425. [CrossRef]
2. Qin, L.; Yao, S.; Zhao, J.; Zhou, C.; Oates, T.W.; Weir, M.D.; Wu, J.; Xu, H.H. Review on development and dental applications of polyetheretherketone-based biomaterials and restorations. *Materials* **2021**, *14*, 408. [CrossRef] [PubMed]
3. Wang, H.; Xu, Z.; Li, Q.; Wu, J. Application of metal-based biomaterials in wound repair. *Eng. Regen.* **2021**, *2*, 137–153. [CrossRef]
4. Markovsky, P. Microstructure and Tensile Properties of Cost-Efficient Thermally Hardenable $\alpha+\beta$ Alloys of Ti-Al-Mo-Fe and Ti-Al-Mo-Cr Systems. *Metallogr. Microstruct. Anal.* **2020**, *9*, 856–872. [CrossRef]
5. Li, C.L.; Ye, W.J.; Mi, X.J.; Hui, S.X.; Lee, D.G.; Lee, Y.T. Development of low cost and low elastic modulus of Ti-Al-Mo-Fe alloys for automotive applications. *Key Eng. Mater. Trans. Tech. Publ.* **2013**, *551*, 114–117. [CrossRef]
6. Lu, J.; Ge, P.; Zhao, Y. Recent development of effect mechanism of alloying elements in titanium alloy design. *Rare Met. Mater. Eng.* **2014**, *43*, 775–779. [CrossRef]
7. Xu, R.; Liu, B.; Yan, Z.; Chen, F.; Guo, W.; Liu, Y. Low-cost and high-strength powder metallurgy Ti–Al–Mo–Fe alloy and its application. *J. Mater. Sci.* **2019**, *54*, 12049–12060. [CrossRef]
8. Li, C.; Lee, D.-G.; Mi, X.; Ye, W.; Hui, S.; Lee, Y. Effect of Al addition on ω precipitation and age hardening of Ti-Al-Mo-Fe alloys. *Metall. Mater. Trans. A* **2016**, *47*, 2454–2461. [CrossRef]
9. Liang, Z.; Miao, J.; Brown, T.; Sachdev, A.K.; Williams, J.C.; Luo, A.A. A low-cost and high-strength Ti-Al-Fe-based cast titanium alloy for structural applications. *Scr. Mater.* **2018**, *157*, 124–128. [CrossRef]
10. Dos Santos, R.F.; Rossi, M.C.; Vidilli, A.L.; Borrás, V.A.; Afonso, C.R.M. Assessment of β stabilizers additions on microstructure and properties of as-cast β Ti–Nb based alloys. *J. Mater. Res. Technol.* **2023**, *22*, 3511–3524. [CrossRef]
11. Ehtemam-Haghighi, S.; Attar, H.; Okulov, I.V.; Dargusch, M.S.; Kent, D. Microstructural evolution and mechanical properties of bulk and porous low-cost Ti–Mo–Fe alloys produced by powder metallurgy. *J. Alloy. Compd.* **2021**, *853*, 156768. [CrossRef]
12. Santos, P.F.; Niinomi, M.; Liu, H.; Cho, K.; Nakai, M.; Trenggono, A.; Champagne, S.; Hermawan, H.; Narushima, T. Improvement of microstructure, mechanical and corrosion properties of biomedical Ti-Mn alloys by Mo addition. *Mater. Des.* **2016**, *110*, 414–424. [CrossRef]
13. Bodunrin, M.O.; Chown, L.H.; Omotoyinbo, J.A. Development of low-cost titanium alloys: A chronicle of challenges and opportunities. *Mater. Today Proc.* **2021**, *38*, 564–569. [CrossRef]
14. Fujii, H.; Fujisawa, K.; Ishiii, M. Development of Low-Cost High-Strength Ti-Fe-O-N Alloy Series. *Nippon Steel Tech. Rep.* **2002**, *85*, 107–112.
15. Senopati, G.; Rahman Rashid, R.A.; Kartika, I.; Palanisamy, S. Recent Development of Low-Cost β-Ti Alloys for Biomedical Applications: A Review. *Metals* **2023**, *13*, 194. [CrossRef]
16. Huang, K.; Marthinsen, K.; Zhao, Q.; Logé, R.E. The double-edge effect of second-phase particles on the recrystallization behaviour and associated mechanical properties of metallic materials. *Prog. Mater. Sci.* **2018**, *92*, 284–359. [CrossRef]
17. Fang, X.; Liu, Y.; Shao, Y.; Xu, H.; Yang, F. Forging Temperature Effects on Crack Tip Creep Behaviour of Hot Hammer Forged Ti-6Al-4V Alloy. *Adv. Mater. Sci. Eng.* **2023**, *2023*, 4414502. [CrossRef]
18. Astarita, A.; Armentani, E.; Ceretti, E.; Giorleo, L.; Mastrilli, P.; Paradiso, V.; Scherillo, F.; Squillace, A.; Velotti, C. Hot stretch forming of a titanium alloy component for aeronautic: Mechanical and modeling. *Key Eng. Mater. Trans. Tech. Publ.* **2013**, *554*, 647–656. [CrossRef]
19. Bortolan, C.C.; Campanelli, L.C.; Mengucci, P.; Barucca, G.; Giguère, N.; Brodusch, N.; Paternoster, C.; Bolfarini, C.; Gauvin, R.; Mantovani, D. Development of Ti-Mo-Fe alloys combining different plastic deformation mechanisms for improved strength-ductility trade-off and high work hardening rate. *J. Alloy. Compd.* **2022**, *925*, 166757. [CrossRef]
20. Xu, S.; Yuan, L.; Xu, W.; Shan, D.; Guo, B. Effects of Forging and Heat Treatment on Martensite Lath, Recrystallization and Mechanical Properties Evolution of 18Ni (250) Maraging Steel. *Materials* **2022**, *15*, 4600. [CrossRef]
21. Luo, S.; Yao, J.; Li, J.; Du, H.; Liu, H.; Yu, F. Influence of forging velocity on temperature and phases of forged Ti-6Al-4V turbine blade. *J. Mater. Res. Technol.* **2020**, *9*, 12043–12051. [CrossRef]

22. Chong, Y.; Bhattacharjee, T.; Shibata, A.; Tsuji, N. Investigation of the grain size effect on mechanical properties of Ti-6Al-4V alloy with equiaxed and bimodal microstructures. *IOP Conf. Ser. Mater. Sci. Eng.* **2017**, *219*, 012013. [CrossRef]
23. Figueiredo, R.B.; Langdon, T.G. Effect of grain size on strength and strain rate sensitivity in metals. *J. Mater. Sci.* **2022**, *57*, 5210–5229. [CrossRef]
24. Ahmed, M.; Savvakin, D.G.; Ivasishin, O.M.; Pereloma, E.V. The effect of cooling rates on the microstructure and mechanical properties of thermo-mechanically processed Ti–Al–Mo–V–Cr–Fe alloys. *Mater. Sci. Eng. A* **2013**, *576*, 167–177. [CrossRef]
25. Welsch, G.; Boyer, R.; Collings, E. *Materials Properties Handbook: Titanium Alloys*; ASM International: Detroit, MI, USA, 1993.
26. Viswanathan, R.; Ramesh, S.; Elango, N.; Kamesh Kumar, D. Temperature Measurement and Optimisation in Machining Magnesium Alloy Using RSM and ANOVA. *Pertanika J. Sci. Technol.* **2017**, *25*, 255–262.
27. Yu, F.; Zhang, Y.; Kong, C.; Yu, H. Microstructure and mechanical properties of Ti–6Al–4V alloy sheets via room-temperature rolling and cryorolling. *Mater. Sci. Eng. A* **2022**, *834*, 142600. [CrossRef]
28. Gao, S.; He, T.; Li, Q.; Sun, Y.; Sang, Y.; Wu, Y.; Ying, L. Anisotropic behavior and mechanical properties of Ti-6Al-4V alloy in high temperature deformation. *J. Mater. Sci.* **2022**, *57*, 651–670. [CrossRef]
29. Cvijović-Alagić, I. Damage and Fracture Resistance of Titanium Based Alloys for Medical Application. Ph.D. Thesis, Faculty of Technology and Metallurgy, University of Belgrade, Beograd, Serbia, 2013. (In Serbian).
30. Etesami, S.A.; Fotovvati, B.; Asadi, E. Heat treatment of Ti-6Al-4V alloy manufactured by laser-based powder-bed fusion: Process, microstructures, and mechanical properties correlations. *J. Alloy. Compd.* **2022**, *895*, 162618. [CrossRef]
31. Okazaki, Y. On the effects of hot forging and hot rolling on the microstructural development and mechanical response of a biocompatible Ti alloy. *Materials* **2012**, *5*, 1439–1461. [CrossRef]
32. Dadras, P.; Thomas, J. Characterization and modeling for forging deformation of Ti-6Ai-2Sn-4Zr-2Mo-0.1 Si. *Metall. Trans. A* **1981**, *12*, 1867–1876. [CrossRef]
33. Yeshanew, S.K.; Bai, C.; Jia, Q.; Xi, T.; Zhang, Z.; Li, D.; Xia, Z.; Yang, R.; Yang, K. Microstructure Evolution and Deformation Mechanisms of As-Cast Antibacterial Ti6Al4V-5Cu Alloy for Isothermal Forging Process. *Materials* **2022**, *15*, 3349. [CrossRef]
34. Zhang, F.-Z.; Bo, W.; Cao, Q.-H.; Zhang, L.; Zhang, H.-Y. Effect of deformation degree on microstructure and mechanical properties evolution of TiBw/Ti60 composites during isothermal forging. *Trans. Nonferrous Met. Soc. China* **2023**, *33*, 802–815. [CrossRef]
35. Yu, W.; Li, M.; Luo, J. Effect of processing parameters on microstructure and mechanical properties in high temperature deformation of Ti-6Al-4V alloy. *Rare Met. Mater. Eng.* **2009**, *38*, 19–24. [CrossRef]
36. Matsumoto, H.; Velay, V. Mesoscale modeling of dynamic recrystallization behavior, grain size evolution, dislocation density, processing map characteristic, and room temperature strength of Ti-6Al-4V alloy forged in the (α+ β) region. *J. Alloy. Compd.* **2017**, *708*, 404–413. [CrossRef]
37. Guo, P.; Zhao, Y.; Zeng, W.; Hong, Q. The effect of microstructure on the mechanical properties of TC4-DT titanium alloys. *Mater. Sci. Eng. A Struct. Mater. Prop. Microstruct. Process.* **2013**, *563*, 106–111. [CrossRef]
38. Hu, Y.; Ao, N.; Wu, S.; Yu, Y.; Zhang, H.; Qian, W.; Guo, G.; Zhang, M.; Wang, G. Influence of in situ micro-rolling on the improved strength and ductility of hybrid additively manufactured metals. *Eng. Fract. Mech.* **2021**, *253*, 107868. [CrossRef]
39. Alluaibi, M.H.I.; Cojocaru, E.M.; Rusea, A.; Şerban, N.; Coman, G.; Cojocaru, V.D. Microstructure and mechanical properties evolution during solution and ageing treatment for a hot deformed, above β-transus, ti-6246 alloy. *Metals* **2020**, *10*, 1114. [CrossRef]
40. Zhang, H.; Yi, J.; Wang, J.; Xiao, H.; Wang, M.; Wang, W. Effect of Direct Rolling Process on Microstructure and Mechanical Properties of the Electron Beam Cold Hearth Melting Ti-6Al-4V Alloy. *Metals* **2022**, *12*, 2018. [CrossRef]
41. Zhu, W.; Sun, Q.; Tan, C.; Li, P.; Xiao, L.; Sun, J. Tensile brittleness and ductility improvement in a novel metastable β titanium alloy with lamella structure. *J. Alloy. Compd.* **2020**, *827*, 154311. [CrossRef]
42. Gornakova, A.; Straumal, B.; Golovin, Y.I.; Afonikova, N.; Pirozhkova, T.; Tyurin, A. Phase Transformations and Mechanical Properties of Two-Component Titanium Alloys after Heat Treatment in the Two-Phase Region (α+ Intermetallic Compound) and High-Pressure Torsion. *J. Surf. Investig. X-Ray Synchrotron Neutron Tech.* **2021**, *15*, 1154–1158. [CrossRef]
43. Shchitsyn, Y.; Kartashev, M.; Krivonosova, E.; Olshanskaya, T.; Trushnikov, D. Formation of structure and properties of two-phase Ti-6Al-4V alloy during cold metal transfer additive deposition with interpass forging. *Materials* **2021**, *14*, 4415. [CrossRef] [PubMed]
44. Filip, R.; Kubiak, K.; Ziaja, W.; Sieniawski, J. The effect of microstructure on the mechanical properties of two-phase titanium alloys. *J. Mater. Process. Technol.* **2003**, *133*, 84–89. [CrossRef]

materials

MDPI

Article

Research on Automatic Identification and Rating of Ferrite–Pearlite Grain Boundaries Based on Deep Learning

Xiaolin Zhu [1,2,*], Yuhong Zhu [1], Cairong Kang [3], Mingqi Liu [4], Qiang Yao [2], Pingze Zhang [1], Guanxi Huang [2], Linning Qian [2], Zhitao Zhang [4] and Zhengjun Yao [1,*]

[1] College of Material Science and Technology, Nanjing University of Aeronautics and Astronautics, Nanjing 211106, China
[2] Jiangsu Product Quality Testing & Inspection Institute, Nanjing 210007, China
[3] Jiangsu Zhongxin Pipe Sci-Tec Co., Ltd., Nanjing 211100, China
[4] Dongying Industrial Product Inspection & Metrology Verification Center, Dongying 257000, China
* Correspondence: zxlin@nuaa.edu.cn (X.Z.); yaozj@nuaa.edu.cn (Z.Y.)

Abstract: Grain size has a significant effect on the mechanical properties of metals. It is very important to accurately rate the grain size number of steels. This paper presents a model for automatic detection and quantitative analysis of the grain size of ferrite–pearlite two-phase microstructure to segment ferrite grain boundaries. In view of the challenging problem of hidden grain boundaries in pearlite microstructure, the number of hidden grain boundaries is inferred by detecting them with the confidence of average grain size. The grain size number is then rated using the three-circle intercept procedure. The results show that grain boundaries can be accurately segmented by using this procedure. According to the rating results of grain size number of four types of ferrite–pearlite two-phase microstructure samples, the accuracy of this procedure is greater than 90%. The grain size rating results deviate from those calculated by experts using the manual intercept procedure by less than Grade 0.5—the allowable detection error specified in the standard. In addition, the detection time is shortened from 30 min of the manual intercept procedure to 2 s. The procedure presented in this paper allows automatic rating of grain size number of ferrite–pearlite microstructure, thereby effectively improving the detection efficiency and reducing the labor intensity.

Keywords: grain size; grain boundary; segmentation; classification; intelligent rating; deep learning

1. Introduction

Grain size is a parameter used to describe the dimensions of grains of polycrystalline materials. It has a significant effect on the properties of metals [1]. When polycrystalline materials are deformed under external force, the dislocation source will continuously release dislocations, which continuously slip inside the crystal. When dislocations reach the grain boundary, they can only continue moving after overcoming the impediment of the grain boundary to the dislocation movement and transfer the deformation from one grain to adjacent ones, and thus yielding the material. Generally, the grain boundaries in polycrystalline materials have large resistance to deformation, and the deformation of a single grain is implicated by adjacent grains, so the strength of polycrystalline materials at room temperature usually increases with the refinement of grains. That is to say, for polycrystalline materials of the same volume, the smaller the grain, the higher the grain boundary content, the more obvious the impediment of grain boundary to dislocation movement, and the higher the yield limit of the material. This is the fine-grained strengthening effect in polycrystalline materials [2,3]. The Hall–Petch equation [4,5] $\sigma_s = \sigma_0 + Kd^{-\frac{1}{2}}$ is usually used to describe the relationship between the grain size and the yield strength. In the equation, σ_s is the yield strength of the material, the yield strength of single crystal of the material, d the average grain size, K the gradient of H-P as a constant. Therefore, grain size is a technical item that needs to be strictly controlled in metal processing, especially in hot working. The rating procedures for grain size number include comparison procedure, the planimetric (or Jeffries) procedure, and the intercept counting procedure [6]. In the circle intercept procedure, the error caused by deviation from equiaxed grain can be automatically compensated for without excessive additional fields of view. It is a common procedure that gives consideration to both accuracy and efficiency. In the circle intercept procedure, a number of concentric circles are used to form a measuring network in the metallographic image, and appropriate magnification is selected and used to make the grain boundary intersection count between 50 and 100. During the calculation of intersections, it is counted as one intersection if the measuring circle is tangent to the grain boundary and as two intersections if it is significantly overlapped with the confluence of three grains. The grain size number is calculated by measuring the average grain intercept distance:

$$G = 6.643856 \lg \frac{MP}{L} - 3.288 \tag{1}$$

where, G is the grain size number; M is the magnification used; P is the grain boundary intersection count; L is the intercept length.

At present, the grain size number in the intercept procedure is mainly rated by manual measurement and calculation. Manual rating has obvious disadvantages. First, it has great subjective influence. Only experienced professionals can accurately identify grain boundaries and judge intersections, and human subjectivity during the measurement affects the rating accuracy. Second, it has low efficiency. For a metallographic image, if the intercept procedure is used, it will take at least 30 min for measurement of hundreds of intersections in multiple fields of view and for the subsequent calculation. Third, it affects health. The rating of grain size number is analyzed under a microscope, and long-term work under high-intensity light will affect the health of testing personnel.

In recent years, the continuous development of artificial intelligence technology has achieved remarkable results in various fields. Deep learning, as the leading artificial intelligence technology, has attracted increasing attention and has achieved excellent performance in medical image classification and text recognition [7–10]. In the field of material tissue analysis, the automatic metallographic analysis of metal materials based on computer vision and machine learning technologies has become the exploring direction of researchers at home and abroad. Zhang [11] et al. used the dual-threshold method and watershed algorithm to segment different metallographic structures. Geng [12] realized threshold segmentation of 2D exponential gray-level entropy through PCA noise reduction based on Gaussian Mixture Model. Zhu [13] proposed an efficient grain boundary division method based on the structured random tree. Patxi [14] et al. reconstructed the austenite grain boundary in martensitic steel using the deep learning algorithm. Wang [15] et al. applied computer vision technology and machine learning to identify the location of fatigue crack initiation in materials. Azimi [16] et al. pixel-wise segmentation based on Full Convolutional Neural Networks (FCNN), using deep learning method to classify and identify the microstructure of mild steel, achieving 93.94% classification accuracy. In Li's paper [17], the machine learning method of Gradient Boosting Decision Tree was used to recognize the boundaries of a lath bainitic microstructure. The optimized machine learning model achieved greater than 88% recognition accuracy for all boundaries in the bainitic microstructure, where the recall score of PAG boundary was 93%. Han [18] proposed a high-throughput characterization method based on deep learning, rapid acquisition techniques and mathematical statistics to identify, segment and quantify the microstructure of weathering steel. The segmentation accuracies of 89.95% and 90.86% for non-metallic inclusions and pearlite phases, respectively, and the detection time are significantly reduced. Tharindu [19] proposes a computational design framework and uses artificial neural networks (ANNs) for generating and predicting the forging response of preform shapes, respectively. This study predicted the average effective plastic strain response in spatially varying regions of the forging to within ±8% of the ground truth.

Currently, scholars have studied the research of using image-processing techniques to identify grain boundaries and calculate grain size [13,14], such as segmentation-based on grayscale thresholding. This technique is an image transformation and analysis based on the gray value of each pixel, which is influenced by factors such as image quality and illumination, and the accuracy of detection is not high. In contrast, using deep learning techniques, the model can be trained and iterated using a large number of calibration samples, and the features of different types of grain boundaries can be extracted, thus achieving accurate segmentation of grain boundaries and accurate calculation of grain size. In this paper, a sample of ferrite–pearlite microstructure (see Figure 1) is used as an example to detect grain boundaries and thus calculate their grain size by using a U-Net-based image segmentation model. These two phases are the most common microstructures in steels. A large number of medium and low-carbon hot-rolled, annealed or normalized steels are ferrite–pearlite microstructures, so it is necessary to study the automatic rating of grain size number of this two-phase microstructure.

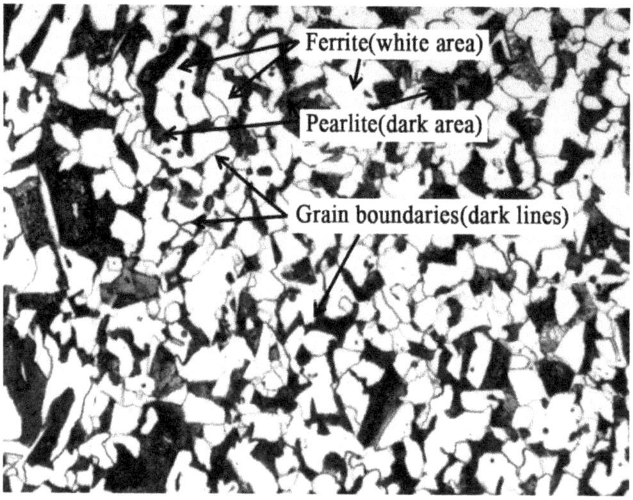

Figure 1. Typical ferrite–pearlite two-phase microstructure.

There are four facts relevant to the grain size testing of ferrite–pearlite in steel mills. (1) Almost all the steel mills use optical microscopes when testing the grain size. (2) Under the optical microscope, the grain boundary in the ferrite has obvious contrast difference from the grain inside, so such grain boundaries can be easily identified with human eyes. However, for pearlite, due to low resolution, the boundaries of pearlite colonies and the slices of ferrite–cementite cannot be seen. Hence, the pearlite grain size (the size of pearlite colonies) cannot be tested under the optical microscope. (3) Therefore, when the steel mills test the actual grain size of ferrite–pearlitic steel, almost all of them test only the grain size of the ferrite, while the pearlite part is ignored, and finally the grain size of the ferrite part is taken as the overall grain size of the sample. (4) The difficulty of testing the grain size using this method (ferrite only) is different for steels with different carbon contents. For steel with less carbon content, the grain size of ferrite can be easily tested directly because ferrite dominates in the sample and the whole sample indicates a great ferrite region with very little pearlite. However, for steel with high carbon content, since pearlite is randomly and approximately uniformly distributed throughout the sample and there is no large pure ferrite area, it is difficult to directly test the grain size of ferrite. In such cases, the presence of pearlite will significantly interfere with the efficiency and accuracy of ferrite grain size testing, and the efficiency of manual calculation of the grain size is extremely low.

Based on the four facts above, when evaluating the grain size of ferrite–pearlite steels by testing the ferrite grain size, in order to solve the interference of pearlite regions to the testing process and results of ferrite, and to improve the testing efficiency, we propose a new and simple method based on artificial intelligence techniques. Since the steel mills ignore the pearlite and consider the grain size of ferrite as the overall grain size of the sample. Then, we assume that all the pearlite areas are ferrite. At this point, the exact size of grains in the pearlite (considered as ferrite) region is unknown, but it can be predicted according to the confidence of the average grain size of ferrite (actual ferrite) using the model (the grain boundaries predicted by the model, which are not real but assumed by the model, are called "hidden grain boundaries"). In this way, we, (1) identify and test the grain boundaries in ferrite, (2) treat pearlite as ferrite, and use our model to predict the "hidden grain boundaries" inside it, and (3) count and calculate the intersections in the actual grain boundaries in ferrite and the predicted "hidden grain boundaries" in pearlite using the three-circle intercept procedure. Then, statistical calculation is made to achieve automatic rating of grain size number.

2. Model and Method

Image segmentation is the first step and also the most important and difficult part of image analysis. Image segmentation is the process of partitioning an image into several disjoint parts by grayscale, color, spatial texture or other characteristics [20], so that these characteristics show consistent differences in adjacent parts, thus achieving the purpose of separating the target from the image. Traditional image segmentation is often carried out with the knowledge of digital image processing, topology, mathematics, etc. Compared with the deep learning-based method, traditional methods have poor ability to deal with impurities, patterns, hidden boundaries, etc. Since the texture of metallographic images is complex [21], and it is difficult and costly to obtain training sets of metallographic images, the image segmentation algorithm of U-Net neural network is used in this paper to extract grain boundaries.

U-Net is a variant of fully convolutional neural network (FCN) [22]. Its architecture is in the shape of a "U", hence the name. This network consists of two parts: the contracting path and the expanding path. The contracting path is generally used to capture the context information in the image, and the expanding path is used to enable precise localization of the parts to be separated from the image. U-Net is an improved model based on FCN, and it can be trained by using small sample datasets to achieve good segmentation effect through data augmentation. It is especially suitable for application scenarios with high cost and great difficulty in sample collection and calibration, such as metallographic image analysis.

In the contracting path of U-Net, each contraction consists of a 2×2 max pooling layer (with stride 2) and two 3×3 unpadded convolutional layers. Each unpadded convolutional layer is followed by a ReLU activation parameter for downsampling of original data. At each downsampling step, the number of feature channels is doubled. In the deconvolution of the expanding path, every step consists of a 2×2 unpadded convolutional layer (also followed by the activation parameter ReLU) and two sets of 3×3 unpadded convolutional layers. At the same time, each deconvolution is followed by an addition of the feature map cropped from the corresponding contracting path (clipped to maintain the same shape). The final layer of the network is a 1×1 unpadded convolutional layer that is used to map the 64-channel feature vector to the desired number of classes. Finally, the U-Net used in this paper has 23 layers in total. The advantage of U-Net is that it has no requirement on the shape and size of the input image, especially when extremely large images are processed. Figure 2 illustrates the U-Net architecture used in this paper.

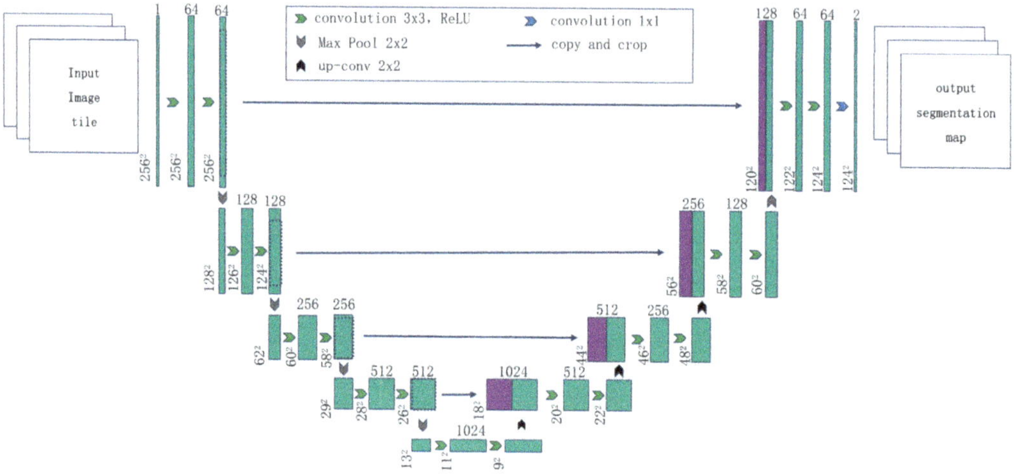

Figure 2. U-Net architecture.

In this paper, cross entropy is used as the loss function of image segmentation, and its principle is as follows:

$$H(y^{(i)} \circ \hat{y}^{(i)} = -\sum_{j=1}^{q} y_j^{(i)} \log \hat{y}_j^{(i)}) \quad (2)$$

where, $y_j^{(i)}$ with a subscript is an element equal to either 0 or 1 in the vector $y^{(i)}$. In the vector $y^{(i)}$, only the $y(i)^{th}$ element, $y_{y(i)}^{(i)}$, equals 1, and all the rest elements equal 0. Therefore, $H\left(y^{(i)} \circ \hat{y}^{(i)} = -\log \hat{y}_{y(i)}^{(i)}\right)$. That is, cross entropy focuses only on the prediction of correct classes. If the value of cross entropy is large enough, it indicates that there is no problem with the classification.

Assuming the number of samples in the training dataset is n, the cross-entropy loss function is defined as:

$$l(\Theta) = \frac{1}{n}\sum_{i=1}^{n} H\left(y^{(i)} \circ \hat{y}^{(i)}\right) \quad (3)$$

where Θ is the model parameter. Similarly, if each sample has only one label, the cross-entropy loss can be expressed as $l(\Theta) = -\frac{1}{n}\sum_{i=1}^{n} log\hat{y}_{y(i)}^{(i)}$. From another perspective, we know that minimizing $l(\Theta)$ is equivalent to maximizing $\exp(-nl(\Theta)) = \prod_{i=1}^{n} \hat{y}_{y(i)}^{(i)}$, that is, minimizing the cross-entropy loss function is equivalent to maximizing the joint prediction probability for all label classes in the training dataset.

The two feature vectors output by U-Net are the category label vector $L = (L_1, \cdots, L_\eta)$ and the correlation confidence degree $F = (F_1, \cdots, F_\eta)$, where $L_i, F_i = F(L_i)$ are the category label of the ith pixel and the corresponding confidence degree respectively. $L_i \in \{c_1, c_2, \cdots, c_T | c_j \in N\}$ is a class label; $0 \leq F_i \leq 1$ is the confidence value for the ith pixel classified as l_i. In the above definition, $1 \leq i \leq \eta$, η is the maximum order value obtained by sequencing the training image pixels in the way of the first method.

3. Experiment

3.1. Dataset

A total of 200 metallographic samples with a diameter of 10 mm were prepared with hot-rolled ribbed bars as the test samples. After grinding, polishing and corrosion, the samples showed clear ferrite–pearlite two-phase microstructures and ferrite grain boundaries.

The LabelMe platform was used to distribute grain size images for experts, and the circle intercept procedure was used for the measurement and statistics of intersections and the calculation of grain size. It should be noted that hidden boundaries do not need to be marked manually, except for visible contour lines. Figure 3 shows an example of a single grain size image and its three-circle intercept calibration. In this paper, 2000 valid grain size images and their annotation files were finally collected, with the image size of 2048 × 1536. The grain sizes of these images were calibrated and classified by experts and then imported into the dataset.

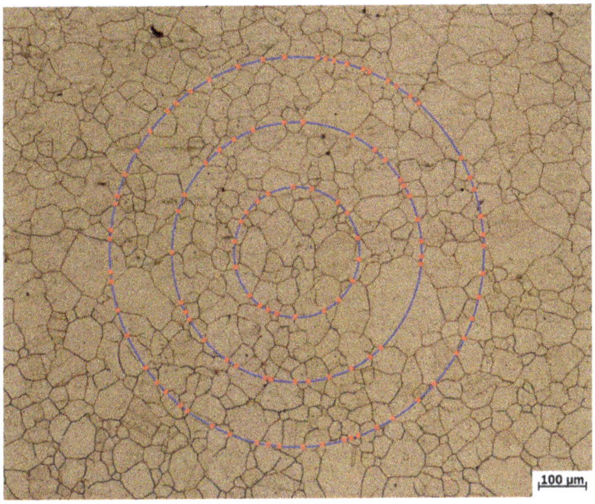

Figure 3. Example of Three-Circle Intercept Calibration.

3.2. Data Augmentation

For deep learning, the size of the dataset has a great impact on the effect of the model [23]. A too-small dataset will lead to overfitting of the model and thus inaccurate prediction results. The dataset used in this paper only included a few metallographic image samples, making it difficult to obtain good results. Therefore, the existing data were augmented to expand the dataset. For this dataset, the data were mainly augmented by random rotation, flipping and scaling, and random erasing. Data augmentation mainly serves two purposes: (1) to increase the data volume for training, so that when the data volume is large, there will be many characteristics, resulting in an improvement in the generalization ability of the model; (2) to eliminate the impact of noise data that increases with the volume of the effective data, so as to improve the robustness of the model. Taking random erasing as an example, the model sometimes shows high consistency with the training data, but does not work very well in data verification, that is to say, the model is overfitting. An important factor to image overfitting is that the training data samples are not rich enough. For example, in actual metallographic images, there may be overexposure, underexposure, ambiguity, occlusion of some areas, etc. If we want to improve the prediction and recognition ability in respect of such images and enhance the generalization ability and robustness of the model, it is necessary to add similar forms of data input during training. This is the function of random erasing.

Random rotation and scaling mean randomly rotating the training image by random angle or scaling by random magnification for data augmentation. For a training sample, the augmented sample obtained by image flipping and image rotation can enable the model to learn features that are not deformed by rotation during training, while the augmented sample obtained by image scaling can better enable the model to realize multi-scale training. Let (x, y) be the pixel coordinates of the original image, (x_1, y_1) be the pixel coordinates after rotation, (x_2, y_2) be the pixel coordinates after scaling, θ be the rotation angle, (a, b) be the rotation center, *(left,top)* be the coordinates of the upper left corner of the rotated image, s_x and s_y be the scaling factors. The rotation and scaling formulas are as follows:

$$\begin{bmatrix} x_1 & y_1 & 1 \end{bmatrix} = \begin{bmatrix} x & y & 1 \end{bmatrix} \begin{bmatrix} 1 & 0 & 0 \\ 0 & -1 & 0 \\ -a & b & 1 \end{bmatrix} \begin{bmatrix} cos\theta & -sin\theta & 0 \\ sin\theta & cos\theta & 0 \\ 0 & 0 & 1 \end{bmatrix} \begin{bmatrix} 1 & 0 & 0 \\ 0 & -1 & 0 \\ left & top & 1 \end{bmatrix}$$

$$\begin{bmatrix} x_2 & y_2 & 1 \end{bmatrix} = \begin{bmatrix} x & y & 1 \end{bmatrix} \begin{bmatrix} s_x & 0 & 0 \\ 0 & s_y & 0 \\ 0 & 0 & 1 \end{bmatrix}$$

The results are shown in Figure 4.

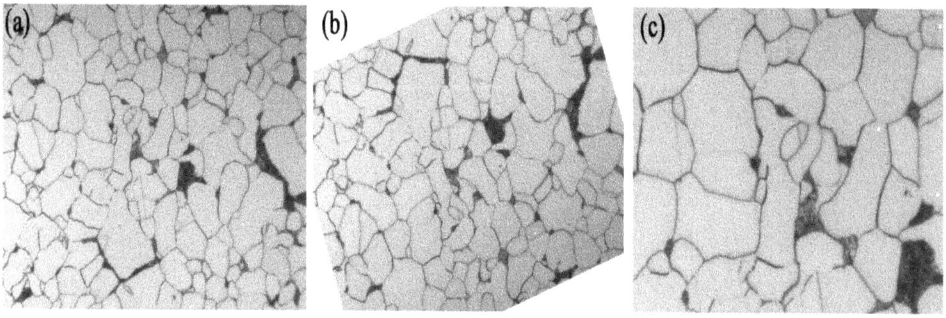

Figure 4. Random rotation and scaling images (Original (**a**), Rotated (**b**), Enlarged (**c**)).

Random erasing is a new data measurement technique, which erases the pixels with random values in the randomly selected rectangular areas in training, so as to obtain more occlusions or erase pixel values and images with different areas for data enhancement. This technique can reduce the risk of fitting and make the model robust to occlusion and erasure. The results are shown in Figure 5.

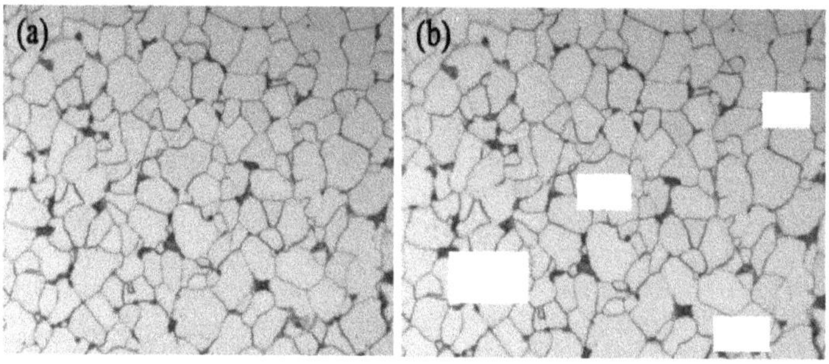

Figure 5. Random erasing images (Original (**a**), Erased (**b**)).

3.3. Algorithm Steps

The algorithm used in this paper falls into five modules: metallographic image acquisition module, contour extraction network module, visible intersection statistics module, hidden intersection statistics module and automatic rating module.

The image acquisition module measures the metallographic samples and calculates their grain size number by the conventional three-circle intercept procedure through experts of equivalent level, labels the images with the LabelMe v5.1.1 software and uploads them to this dataset for image acquisition. The module contains 2000 effective images of grain size numbers and their labeling files, which are used as the training set of the contour extraction network module.

The contour extraction network module is mainly for image preprocessing and grain boundaries contour extraction. One of the most critical steps is the training of the model, which is specifically divided into the following substeps: (1) Calibrate the original data of image size 2048 × 1536 with LabelMe, mark the visible boundary and fill the prolongable boundary. Name the images with id_magnification used_level.png (See Figure 6). (2) The Labeled image of Figure 6b is binarized and then a random mask is applied to form Figure 7a. Segment the labeled images into images of size 256 × 256 with a step size of 128 and segment the labels in the same way (See Figure 7). (3) Input the dataset into U-Net for training, use the PyTorch framework for model building and training, and save the trained model files.

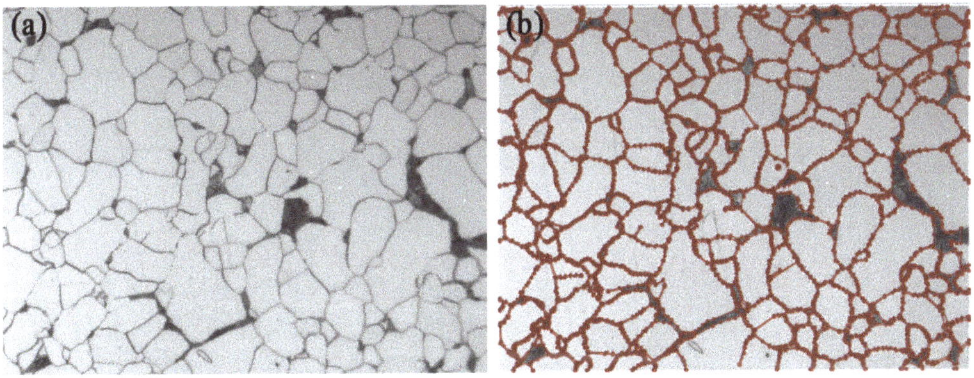

Figure 6. Original image (**a**) and Labeled image (**b**).

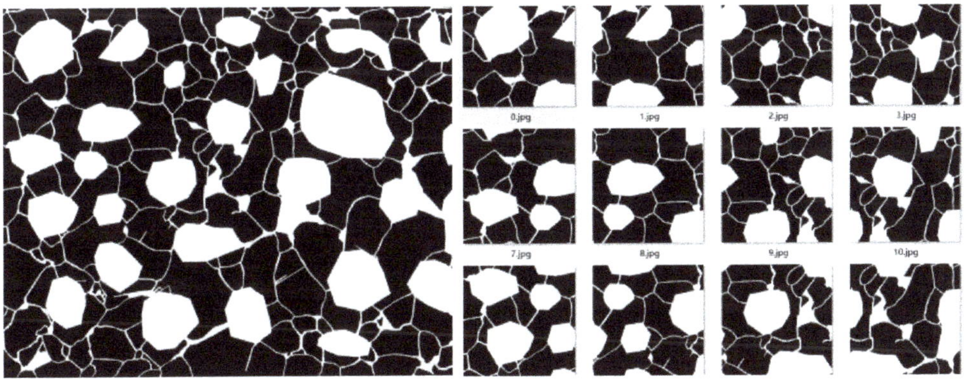

Figure 7. Original Labeled image (**Left**) and Segment image (**Right**).

Start the test after the model is trained. Enter a metallographic image of size 2048 × 1536. The program first segments the image, runs the contour extraction model for model extraction, and finally splices the segmented images, and processes the overlapping part with the mean value. In this way, a complete contour extraction image will be obtained. Both model training and prediction are carried out in an end-to-end manner. Extraction of contours by sections greatly reduces the complexity of model training and improves the running speed of the model. The formula for dealing with overlapping parts is as follows:

$$x_{avg} = average(x_1, x_2 \cdots x_n)\ ,\ x_n \in \{0, 1\}$$

$$x_{new} = \begin{cases} 1\ , x_{avg} > \alpha \\ 0, x_{avg} \leq \alpha \end{cases}, \alpha \in (0, 1)$$

where, n represents the number of overlaps, x_n represents the value of the nth overlap prediction, and the value is 0 or 1,0 represents that this pixel is not predicted to be the contour, 1 represents that this pixel is predicted to be the contour, average is the mean operation, the result is x_{avg}, α is the classification threshold, and the value in this paper is 0.8, x_{new} represents the type of the overlap region after processing. Contour extraction results are as Figure 8:

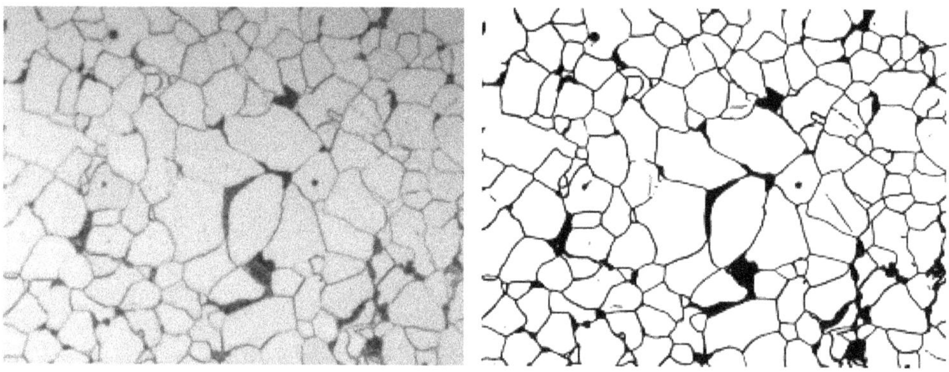

(**a**) (**b**)

Figure 8. Grain-boundary contours identified by the model. Original image (**a**) and image of model predictions (**b**).

As for the visible intersection count statistics module, the binary image of edge contour output by the contour extraction network is input, and based on the three-circle truncation method, the edge contour line is judged to intersect the circle, and the position of the pixel where the intersection point is recorded, so as to conduct preliminary statistics of the truncation point, so the number of visible cut-off points P is calculated.

As for the hidden intersection count statistics module, the binary image of the grain boundary contour output by the contour extraction network is input, and the intersection count of the pearlite area is estimated based on confidence. The calculation of grain size number is a description of the overall ferrite properties. It may be assumed that the size of each grain boundary in a ferrite section is approximately normally distributed after the hidden boundary is delineated. The specific steps are as follows: After the grain boundary segmentation of the first part, we obtain a set of grain boundary sizes $X = \{x1, x2, x3..xn\}$, x_n represents the Manhattan distance between two adjacent grain boundaries. For some steel, we solve $\overline{X} = Averge\ \{x_i | x_a < x_i < x_b\}$, where x_a, x_b meet $Pr(x_a < \mu < x_b) = \alpha$, which means the calculated average of the confidence interval of the distribution serves as the statistical magnitude describing the overall grain size number of the steel. α varies in different types of ferrites (with different carbon contents), and parameters need to be adjusted for optimization. After computing, $\mathrm{x_n}$ contains hidden grain boundary cutoff points x_{new} $(x_n \in \{x_i | x_b < x_i, i = 0, 1 \cdots n\})$, finally, present $x_{new} = x \div \overline{x} + 1$ to calculate processed grain boundary points (including the number of hidden grain boundary).

The automatic rating module accumulates the visible and hidden intersection count to obtain the final intersection count and calculates the grain size number by Formula (1).

The model architecture is shown in Figure 9. Capture the target field of view to obtain the image to be rated as the input of the system. Slice the image and input the sliced images into the contour extraction network module for processing, and then integrate them to obtain the contour extraction image. Input the obtained contour into the visible intersection statistics module and the hidden intersection statistics module respectively for calculation. Then, input the results into the automatic rating module for processing to obtain the final rating results and the labeling results by the three-circle intercept procedure.

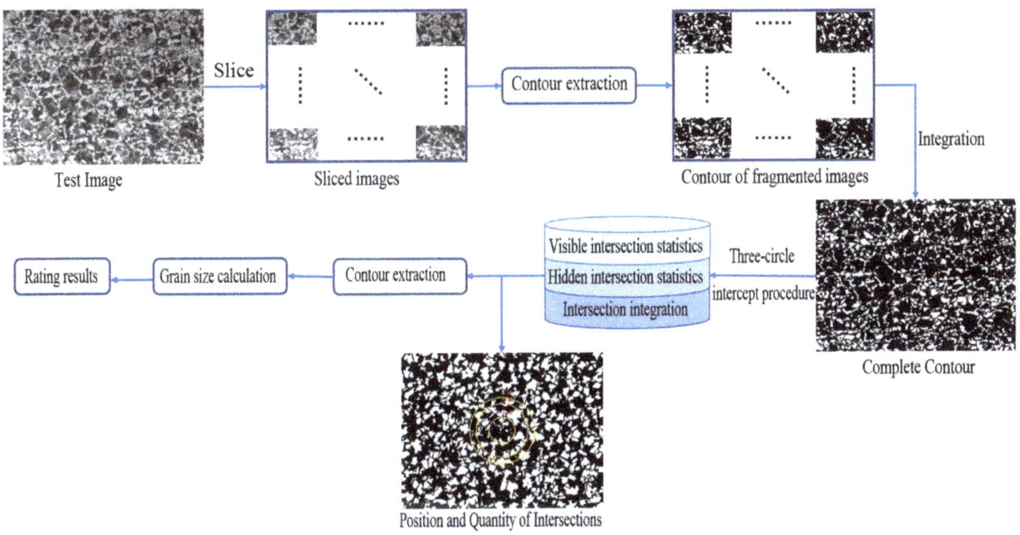

Figure 9. Model architecture.

4. Results and Discussion

The training datasets in this paper are all about hot rolled ribbed steel bar samples (HRB400). A trained model is used to test the grain sizes of four different grades of steel, namely SWRCH6A (with a carbon content of 0.05%, close to that of industrial pure iron), HRB400E (with a carbon content $\leq$ 0.23%), 35K (with a carbon content of 0.32%) and S45C (with a carbon content of 0.45%). With different chemical compositions, they are all of a ferrite–pearlite microstructure. The two-phase ratio of samples with different carbon contents is different. Generally, the higher the carbon content, the higher the pearlite ratio.

4.1. Grain Boundary Segmentation Results

The original image size of the datasets used in this paper is 2048 $\times$ 1536, and the sub-images of 256 $\times$ 256 are cut out with a step size of 128, which fall into the training dataset, verification dataset and test dataset, accounting for 70%, 20% and 10% of the total data respectively. On top of that, the grain size number of three types of steels for which no training data collected are tested.

During model training, the authors adjust the number of iteration rounds, learning rate, batch size and other parameters. The results of grain boundary segmentation of various types of steels are shown in Figure 10. It can be seen that the images are well segmented and spliced, showing no obvious splicing joints compared with the original. In different sample images, the grain boundaries inside the ferrite and the ferrite–pearlite interface are all correctly identified and segmented into grain boundary contours as a whole. The pearlite does not show internal grain boundaries, it is treated as a grayscale part as a whole, and its internal hidden grain boundaries are predicted as ferrite.

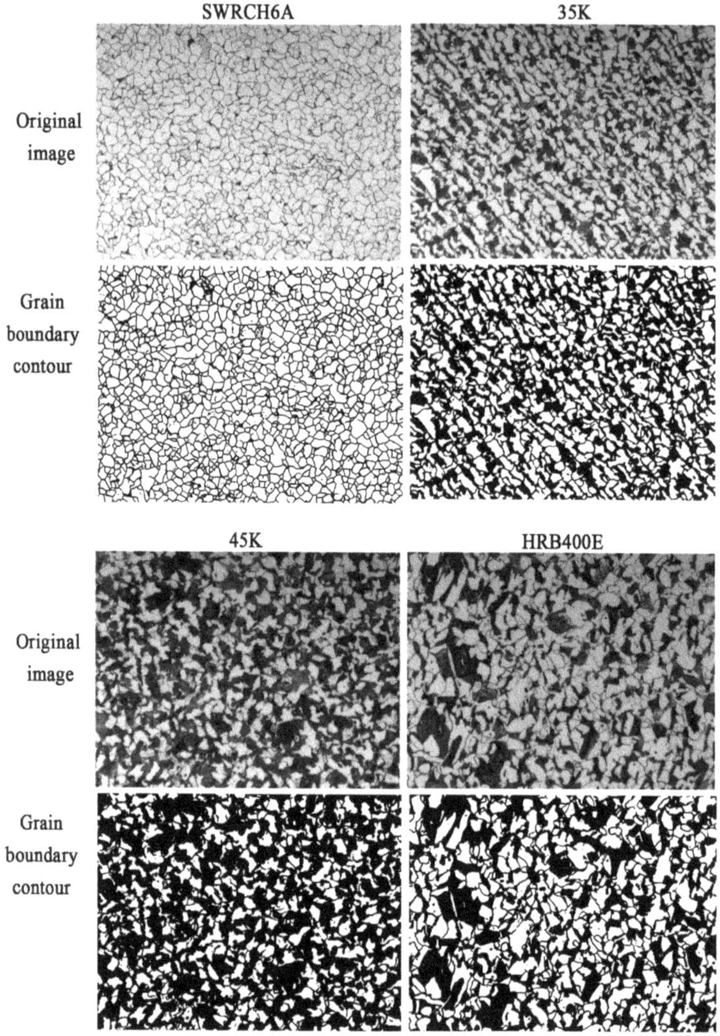

Figure 10. Grain-boundaries segmentation results.

4.2. *Rating of Grain Size Number by Circle Intercept Procedure*

A three-circle intercept procedure is used to count and calculate the grain boundary intersections in the segmented grain boundary images. Three concentric, isometric circles with a total perimeter of 500 mm are used to form a measurement grid, and the point of intersection between the grid and the grain boundary is the intersection. The number of visible intersections in the ferrite microstructure is obtained by counting the binary images of the edge contour with the three-circle intercept procedure. The intersections of the four materials are shown in Figure 11. It can be seen that the intersections of ferrite grain boundaries among the four materials are accurate, and the intersection areas of circumference and ferrite grain boundaries and ferrite–pearlite boundaries are identified as intersections.

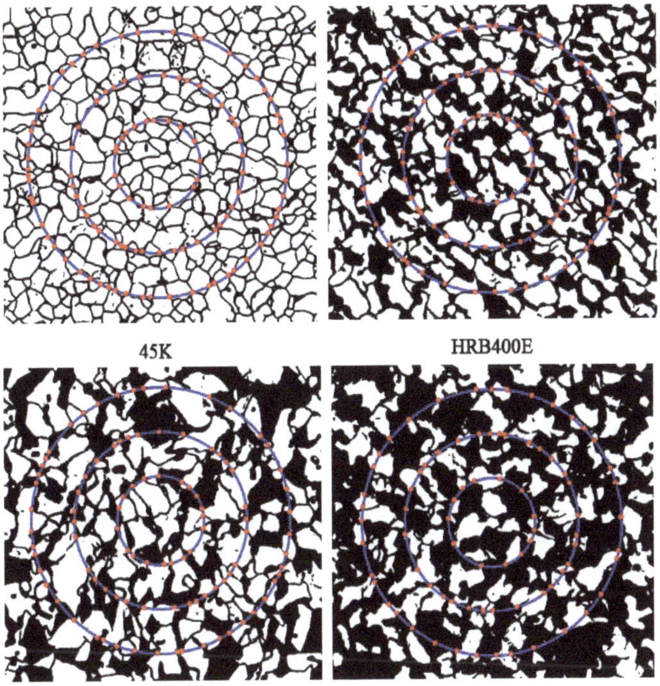

Figure 11. Processing results of four types of steels.

In the end, the automatic rating module integrates the visible intersections and hidden intersections to output the rating results and the labeled image of the three-circle intercept procedure. Then, calculate the average grain size number of the samples according to Formula (1). The rating results of the grain size number of the four samples are shown in Table 1. Compared with the manual rating results of experts, the average error is lower than Grade 0.5, meeting the requirements of the allowable error in the detection standard.

Table 1. Detection Results of Various Types of Steel.

Material	Average Error	Accuracy Rate
SWRCH6A	0.3064	96.67%
HRB400E	0.3160	90.00%
35K	0.2843	96.70%
45K	0.2897	93.33%

The average error and accuracy are defined as follows:

$$e_{i,j} = \left|s_{i,j} - r_{i,j}\right|$$

where *i* represents the sample number and *j* represents the sample category. $e_{i,j}$ represents the error of the ith sample of class *j*; $s_{i,j}$ represents the manual rating result of the *ith* sample of class *j* samples; $r_{i,j}$ represents the model rating result of the ith sample of class j samples.

$$\overline{e_j} = \frac{\sum_{i=1}^{n_j} e_{i,j}}{n_j}$$

$\overline{e_j}$ is the average error of the class j. n_j represents the total number of class-j samples.

$$A_j = \frac{n_{0.5,j}}{n_j}$$

A_j represents the accuracy of class-j samples and $n_{0.5,j}$ represents the number of samples whose error is less than 0.5.

This paper is based on the model training, prediction and data processing of metallographic images, which inevitably relies on the quality of images. An image will be more representative in characteristics if it has fewer interference items and richer data, and as a result, the model training effect will be better. The original metallographs do have such defects as scratches, foreign contaminants and holes, and they all are interference items for the training and prediction of the deep learning model. In this paper, two methods are mainly used to eliminate the negative impact of these interference items on the model as far as possible. Method 1: accurate data calibration. Scratches, foreign contaminants and holes are different from grain boundaries in form. For example, scratches are mainly long straight lines, while foreign contaminants and holes are mostly points rather than curves. They are obviously different from the grain boundaries that we need to identify. In the preliminary data calibration work, we arranged senior metallographic analysis experts to identify and calibrate the grain boundaries and would not mark these interference items as inclusions. The accurate calibration ensures that, in subsequent training and identification, the model algorithm is unlikely to identify these interference items in the verification sample as grain boundaries. Method 2: Identification through the algorithm. As mentioned earlier, interference items differ from grain boundaries in size, grayscale or shape, and we can avoid misjudgment as far as possible by identifying these differences through the algorithm. For example, in terms of size, grain boundaries are curve segments, so we can directly use the algorithm to delete particles whose length and width are less than a certain value (e.g., 3 μm). In terms of grayscale, the grain boundaries are gray or nearly black. Items with lighter or darker colors are generally foreign contaminants, holes, corrosive water stains, or other interference items. We set a grayscale threshold in the algorithm, and this can also exclude interference items to some extent.

This testing process is an approximate, simple and fast method, suitable for mass production inspection in factories, and cannot be used for accurate detection of grain size. If the true and accurate grain size of the sample is required, it is also necessary to detect the grain size of pearlite (the size of pearlite colonies) besides that of ferrite. In this case, it is necessary to use a scanning electron microscope to obtain a metallographic image with higher resolution, which can clearly display the boundaries of pearlite colonies, and thus the grain size of pearlite can be tested. As we know, the pearlite microstructure is in colonies due to different orientations of ferrite–cementite lamellae, and the boundaries between microstructures of different orientations can be regarded as the grain boundaries of pearlite colonies. Unlike solid boundaries such as ferrite grain boundaries, such boundaries are not solid boundaries formed by lines or specific grayscales or colors, but virtual boundaries formed only due to different orientations of lamellae. The areas on both sides of such virtual boundaries have no difference, except the difference in the orientations of lamellae. Unlike the hidden boundary described earlier as well, this virtual boundary is objectively present and is not visible to the human eyes simply because of the low resolution in the optical images. Such boundaries are easy to identify for man, but difficult for computer. We are studying this aspect and have achieved good results, which will be presented in the next paper.

5. Conclusions

This paper presents a model for automatic detection and quantitative analysis of the grain size number of ferrite–pearlite two-phase microstructure. A U-Net neural network model is designed to segment ferrite grain boundaries. In view of the hidden grain boundaries in pearlite microstructure, the number of hidden grain boundaries can be inferred based on the confidence of average grain size. The grain size number is rated using the three-circle intercept procedure. The results show that the accuracy of this procedure is greater than 90% in the grain boundary segmentation for the four types of ferrite–pearlite two-phase microstructure samples. All of these results deviate from those calculated by experts using the manual intercept procedure by less than Grade 0.5, the allowable detection error specified in the standard. In addition, the detection time is shortened to 2 s from 30 min consumed in the manual intercept procedure, greatly improving the detection efficiency and reducing labor intensity. Accordingly, the method proposed in this paper can be effectively used for the automatic rating of grain size number of ferrite–pearlite microstructure. The results obtained can be applied in steelworks laboratories to control the grain size of rolled ferrite–pearlitic steel.

Author Contributions: Conceptualization, Z.Y. and X.Z.; methodology, Y.Z.; validation, P.Z., L.Q. and M.L.; investigation, X.Z.; data curation, C.K. and Z.Z.; writing—original draft, X.Z.; writing—review and editing, L.Q. and G.H.; supervision, Q.Y.; Resources, Q.Y., project administration, Y.Z.; funding acquisition, X.Z. and Y.Z. All authors have read and agreed to the published version of the manuscript.

Funding: This project was supported by Science and Technology Program of Jiangsu Provincial Administration for Market Regulation (No. KJ204115, KJ21125122, KJ2022062).

Institutional Review Board Statement: Not applicable.

Informed Consent Statement: Not applicable.

Data Availability Statement: The data presented in this study are available on request from the corresponding author.

Conflicts of Interest: The authors declare no conflict of interest.

References

1. Callister, W.D.; Rethwisch, D.G. *Materials Science and Engineering: An Introduction*; Wiley: New York, NY, USA, 2018; pp. 2–9.
2. Zhao, H.L.; Jia, Z.Q.; Yi, N. Development of Quantitative Model of Grain Size and Yield Strength of Niobium Sheel. *Rare Met. Cem. Carbides* **2017**, *45*, 34–36.
3. Huang, C.X.; Hu, W.; Yang, G.; Zhang, Z.F.; Wu, S.D.; Wang, Q.Y.; Gottstein, G. The effect of stacking fault energy on equilibrium grain size and tensile properties of nano structured coper and copper-aluminum alloys processed by equal channel angular pressing. *Mater. Sci. Eng. A* **2012**, *556*, 638–647. [CrossRef]
4. Hall, E.O. The deformation and ageing of mild steel. *Proc. Hysical Soc. Sect. B* **1951**, *64*, 747–753. [CrossRef]
5. Petch, N.J. The cleavage strength of polycrystals. *J. Iron Steel Inst.* **1953**, *174*, 25–28.
6. *ASTM E112-2013*; Standard Test Methods for Determining Average Grain Size. Standard. ASTM: West Conshohocken, PA, USA, 2021; pp. 8–12.
7. Haboubi, S.; Guesmi, T.; Alshammari, B.M.; Alqunun, K.; Alshammari, A.S.; Alsaif, H.; Amiri, H. Improving CNN-BGRU Hybrid Network for Arabic Handwritten Text Recognition. *Comput. Mater. Contin.* **2022**, *73*, 5385–5397. [CrossRef]
8. Vaiyapuri, T.; Srinivasan, S.; Sikkandarn, M.Y.; Balaji, T.S.; Kadry, S.; Meqdad, M.N.; Nam, Y.Y. Intelligent Deep Learning Based Multi-Retinal Disease Diagnosis and Classification Framework. *Comput. Mater. Contin.* **2022**, *73*, 5543–5557. [CrossRef]
9. Valença, J.; Mukhandi, H.; Araújo, A.G.; Couceiro, M.S.; Júlio, E. Benchmarking for Strain Evaluation in CFRP Laminates Using Computer Vision: Machine Learning versus Deep Learning. *Materials* **2022**, *15*, 6310. [CrossRef] [PubMed]
10. Ren, Z.; Ren, G.; Wu, D. Deep Learning Based Feature Selection Algorithm for Small Targets Based on mRMR. *Micromachines* **2022**, *13*, 1765. [CrossRef] [PubMed]
11. Zhang, H.J.; Wang, C.G.; Yu, Z.H. Image Threshold Segmentation based on Simulated Annealing Algorithms in Wavelet Field. *J. Inn. Mong. Agric. Univ.* **2008**, *29*, 161–164.
12. Geng, S. The Application of Digital Image Processing Tometallographic Analysis. Master's Thesis, Nanjing University of Aeronautics and Astronautics, Nanjing, China, 2018.
13. Zhu, J.D. Research on the Algorithm Used for Grain Boundary Extraction from Metallographic Images and Design of Intelligent Rating Software. Master's Thesis, Jiangsu University, Zhenjiang, China, 2018.
14. Patxi, F.Z.; Andrés, M.R.; Quinn, C. Austenitic Parent Grain Reconstruction in Martensitic Steel Using Deep Learning. *Mater. Charact.* **2022**, *185*, 111759. [CrossRef]
15. Wang, S.Y.; Zhang, P.Z.; Zhou, S.Y. A Computer Vision Based Machine Learning Approach for Fatigue Crack Initiation Sites Recognition. *Comput. Mater. Sci.* **2020**, *171*, 109259. [CrossRef]

16. Azimi, S.M.; Britz, D.; Engstler, M.; Fritz, M.; Mücklich, F. Advanced Steel Microstructural Classifcation by Deep Learning Methods. *Sci. Rep.* **2018**, *8*, 2128. [CrossRef] [PubMed]
17. Li, X.C.; Zhao, J.X.; Cong, J.H.; Misra, R.D.K.; Wang, X.M.; Wang, X.L.; Shang, C.K. Machine Learning Guided Automatic Recognition of Crystal Boundaries in Bainitic/Martensitic Alloy and Relationship Between Boundary Types and Ductile-to-Brittle Transition Behavior. *J. Mater. Sci. Technol.* **2021**, *84*, 49–58. [CrossRef]
18. Han, B.; Wan, W.H.; Sun, D.D.; Dong, C.C.; Zhao, L.; Wang, H.Z. A deep learning-based method for segmentation and quantitative characterization of microstructures in weathering steel from sequential scanning electron microscope images. *J. Iron Steel Res. Int.* **2022**, *29*, 836–845. [CrossRef]
19. Tharindu, K.; Stephan, L.; Arash, A. Data-driven prediction of forging outcome: Effect of preform shape on plastic strain in a magnesium alloy forging. *Mater. Today Commun.* **2022**, *31*, 03210. [CrossRef]
20. Yu, H.S.; Yang, Z.G.; Tan, L.; Wang, Y.N.; Sun, W.; Sun, M.G.; Tang, Y.D. Methods and datasets on semantic segmentation: A review. *Neurocomputing* **2018**, *304*, 82–103. [CrossRef]
21. Zhang, J.N. Automatic Grain Size Rating of Metallographic Based on Deep Learning. Master's Thesis, Automation Research and Design Institute of Metallurgical Industry Co., Ltd., Beijing, China, 2021.
22. Ronneberger, O.; Fischer, P.; Brox, T. U-Net: Convolutional Networks for Biomedical Image Segmentation. In *Medical Image Computing and Computer-Assisted Intervention*; Springer: Munich, Germany, 2015. [CrossRef]
23. Bloice, M.D.; Roth, P.M.; Holzinger, A. Biomedical image augmentation using Augmentor. *Bioinformatics* **2019**, *35*, 4522–4524. [CrossRef]

materials

Article

Forming Limit Analysis of Thin-Walled Extruded Aluminum Alloy Tubes under Nonlinear Loading Paths Using an Improved M-K Model

Haihui Zhu [1], Yanli Lin [1,*], Kelin Chen [2] and Zhubin He [1,*]

[1] School of Mechanical Engineering, Dalian University of Technology, Dalian 116024, China
[2] Department of Integrated Systems Engineering, The Ohio State University, Columbus, OH 43210, USA
* Correspondence: linyanli0616@163.com (Y.L.); hezb@dlut.edu.cn (Z.H.)

Abstract: To meet the requirement of lighter weight and better performance in tube hydroforming, one of the most important tasks is to accurately predict the forming limit of thin-walled tubes under nonlinear loading paths. This work established the M-K+DF2012 model, a combination of the M-K model and the DF2012 ductile fracture criterion, for the forming limit prediction of thin-walled tubes under nonlinear loading paths. In this model, the failure of the groove is determined by the DF2012 criterion, and the corresponding strains in the uniform region are the limit strains. The limit strains of an AA6061 aluminum alloy tube under a set of linear loading paths and two typical nonlinear loading paths were tested. Parameter values of the M-K+DF2012 model for the tube were determined based on the experimental limit strains under linear loading paths, and the limit strains under the two nonlinear loading paths were predicted. Then the strain-based forming limit diagram (ε-FLD) and the polar effective plastic strain FLD (PEPS-FLD) of the tube under different pre-strains were predicted and discussed. The results show that the limit strains of the tube are obviously path-dependent, and the M-K+DF2012 model can reasonably capture the limit strains of the tube under both linear and nonlinear loading paths. The predicted ε-FLD shows a strong dependence on the pre-strain, while the predicted PEPS-FLD is weakly strain path-dependent and almost path-independent on the right-hand side for the AA6061 tube.

Keywords: forming limit; nonlinear loading path; M-K+DF2012 model; ductile fracture criterion; extruded aluminum alloy tube

1. Introduction

Tube hydroforming is one of the advanced manufacturing technologies to form lightweight tubular parts and meet the lightweight requirement in industries such as the aircraft, aerospace, and automobile fields [1–3]. Tube hydroforming processes often involve complex contact conditions and combinations of internal pressure and axial force [4], and even include several pre-deformation steps before the final bulging. Therefore, the deformation history during tube hydroforming is complex, and the strain path is nonlinear even in a simple tube hydro-bulging test with no axial feed [5,6].

As a type of lightweight material, aluminum alloy has been widely used in key structural parts and is being increasingly applied, especially in the automobile industry. However, the ductility of the aluminum alloy at room temperature is relatively low, and the most common challenge in the hydroforming of an aluminum alloy tube is fracture, which limits the development of the product for lighter weight and better performance. Therefore, it is important to accurately predict the forming limit of aluminum alloy tubes under complex loading conditions and avoid fracture by optimizing the loading path [7,8]. Considering that the occurrence of necking before fracture usually reflects the failure of the material, the concept of forming limit strain in this paper refers to the necking strain. The experimental necking strains of a tube can be defined with different methods, such as the strain-rate method [9], the maximum pressure method [10], and the time-dependent method [11].

The forming limit diagram (FLD) is an important tool to evaluate the formability of thin-walled metals. The traditional strain-based FLD (ε-FLD) proposed by Keeler et al. [12] and Goodwin [13] is determined under linear loading paths. However, the ε-FLD has been shown to be path-dependent [14–16]. Thus, the ε-FLD obtained under linear loading paths are not suitable for the nonlinear loading conditions. In order to overcome the influence of the load path, several stress-based FLDs have been developed, for example, the forming limit stress diagram (FLSD) [17,18] and the extended stress FLD (XSFLD) [19]. Experimental results show that the path sensitivity of these FLDs is significantly reduced. However, due to the reduction of the slope of the true stress–strain curve as the strain increases, a small change in stress close to the necking limit may lead to a large change in strain. Moreover, the stress states may not be accurately measured directly. As a result, the application of these stress-based FLDs is limited.

Stoughton and Yoon [20] established the polar effective plastic strain FLD (PEPS-FLD) using the effective plastic strain to represent formability. The path in the PEPS-FLD is determined based on the magnitude of the effective plastic strain radius and the direction of the strain increment in the conventional strain diagram. Nguyen et al. [21] predicted necking in free-expansion hydroforming of the tube with tensile pre-straining in the axial direction and demonstrated that PEPS-FLD captured the forming limit accurately. The main benefits of the PEPS-FLD are (a) its weak strain path-dependency, (b) its similar shape to ε-FLD, and its ease of understanding [22]. However, experimental studies showed that the path dependence of PEPS-FLD depends on the specific material [23]. For some materials, the influence of pre-strains on the PEPS-FLD cannot be ignored. In this case, the PEPS-FLD determined under linear loading paths is not an appropriate criterion for limit strain prediction under complex loading conditions [23].

On the other hand, the theoretical prediction of the forming limit is also critical in the design and analysis of forming processes. At present, many theoretical models for forming limit prediction have been proposed. The M-K model is one of the most widely used models in the prediction of forming limit of thin-walled metals [24,25]. The model can be combined with any anisotropic yield criterion and hardening law [26,27] and is suitable for nonlinear loading conditions. In the traditional M-K model, the limit state is determined by the ratio of major strain increment in the groove to that in the uniform region. However, it is rather empirical and physically unsound to use a constant critical strain increment ratio as the failure criterion, considering that the major strain incremental ratio at failure may be influenced by the stress state [28].

In order to accurately predict the forming limit, the M-K model can be combined with a ductile fracture criterion (DFC), for example, the well-known Gurson model and Gurson–Tvergaard–Needleman model [29–31]. This type of M-K+DFC model has been discussed briefly in the reviews of Banabic et al. [25,32] and Zhang et al. [33]. However, the computational cost of a combination of the M-K model and a physical DFC is much more expensive than the traditional M-K model [34]. It is worth noting that a phenomenological DFC is generally a function of stress components and strain increments, which can predict forming limits with much lower computational costs and take the nonlinear loading path into account through incremental calculation [35–37]. Therefore, the combination of the M-K model and a phenomenological DFC is worth studying in the prediction of the forming limit under nonlinear loading paths.

For some metals, the fracture forming limit curve (FFLC) monotonically decreases from uniaxial tension to equibiaxial tension, which can be captured by classic DFCs such as the Clift criterion [38], the Cockcroft–Latham criterion [39], and the McClintock criterion [40]. However, the FFLCs of many materials may have more complex shapes, for example, a shape of "V" [41]. Lou et al. [42] proposed the DF2012 criterion, and both monotonically decreasing and V-shaped curves can be predicted by the DF2012 criterion. In our previous work, the FLCs predicted by different M-K+DFC models under linear loading conditions were compared, and the M-K+DF2012 model gave the best prediction [43]. However, the performance of an M-K+DFC model under nonlinear loading paths is also very important and has not been studied systematically. Therefore, it is necessary to carry out further studies on the combined model under nonlinear loading conditions.

In this work, an improved M-K+DF2012 model will be established first to predict the forming limit at the necking of an AA6061 aluminum alloy tube under nonlinear loading paths. Then, the limit strains of the tube under linear and nonlinear loading paths will be tested through tube-controllable biaxial loading experiments. After that, the validity of the M-K+DF2012 model will be validated by the experimental results. Finally, the effect of pre-strains on the ε-FLD and the PEPS-FLD predicted by the M-K+DF2012 model will be discussed.

2. M-K+DF2012 Model for Tubes

2.1. Fundamental Assumptions

In tube hydroforming, due to the action of internal pressure, the hoop strain is usually the largest principal strain and the crack is often formed along the axial direction. Meanwhile, the thickness distribution in the hoop direction of an extruded tube is not homogenized because of the eccentricity of the extrusion mandrel [44]. Therefore, a groove along the axial direction is assumed to represent the thickness inhomogeneity of the tube, as shown in Figure 1. The tube specimen is divided into a uniform region (region A) with thickness t_A and a groove (region B) with thickness t_B. The initial coefficient of non-homogeneity f_0 is defined as the ratio of the initial thicknesses of regions B and A:

$$f_0 = t_{B0}/t_{A0} \tag{1}$$

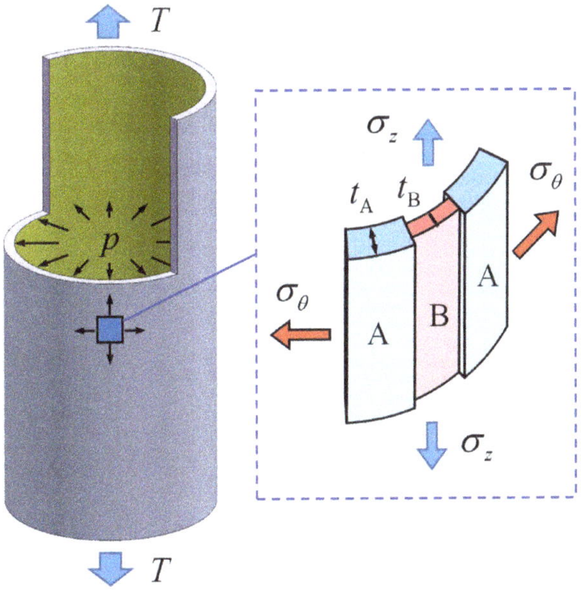

Figure 1. Schematic diagram of the axial groove in tubes.

It is important to point out that the normal stress σ_n is usually much smaller than the axial stress σ_z and hoop stress σ_θ in the hydroforming of a thin-walled tube. Thus, the normal stress can be ignored, and the stress condition can be simplified to the plane stress state. At the same time, the force equilibrium condition along the hoop direction between regions A and B must be satisfied:

$$\sigma_{\theta A} t_A = \sigma_{\theta B} t_B \tag{2}$$

where $\sigma_{\theta A}$ and $\sigma_{\theta B}$ are the hoop stresses in regions A and B, respectively.

On the other hand, the axial strain increments in regions A and B should be equal according to the requirement of coordinated deformation:

$$d\varepsilon_{zA} = d\varepsilon_{zB} \tag{3}$$

In the traditional M-K model, when the ratio of the major strain increment becomes very large (for example, 10), the strains in region A are the limit strains. Differently, in the M-K+DF2012 model, the onset of fracture in region B is determined by the DF2012 criterion, and the strains in region A at the moment are defined as the limit strains, and the DF2012 criterion can be expressed as [42]:

$$\int_0^{\bar{\varepsilon}_f} \left(\frac{2\tau_{\max}}{\bar{\sigma}}\right)^{C_1} \left(\frac{\langle 1+3\eta \rangle}{2}\right)^{C_2} d\bar{\varepsilon} = C_3,\ \langle x \rangle = \begin{cases} x \text{ when } x \geq 0 \\ 0 \text{ when } x < 0 \end{cases} \tag{4}$$

where $\eta = \sigma_m/\bar{\sigma}$ is the stress triaxiality, $\bar{\sigma}$ is the equivalent stress, and $\sigma_m = (\sigma_1 + \sigma_2 + \sigma_3)/3$ is the hydrostatic stress; σ_1, σ_2 and σ_3 are the three principal stress components, and $\sigma_1 > \sigma_2 > \sigma_3$; $\tau_{\max} = (\sigma_1 - \sigma_3)/2$ is the maximum shear stress; C_1, C_2, and C_3 are material constants. For a thin-walled tube subjected to a combined load of internal pressure and axial tension, the typical stress states are $\sigma_1 = \sigma_\theta$, $\sigma_2 = \sigma_z$, and $\sigma_3 = \sigma_t = 0$.

Considering that both the M-K model and the DF2012 criterion can be used with anisotropic plastic constitutive models and nonlinear loading paths, the M-K+DF2012 model is also suitable for predicting the forming limits of anisotropic materials under nonlinear loading conditions.

2.2. Prediction Process

The process of applying the M-K+DF2012 model to predict the limit strain under a nonlinear strain path is shown in Figure 2. Firstly, the strain path in region A should be discretized into sufficiently small incremental steps. According to the associated flow rule, the plastic potential equals the yield function ϕ, and there is a relationship between the strain increment ratio β and the stress state as follows:

$$\beta = \frac{d\varepsilon_z}{d\varepsilon_\theta} = \frac{\partial\phi/\partial\sigma_z}{\partial\phi/\partial\sigma_\theta} \tag{5}$$

where $d\varepsilon_\theta$ and $d\varepsilon_z$ are the hoop and axial strain increments, respectively.

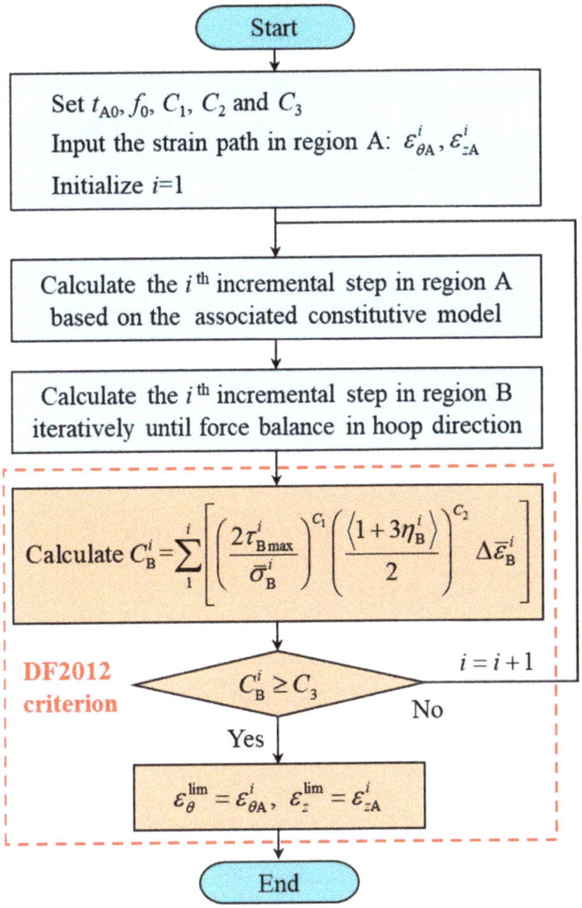

Figure 2. Prediction process for the limit strain under a nonlinear loading path using the M-K+DF2012 model.

Hence, the stress components in region A at step i can be calculated based on the strain increments and the associated constitutive model. Then, the strain increments and stress components in region B at step i can be calculated as follows: The axial strain increment in region B, $d\varepsilon_{zB}$, equals that in region A; and the hoop strain increment $d\varepsilon_{\theta B}$, the axial stress σ_{zB}, and the hoop stress $\sigma_{\theta B}$ can be obtained by numerical iteration using force balance in the hoop direction, as seen in Equation (2). In the meantime, the equivalent strain increment $d\bar{\varepsilon}$ can be obtained based on the rule of the equal plastic work rate [45]:

$$dW = \sigma_\theta d\varepsilon_\theta + \sigma_z d\varepsilon_z = \bar{\sigma}(\bar{\varepsilon})d\bar{\varepsilon} \tag{6}$$

Therefore, the stress and strain components in regions A and B can be calculated step by step through the incremental method until the deformation in region B satisfies the DF2012 criterion. The strain in region A at that moment is the limit strain at necking under the given loading path.

2.3. Constitutive Model

A reasonable constitutive model is critical for the prediction of the forming limits. Here, an associated constitutive model with the Yld2000-2d anisotropic yield criterion and the power hardening law is adopted. The Yld2000-2d criterion is an advanced and commonly used yield criterion for anisotropic aluminum alloys [46], which is defined as:

$$\phi = \left|X_1' - X_2'\right|^k + \left|X_1'' + 2X_2''\right|^k + \left|2X_1'' + X_2''\right|^k = 2\bar{\sigma}^k \tag{7}$$

where X_i' and X_i'' (i = 1, 2) are the principal values of tensors $\mathbf{X}'$=$\mathbf{L}' \cdot \boldsymbol{\sigma}$ and $\mathbf{X}''$=$\mathbf{L}'' \cdot \boldsymbol{\sigma}$, respectively. $\boldsymbol{\sigma}$, $\mathbf{L}'$, and $\mathbf{L}''$ can be written as:

$$[\boldsymbol{\sigma}] = \begin{bmatrix} \sigma_\theta & \sigma_z & \sigma_{\theta z} \end{bmatrix}^{\mathrm{T}} \tag{8}$$

$$\left[\mathbf{L}'\right] = \begin{bmatrix} L_{11}' & L_{12}' & 0 \\ L_{21}' & L_{22}' & 0 \\ 0 & 0 & L_{66}' \end{bmatrix} = \frac{1}{3}\begin{bmatrix} 2\alpha_1 & -\alpha_1 & 0 \\ -\alpha_2 & 2\alpha_2 & 0 \\ 0 & 0 & \alpha_7 \end{bmatrix} \tag{9}$$

$$\left[\mathbf{L}''\right] = \begin{bmatrix} L_{11}'' & L_{12}'' & 0 \\ L_{21}'' & L_{22}'' & 0 \\ 0 & 0 & L_{66}'' \end{bmatrix} = \frac{1}{9}\begin{bmatrix} 8\alpha_5 - 2\alpha_3 - 2\alpha_6 + 2\alpha_4 & 4\alpha_6 - 4\alpha_4 - 4\alpha_5 + \alpha_3 & 0 \\ 4\alpha_3 - 4\alpha_5 - 4\alpha_4 + \alpha_6 & 8\alpha_4 - 2\alpha_6 - 2\alpha_3 + 2\alpha_5 & 0 \\ 0 & 0 & 9\alpha_8 \end{bmatrix} \tag{10}$$

in which α_i (i = 1, 2 ... 8) represents pending coefficients.

The principal values of **X** are

$$\begin{cases} X_1 = \frac{1}{2}\left(X_{11} + X_{22} + \sqrt{(X_{11} - X_{22})^2 + 4X_{12}^2}\right) \\ X_2 = \frac{1}{2}\left(X_{11} + X_{22} - \sqrt{(X_{11} - X_{22})^2 + 4X_{12}^2}\right) \end{cases} \tag{11}$$

On the other hand, the widely used power-hardening law is as follows:

$$\bar{\sigma} = K\bar{\varepsilon}^n \tag{12}$$

where K and n are the hardening coefficient and strain-hardening exponent, respectively.

3. Experiments

3.1. Experimental Principle and Setup

In this paper, the limit strains of the tube were tested in tube-controllable biaxial loading experiments, of which the principle and setup are shown in Figure 3. The tube specimen is bulged by the combined load of internal pressure p and axial load T. The deformation process of the specimen can be measured and recorded by the digital image correlation (DIC) system in real time, and the two principal stresses σ_z and σ_θ at the central point of the specimen can be controlled in real time by controlling p and T.

The stress state of the tube can be analyzed with the aid of the membrane theory if the ratio of thickness to diameter is less than 1/20 [47]. The axial stress σ_z and hoop stress σ_θ can be calculated by the following equations [48].

$$\begin{cases} \sigma_z = \frac{p\pi(\rho_\theta - t)^2 + T}{\pi(2\rho_\theta - t)t} \\ \sigma_\theta = \frac{2(\rho_z - t)(\rho_\theta - t)}{(2\rho_z - t)t}p - \frac{2\rho_\theta - t}{2\rho_z - t}\sigma_z \end{cases} \tag{13}$$

where t, ρ_θ and ρ_z are the instantaneous thickness, hoop, and axial radii of curvature at point P, respectively.

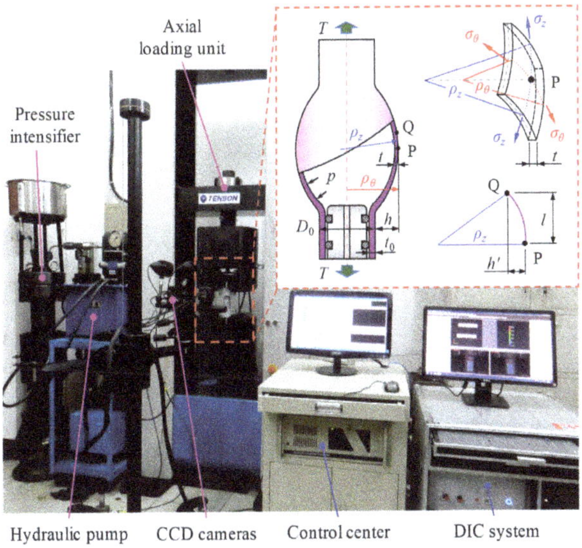

Figure 3. Principle and setup of tube controllable biaxial loading test.

According to the geometric relationship, the values of ρ_θ and ρ_z can be calculated by Equations (14) and (15), respectively.

$$\rho_\theta = \frac{D_0}{2} + h \tag{14}$$

$$\rho_z = \frac{h'^2 + l^2}{2h'} \tag{15}$$

where D_0 is the initial diameter of the tube, h is the bulging height of point P, l and h' are the axial distance and the radial distance between points P and Q, and the axial profile of the tube specimen is assumed to be circular, as can be seen in Figure 3.

In addition, the instantaneous thickness t at point P can be calculated by

$$t = t_0 \exp(-\varepsilon_z - \varepsilon_\theta) \tag{16}$$

where t_0 is the initial thickness of the tube.

The setup shown in Figure 3 consists of four units: Axial loading, internal pressure, displacement and strain measurement, and the control center. The axial loading unit acting on the ends of the tube is an electronic universal testing machine with a load capacity of ±200 kN. The internal pressure unit is a pressure intensifier with a maximum pressure of 40 MPa, which provides the internal pressure required for bulging. The displacement and strain measurement unit is a three-dimensional DIC system developed by the Institute of Mould and Advanced Forming Technology of Xi'an Jiaotong University. The control center unit is established using an industrial computer to achieve the simultaneous functioning of the above three units.

In this work, an extruded aluminum alloy seamless tube of AA6061 with a diameter of 40 mm and a thickness of 1.2 mm was used. As shown in Figure 4, the initial gauge length L_0 of the tube specimen was designed to be 80 mm, which was twice the initial outer diameter of the tube. Before testing, a speckle pattern was sprayed on the surface of the specimen to record the deformation process by DIC technology. In the meantime, the central point P and another point Q were marked with an initial axial distance of l_0 = 15 mm.

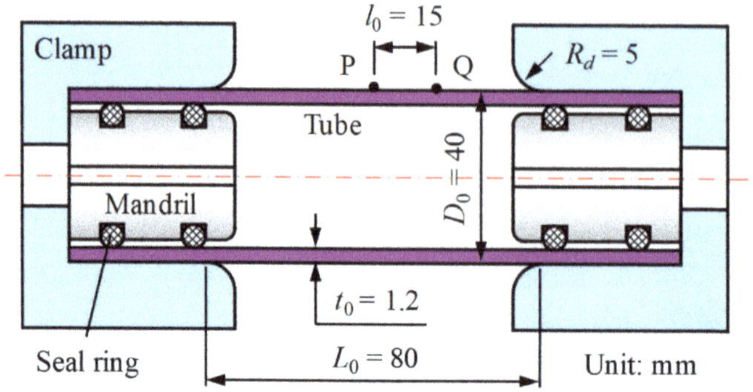

Figure 4. Schematic diagram of the tube specimen.

3.2. Scheme of Loading Paths

In this paper, the ε-FLD of the tube under linear loading conditions was tested first because it serves as the reference for discussing the influence of loading paths on the limit strains. In order to obtain the forming limit strains from hoop uniaxial tension to equibiaxial tension, nine linear loading tests were carried out with the axial–hoop stress ratios σ_z/σ_θ = 0, 0.125, 0.25, 0.375, 0.5, 0.625, 0.75, 0.875, and 1.0.

Pre-bulging and bending are two important pre-deformation forms in tube hydroforming, of which the strain states are approximately hoop plane strain and axial plane strain, respectively. In order to analyze the influence of plane strain pre-deformation and verify the validity of the M-K+DF2012 model under nonlinear loading paths, two nonlinear stress loading paths with plane strain pre-deformation were designed, as shown in Figure 5. In the first stage of path A, the stress ratio is prescribed as σ_z/σ_θ = 0.5, while the hoop stress remains constant in the second stage, and the turning point is located at (130, 260) MPa. For path B, the stress state in the first stage is σ_z/σ_θ = 2.0, the axial stress remains constant in the second stage, and the turning point is located at (225, 112.5) MPa. The stress states σ_z/σ_θ = 0.5 and 2.0 correspond to the hoop plane strain tension and axial plane strain tension for an isotropic tube, respectively.

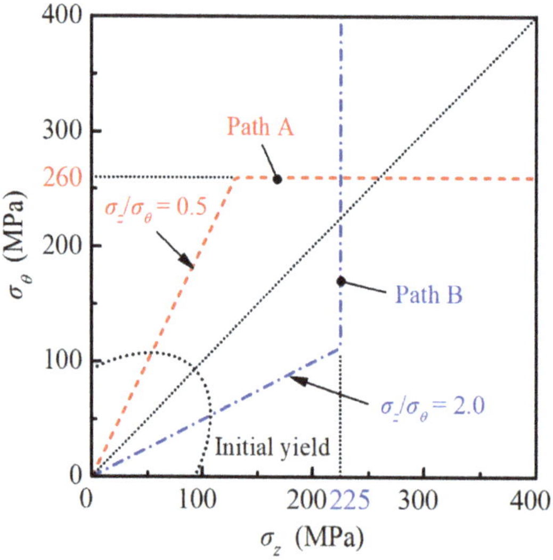

Figure 5. Designed continuous nonlinear loading paths for the AA6061 tube.

4. Results

4.1. Material Properties

Three tests were carried out to test the mechanical properties of the AA6061 tube, i.e., an axial uniaxial tensile test on a strip-shape specimen cut from the tube (refer to ASTM: E8), a controllable biaxial loading test with $\sigma_z/\sigma_\theta = 0$ as the alternative to the hoop uniaxial tension, and an equibiaxial tensile test with $\sigma_z/\sigma_\theta = 1.0$. Anisotropic parameters of the tube deduced from the experimental results are listed in Table 1. $r_b = \varepsilon_z/\varepsilon_\theta$ is the biaxial anisotropy coefficient under the loading condition of $\sigma_z/\sigma_\theta = 1.0$. Here, ε_θ and $\sigma_{\theta 0}$ were chosen as the denominators considering that the major limit strain is usually in the hoop direction.

Table 1. Mechanical properties and the Yld2000-2d yield criterion of the AA6061 tube (Adapted from [43]).

Parameters	Anisotropic Parameters					Power Hardening		Coefficients of the Yld2000-2d Yield Criterion					
	r_z	r_θ	r_b	$\sigma_{z0}/\sigma_{\theta 0}$	$\sigma_{b0}/\sigma_{\theta 0}$	K (MPa)	n	α_1	α_2	α_3	α_4	α_5	α_6
Values	0.454	0.927	2.80	0.988	0.959	391.7	0.2636	1.0072	0.9436	1.1625	1.0764	0.9725	0.8519

Meanwhile, the power hardening law of the tube was obtained by fitting the hoop true stress–plastic strain curve, the obtained values of K and n are also given in Table 1. As shown in Figure 6, the fitted hardening curve is in good agreement with the experimental stress–strain curve of the tube. The coefficients α_1 to α_6 of the Yld2000-2d yield criterion for the AA6061 tube are determined iteratively according to the r-values and stress ratios in Table 1. During the calculation, the exponent k was set to 8 considering that the aluminum alloy is an FCC material, and the coefficients α_7 and α_8 were set to 1.0 because the two coefficients are related to the shear stress components and do not influence the prediction of the tube controllable biaxial loading process [49]. The obtained values of α_1 to α_6 are also listed in Table 1.

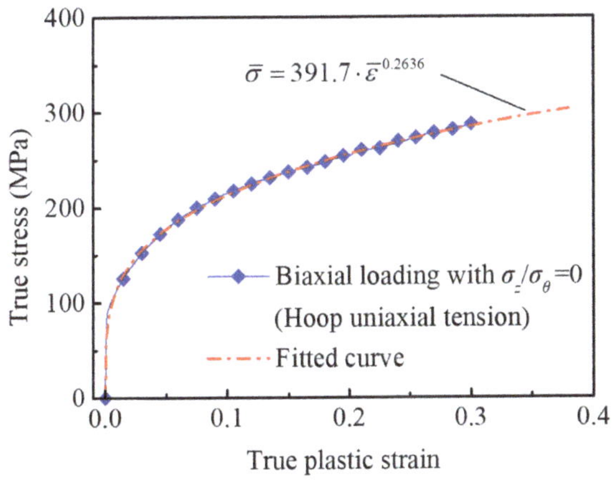

Figure 6. Stress–strain curve of the AA6061 tube in hoop direction.

4.2. Experimental Limit Strains

According to the time-dependent method proposed by Martínez-Donaire et al. [11], the onset of necking corresponds to the occurrence of a maximum of the first derivative of ε_θ at the boundary of the instability region. The limit strains at necking and corresponding strain paths at the central point of tube specimens under loading conditions of σ_z/σ_θ = 0, 0.125, 0.25, 0.375, 0.50, 0.625, 0.75, and 0.875 are shown in Figure 7. The necking point under the condition of σ_z/σ_θ = 1.0 is not given because the crack is along the hoop direction, which means that the assumption of the groove along the axial direction is not satisfied for this case. It can be seen from Figure 7 that these strain paths are all approximately linear and the experimental forming limit curve (FLC) is V-shaped.

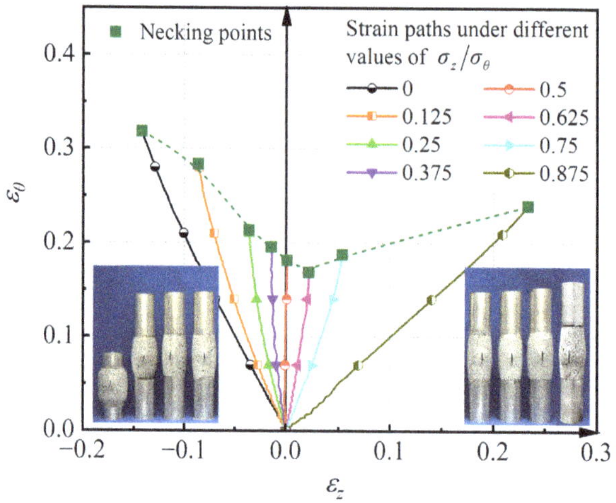

Figure 7. Necking points and strain paths of the AA6061 tube under linear loading paths.

Figure 8 shows the experimental limit strain points and corresponding strain paths of the AA6061 tube under nonlinear loading paths A and B. Both the two strain paths in the first stage are linear and close to the state of plane strain. While in the second stage, the strain paths gradually deviate from plane strain. Finally, both nonlinear strain paths converge to the plane strain state of $d\varepsilon_z = 0$ after necking, which leads to the axial cracks in the two specimens. It can be also noted that the limit strain point under the nonlinear loading path A is obviously higher than the experimental FLC in Figure 7 under linear loading conditions, but the limit strain point under path B is lower. It indicates that the formability of the AA6061 tube is significantly affected by the loading path.

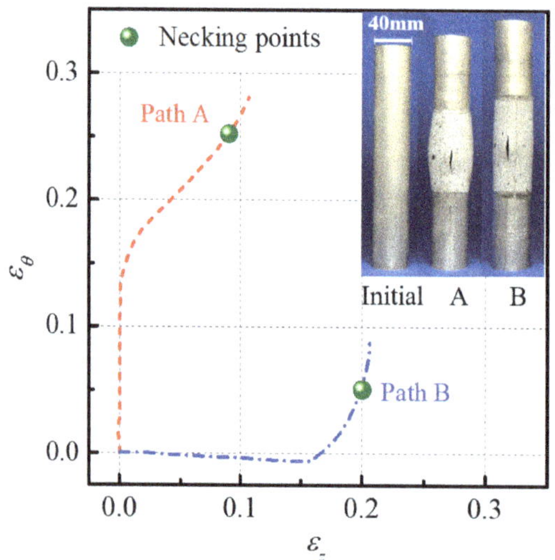

Figure 8. Necking points and strain paths of the AA6061 tube under nonlinear loading paths.

4.3. Prediction of Limit Strains

In order to predict the forming limit of the AA6061 tube using the M-K+DF2012 model, material constants C_1, C_2, and C_3 in the DF2012 criterion and the initial coefficient of non-homogeneity f_0 of the tube must be determined first. The values of these four parameters are listed in Table 2, which were determined by an optimization method aiming at the minimum root mean squared error between the experimental hoop limit strains shown in Figure 7 and the corresponding predicted values. The experimental data under nonlinear loading paths in Figure 8 are not included in the optimization.

Table 2. Parameter values of the M-K+DF2012 model for the AA6061 tube.

Parameters	C_1	C_2	C_3	f_0
Values	10.77	−2.814	0.4114	0.9852

In Figure 9, the limit strains of the tube predicted by the M-K+DF2012 model are compared with the experimental strain points. The M-K+DF2012 model provides accurate predictions under all eight linear loading paths and the two nonlinear loading paths. It indicates that the M-K-DF2012 model determined by only the data of linear loading can reasonably predict the influence of the nonlinear loading path on the forming limit of the AA6061 tube.

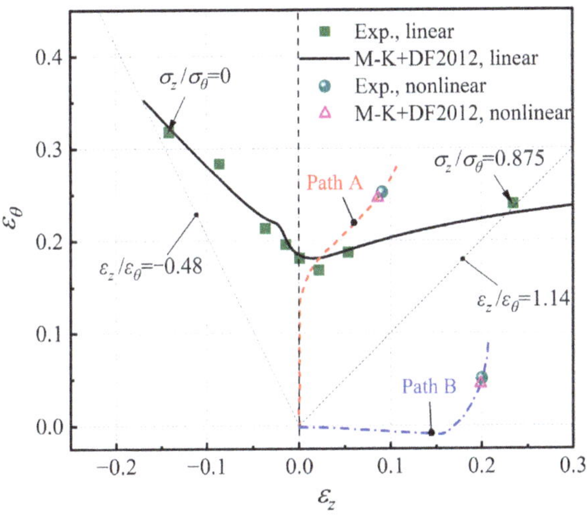

Figure 9. Comparison of the predicted and experimental limit strains of the AA6061 tube.

5. Discussion

5.1. Effect of Pre-Strain on Predicted ε-FLD

In tube hydroforming, proper pre-deformation steps are usually required, and the strain state of the pre-deformation is primarily selected in the range of hoop uniaxial pre-straining to axial plane strain pre-straining. In order to analyze the influence of pre-strains on the limit strains, the ε-FLCs of the AA6061 tube under four different pre-straining modes, namely, hoop uniaxial tensile pre-straining (UT-θ), hoop plane strain tensile pre-straining (PT-θ), biaxial tensile pre-straining with $\varepsilon_z = \varepsilon_\theta$ (BT), and axial plane strain tensile pre-straining (PT-z), were predicted using the M-K+DF2012 model.

The nonlinear loading path is the two subsequent linear loading steps described by the following relation:

$$\begin{cases} \beta = \frac{d\varepsilon_z}{d\varepsilon_\theta} = \beta_1, \text{ If } \bar{\varepsilon} \leq \bar{\varepsilon}^* \\ \beta = \frac{d\varepsilon_z}{d\varepsilon_\theta} = \beta_2, \text{ If } \bar{\varepsilon} > \bar{\varepsilon}^* \end{cases} \tag{17}$$

where β_1 and β_2 are the strain increment ratios related to the pre-strain step and the second loading step, respectively, and $\bar{\varepsilon}^*$ is the equivalent strain of the pre-deformation in region A.

Figure 10 shows ε-FLCs of the tube predicted by the M-K+DF2012 model under different pre-strain conditions, where the equivalent pre-strains are all set to $\bar{\varepsilon}^* = 0.1$. It can be easily found that ε-FLCs are obviously affected by pre-strains. Compared with the ε-FLC under linear loading conditions, the ε-FLC after UT-θ pre-straining shifts towards the tension-compression side. The ε-FLC after PT-θ pre-straining increases slightly in the tension-compression side and increases obviously in the tension-tension side but remains unchanged in the hoop plane strain tensile state. Meanwhile, the ε-FLCs after BT pre-straining and PT-z pre-straining shift towards the tension-tension side with a significant reduction of hoop strain. The moving direction of the ε-FLC predicted under different pre-strains is approximately consistent with the experimental results in the literature [23].

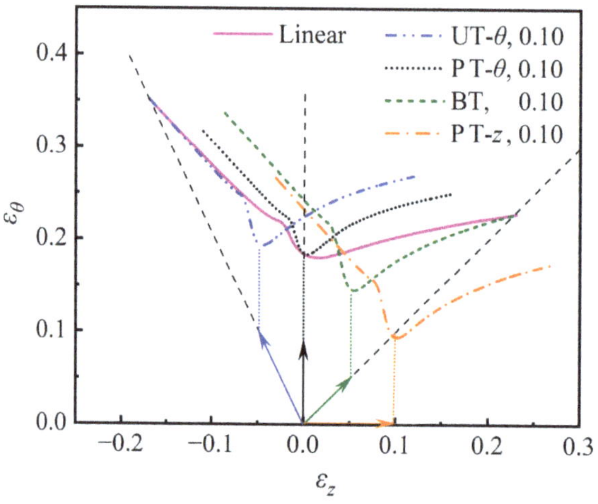

Figure 10. ε-FLCs of the AA6061 tube predicted by the M-K+DF2012 model under different pre-strain conditions.

5.2. Effect of Pre-Strain on Predicted PEPS-FLD

In order to investigate the effect of the loading path on the PEPS-FLD predicted by the M-K+DF2012 model, the ε-FLCs in Figure 10 were converted to PEPS-FLCs as shown in Figure 11. It was observed that the PEPS-FLD is weakly strain path-dependent. On the right-hand side of the FLD, all PEPS-FLCs almost fall on a single curve. On the left-hand side of the FLD, the equivalent limit strain at necking increases slightly after pre-straining such as UT-θ, BT, and PT-z, while the effect of the PT-θ pre-straining on the equivalent limit strain is small. Therefore, if the final deformation state of the AA6061 tube is located on the right-hand side of the PEPS-FLD, e.g., a tension–tension strain state, the forming limit can be evaluated through the PEPS-FLD regardless of the loading paths.

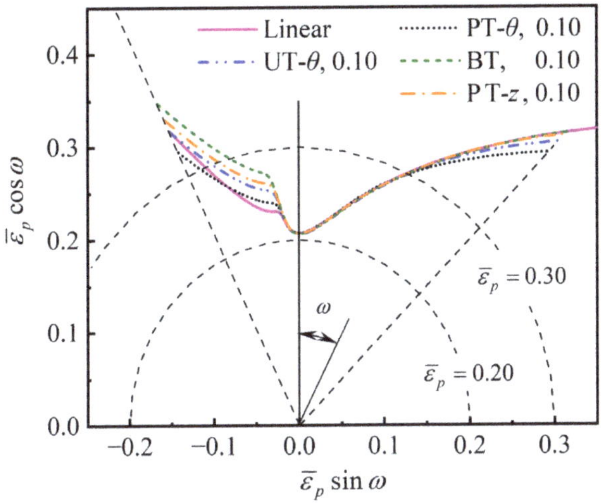

Figure 11. Predicted PEPS-FLD of the AA6061 tube under different pre-strain conditions.

Then, the PEPS-FLC predicted by the M-K+DF2012 model under linear loading conditions is compared with the experimental results under linear paths and nonlinear paths A and B, as shown in Figure 12. The two experimental points of nonlinear loading are on the right-hand side, and all the points on the right-hand side roughly follow the same curve. It indicates that the path dependence of the PEPS-FLD of the AA6061 tube on the right-hand side is weak. In other words, the small difference between the predicted curves on the right-hand side of Figure 11 is consistent with the experimental phenomenon.

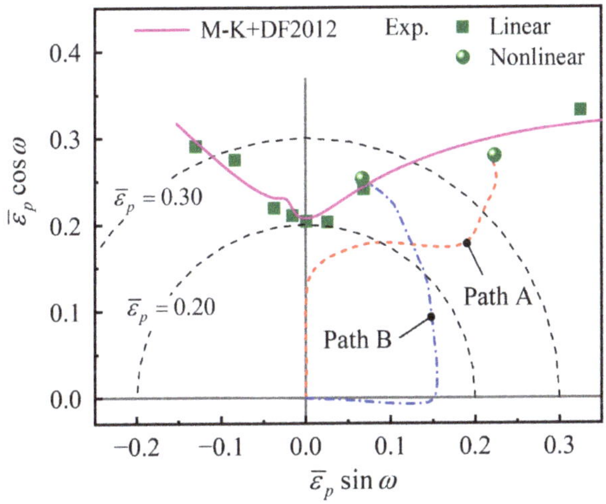

Figure 12. Comparison of predicted PEPS-FLC and experimental limit strains of the AA6061 tube.

6. Conclusions

In this paper, the use of the M-K+DF2012 model for the prediction of forming limits under nonlinear loading paths was established, the limit strains of the AA6061 aluminum alloy tube under a set of linear and two typical nonlinear loading paths were tested, the validity of the M-K+DF2012 model was validated by the experimental results, and the ε-FLD and PEPS-FLD under different pre-strains were discussed. The following conclusions were obtained:

(1) The forming limit of the AA6061 tube is significantly affected by the loading path. The limit strain in the tension–tension strain state increases obviously after a hoop plane strain tensile pre-straining.

(2) The M-K+DF2012 model can reasonably predict the limit strains of the AA6061 aluminum alloy tube under both linear and nonlinear loading paths.

(3) The ε-FLD predicted by the M-K+DF2012 model under different pre-strain conditions shows a strong dependence on the pre-strain, and the direction in which the FLC shifts relative to the conventional FLC under linear loading conditions depends on the types of pre-strain.

(4) The PEPS-FLD predicted by the M-K+DF2012 model is weakly strain path dependent, and almost path-independent on the right-hand side for the AA6061 tube. The curve predicted under linear loading conditions is in good agreement with the experimental limit strain points of the AA6061 tube.

In the future, it is worth applying the M-K+DF2012 model to different materials such as Magnesium alloy and Titanium alloy and improving the model predictability according to the characteristics of the materials. On the other hand, more influential factors should be introduced into the model in order to predict the forming limit under more complex deformation conditions. The influential factor could be the angle of the groove, the strain rate sensitivity, and the stress in the thickness direction.

Author Contributions: Conceptualization, H.Z. and Z.H.; methodology, H.Z. and Y.L.; validation, Y.L. and K.C.; formal analysis, H.Z., Y.L. and K.C.; investigation, H.Z. and Z.H.; resources, Z.H.; data curation, H.Z. and Y.L.; writing—original draft preparation, H.Z., Y.L. and K.C.; supervision, Z.H.; project administration, Z.H. All authors have read and agreed to the published version of the manuscript.

Funding: This research was funded by the National Natural Science Foundation of China, grant numbers: 52075075, 51575131.

Institutional Review Board Statement: Not applicable.

Informed Consent Statement: Not applicable.

Data Availability Statement: Not applicable.

Acknowledgments: This work was supported by the National Natural Science Foundation of China (grant numbers: 52075075, 51575131). The authors would like to express their sincere appreciation to the fund.

Conflicts of Interest: The authors declare no conflict of interest.

References

1. Alaswad, A.; Benyounis, K.; Olabi, A.G. Tube hydroforming process: A reference guide. *Mater. Des.* **2012**, *33*, 328–339. [CrossRef]
2. Korkolis, Y.P.; Kyriakides, S. Hydroforming of anisotropic aluminum tubes: Part I experiments. *Int. J. Mech. Sci.* **2011**, *53*, 75–82. [CrossRef]
3. Manabe, K.-I.; Fuchizawa, S. Further development on tube hydroforming. In *60 Excellent Inventions in Metal Forming*; Tekkaya, A.E., Homberg, W., Brosius, A., Eds.; Springer: Berlin/Heidelberg, Germany, 2015; pp. 387–393. [CrossRef]
4. Mirzaali, M.; Seyedkashi, S.M.H.; Liaghat, G.H.; Moslemi Naeini, H.; Shojaee G., K.; Moon, Y.H. Application of simulated annealing method to pressure and force loading optimization in tube hydroforming process. *Int. J. Mech. Sci.* **2011**, *55*, 78–84. [CrossRef]
5. Boudeau, N.; Malécot, P. A simplified analytical model for post-processing experimental results from tube bulging test: Theory, experimentations, simulations. *Int. J. Mech. Sci.* **2012**, *65*, 1–11. [CrossRef]

6. Zhu, H.; He, Z.; Lin, Y.; Zheng, K.; Fan, X.; Yuan, S. The development of a novel forming limit diagram under nonlinear loading paths in tube hydroforming. *Int. J. Mech. Sci.* **2020**, *172*, 105392. [CrossRef]
7. Hashemi, R.; Assempour, A.; Abad, E.M.K. Implementation of the forming limit stress diagram to obtain suitable load path in tube hydroforming considering M–K model. *Mater. Des.* **2009**, *30*, 3545–3553. [CrossRef]
8. Hashemi, R.; Madoliat, R.; Afshar, A. Prediction of forming limit diagrams using the modified M-K method in hydroforming of aluminum tubes. *Int. J. Mater. Form.* **2014**, *9*, 297–303. [CrossRef]
9. Marron, G.; Moinier, L.; Patou, P.; Céleski, J. A new necking criterion for the forming limit diagrams. *Metall. Res. Technol.* **1997**, *94*, 837–845.
10. Scales, M.; Chen, K.; Kyriakides, S. Response, Localization, and Rupture of Anisotropic Tubes under Combined Pressure and Tension. *J. Appl. Mech.* **2021**, *88*, 011008. [CrossRef]
11. Martínez-Donaire, A.J.; García-Lomas, F.J.; Vallellano, C. New approaches to detect the onset of localised necking in sheets under through-thickness strain gradients. *Mater. Des.* **2014**, *57*, 135–145. [CrossRef]
12. Keeler, S.P.; Backofen, W.A. Plastic instability and fracture in sheets stretched over rigid punches. *Trans. Am. Soc. Met.* **1963**, *56*, 25–48.
13. Goodwin, G.M. Application of Strain Analysis to Sheet Metal Forming Problems in the Press Shop. *Soc. Automot. Eng.* **1968**, *77*, 380–387. [CrossRef]
14. Kleemola, H.J.; Pelkkikangas, M.T. Effect of pre-deformation and strain path on the forming limits of steel, copper and brass. *Sheet Met. Ind.* **1977**, *63*, 591–596.
15. Graf, A.; Hosford, W. The influence of strain-path changes on forming limit diagrams of A1 6111 T4. *Int. J. Mech. Sci.* **1994**, *36*, 897–910. [CrossRef]
16. Wang, H.; Yan, Y.; Han, F.; Wan, M. Experimental and theoretical investigations of the forming limit of 5754O aluminum alloy sheet under different combined loading paths. *Int. J. Mech. Sci.* **2017**, *133*, 147–166. [CrossRef]
17. Arrieux, R.; Boivin, M.; Le Maître, F. Determination of the Forming Limit Stress Curve for Anisotropic Sheets. *CIRP Ann.* **1987**, *36*, 195–198. [CrossRef]
18. Arrieux, R. Determination and use of the forming limit stress diagrams in sheet metal forming. *J. Mater. Process. Technol.* **1995**, *53*, 47–56. [CrossRef]
19. Simha, C.H.M.; Gholipour, J.; Bardelcik, A.; Worswick, M.J. Prediction of Necking in Tubular Hydroforming Using an Extended Stress-Based Forming Limit Curve. *J. Eng. Mater. Technol.* **2007**, *129*, 36–47. [CrossRef]
20. Stoughton, T.B.; Yoon, J.W. Path independent forming limits in strain and stress spaces. *Int. J. Solids Struct.* **2012**, *49*, 3616–3625. [CrossRef]
21. Nguyen, N.T.; Lee, E.; Lee, M.-G.; Kim, H.J.; Kim, H.-Y. Hydroformability assessment of AA6063 tubes using the polar effective plastic strain diagram. *Proc. Inst. Mech. Eng. Part B J. Eng. Manuf.* **2015**, *229*, 647–653. [CrossRef]
22. Paul, S.K. Path independent limiting criteria in sheet metal forming. *J. Manuf. Process.* **2015**, *20*, 291–303. [CrossRef]
23. Basak, S.; Panda, S.K. Necking and fracture limit analyses of different pre-strained sheet materials in polar effective plastic strain locus using Yld2000-2d yield model. *J. Mater. Process. Technol.* **2019**, *267*, 289–307. [CrossRef]
24. Marciniak, Z.; Kuczyński, K. Limit strains in the processes of stretch-forming sheet metal. *Int. J. Mech. Sci.* **1967**, *9*, 609–620. [CrossRef]
25. Banabic, D.; Kami, A.; Comsa, D.-S.; Eyckens, P. Developments of the Marciniak-Kuczynski model for sheet metal formability: A review. *J. Mater. Process. Technol.* **2021**, *287*, 116446. [CrossRef]
26. Kuroda, M.; Tvergaard, V. Effect of strain path change on limits to ductility of anisotropic metal sheets. *Int. J. Mech. Sci.* **2000**, *42*, 867–887. [CrossRef]
27. Hiwatashi, S.; Van Bael, A.; Houtte, P.A.; Teodosiu, C. Prediction of forming limit strains under strain-path changes: Application of an anisotropic model based on texture and dislocation structure. *Int. J. Plast.* **1998**, *14*, 647–669. [CrossRef]
28. Yang, X.-Y.; Lang, L.-H.; Liu, K.-N.; Guo, C. Modified MK model combined with ductile fracture criterion and its application in warm hydroforming. *Trans. Nonferrous Met. Soc. China* **2015**, *25*, 3389–3398. [CrossRef]
29. Hosseini, M.E.; Hosseinipour, S.J.; Bakhshi-Jooybari, M. Theoretical FLD Prediction Based on M-K Model using Gurson's Plastic Potential Function for Steel Sheets. *Procedia Eng.* **2017**, *183*, 119–124. [CrossRef]
30. Shahzamanian, M.M.; Wu, P.D. Study of forming limit diagram (FLD) prediction of anisotropic sheet metals using Gurson model in M-K method. *Int. J. Mater. Form.* **2021**, *14*, 1031–1041. [CrossRef]
31. Hu, P.; Liu, W.; Ying, L.; Zhang, J.; Wang, D. A thermal forming limit prediction method considering material damage for 22MnB5 sheet. *Int. J. Adv. Manuf. Technol.* **2017**, *92*, 627–638. [CrossRef]
32. Banabic, D.; Comsa, D.-S.; Eyckens, P.; Kami, A.; Gologanu, M. Advanced models for the prediction of forming limit curves. In *Multiscale Modelling in Sheet Metal Forming*; Banabic, D., Ed.; Springer: Cham, Switzerland, 2016; pp. 205–300.
33. Zhang, R.; Shao, Z.; Lin, J. A review on modelling techniques for formability prediction of sheet metal forming. *Int. J. Lightweight Mater. Manuf.* **2018**, *1*, 115–125. [CrossRef]
34. Zadpoor, A.A.; Sinke, J.; Benedictus, R. Formability prediction of high strength aluminum sheets. *Int. J. Plast.* **2009**, *25*, 2269–2297. [CrossRef]
35. Zhalehfar, F.; Hosseinipour, S.J.; Nourouzi, S.; Gorji, A.H. Numerical Investigation into the Effect of Uniaxial and Biaxial Pre-Strain on Forming Limit Diagram of 5083 Aluminum Alloy. *AIP Conf. Proc.* **2011**, *1383*, 165–172. [CrossRef]

36. Hu, Q.; Zhang, F.; Li, X.; Chen, J. Overview on the Prediction Models for Sheet Metal Forming Failure: Necking and Ductile Fracture. *Acta Mech. Solida Sin.* **2018**, *31*, 259–289. [CrossRef]
37. Li, H.; Hou, H.; Li, Z.; He, L.; Wang, X. Numerical simulation of hot blanking for boron steel B1500HS based on ductile fracture criterion. *Int. J. Adv. Manuf. Technol.* **2018**, *97*, 3505–3522. [CrossRef]
38. Clift, S.; Hartley, P.; Sturgess, C.; Rowe, G. Fracture prediction in plastic deformation processes. *Int. J. Mech. Sci.* **1990**, *32*, 1–17. [CrossRef]
39. Cockcroft, M.G.; Latham, D.J. Ductility and the workability of metals. *J. Inst. Met.* **1968**, *96*, 33–39.
40. McClintock, F.A. A Criterion for Ductile Fracture by the Growth of Holes. *J. Appl. Mech.* **1968**, *35*, 363–371. [CrossRef]
41. Han, H.N.; Kim, K.-H. A ductile fracture criterion in sheet metal forming process. *J. Mater. Process. Technol.* **2003**, *142*, 231–238. [CrossRef]
42. Lou, Y.; Huh, H.; Lim, S.; Pack, K. New ductile fracture criterion for prediction of fracture forming limit diagrams of sheet metals. *Int. J. Solids Struct.* **2012**, *49*, 3605–3615. [CrossRef]
43. Zhu, H.; Lin, Y.; He, Z.; Yuan, S. An improved approach for forming limit prediction of strongly anisotropic thin-walled tube combining M-K model and ductile fracture criterion. *J. Manuf. Sci. Eng.* **2023**, *145*, 061003. [CrossRef]
44. He, Z.; Wang, Z.; Lin, Y.; Zhu, H.; Yuan, S. A modified Marciniak–Kuczynski model for determining the forming limit of thin-walled tube extruded with initial eccentricity. *Int. J. Mech. Sci.* **2019**, *151*, 715–723. [CrossRef]
45. Hora, P.; Tong, L.; Berisha, B. Modified maximum force criterion, a model for the theoretical prediction of forming limit curves. *Int. J. Mater. Form.* **2013**, *6*, 267–279. [CrossRef]
46. Barlat, F.; Brem, J.C.; Yoon, J.W.; Chung, K.; Dick, R.E.; Lege, D.J.; Pourboghrat, F.; Choi, S.-H.; Chu, E. Plane stress yield function for aluminum alloy sheets—Part 1: Theory. *Int. J. Plast.* **2003**, *19*, 1297–1319. [CrossRef]
47. Timoshenko, S.; Woinowsky-Krieger, S. *Theory of Plates and Shell*, 2nd ed.; McGraw-Hill Book Co., Inc.: New York, NY, USA, 1959.
48. Zhang, K.; He, Z.; Zheng, K.; Yuan, S. Experimental verification of anisotropic constitutive models under tension-tension and tension-compression stress states. *Int. J. Mech. Sci.* **2020**, *178*, 105618. [CrossRef]
49. Korkolis, Y.P.; Kyriakides, S. Inflation and burst of anisotropic aluminum tubes for hydroforming applications. *Int. J. Plast.* **2008**, *24*, 509–543. [CrossRef]

materials

MDPI

Article

A Homogeneous Anisotropic Hardening Model in Plane Stress State for Sheet Metal under Nonlinear Loading Paths

Haihui Zhu [1], Yanli Lin [1,*], Kelin Chen [2], Zhubin He [1] and Shijian Yuan [3]

[1] School of Mechanical Engineering, Dalian University of Technology, Dalian 116024, China
[2] Department of Integrated Systems Engineering, The Ohio State University, Columbus, OH 43210, USA
[3] National Key Laboratory for Precision Hot Processing of Metals, Harbin Institute of Technology, Harbin 150001, China
* Correspondence: linyanli0616@163.com

Abstract: In sheet metal forming, the material is usually subjected to a complex nonlinear loading process, and the anisotropic hardening behavior of the material must be considered in order to accurately predict the deformation of the sheet. In recent years, the homogeneous anisotropic hardening (HAH) model has been applied in the simulation of sheet metal forming. However, the existing HAH model is established in the second-order stress deviator space, which makes the calculation complicated and costly, especially for a plane stress problem such as sheet metal forming. In an attempt to reduce the computational cost, an HAH model in plane stress state is proposed, and called the HAH-2d model in this paper. In the HAH-2d model, both the stress vector and microstructure vector contain only three in-plane components, so the calculation is significantly simplified. The characteristics of the model under typical nonlinear loading paths are analyzed. Additionally, the feasibility of the model is verified by the stress–strain responses of DP780 and EDDQ steel sheets under different two-step uniaxial tension tests. The results show that the HAH-2d model can reasonably reflect the Bauschinger effect and the permanent softening effect in reverse loading, and the latent hardening effect in cross loading, while the predictive accuracy for cross-loading softening remains to be improved. In the future, the HAH-2d model can be further modified to describe more anisotropic hardening behaviors and applied to numerical simulations.

Keywords: homogeneous anisotropic hardening model; sheet metal; nonlinear loading path; plane stress

1. Introduction

With increasing requirements for lightweight materials in the aerospace, aviation, and automotive industries, integrated thin-walled components with complex shapes are finding broader applications [1,2]. During the forming of an integrated component, the material often undergoes complex nonlinear loading paths, even including multi-step pre-deformation. The deformation behavior of sheet metal often exhibits strong path dependence [3–5] and pronounced anisotropic hardening behaviors in nonlinear loading paths; these may be, for example, the Bauschinger effect, the permanent softening effect and the latent hardening effect [6,7].

The Bauschinger effect is the phenomenon wherein the yield stress under reverse loading decreases after a certain amount of pre-deformation [8,9]; the permanent softening effect is another phenomenon during reverse loading wherein the reloading flow stress remains lower than the flow stress of monotonic loading [10]. Meanwhile, there is a wide range of strain path changes between monotonic loading and reverse loading. For some materials, the reloading flow stress at an angle with respect to the first loading may overshoot the monotonic stress–strain curve; this phenomenon is referred to the latent hardening effect [11,12].

Practically, the anisotropic hardening behavior significantly affects the spring-back and forming limit [13–15], and the reduction of spring-back and the avoidance of necking or fracture are of paramount importance in the forming of an integrated component. In order to guide process formulation and reduce the cost of process development, it is necessary to predict the deformation process and optimize the forming process by simulation. At present, the isotropic hardening assumption is commonly used in general simulations of metal forming, i.e., the subsequent yield surface expands uniformly based on the initial yield surface. However, an isotropic hardening model cannot reasonably capture the anisotropic hardening behavior under nonlinear loading paths [16,17]. Therefore, establishing advanced anisotropic hardening models to accurately predict the plastic deformation behavior of thin-walled metals under nonlinear loading conditions is an important research direction [18].

A variety of anisotropic hardening models have been proposed [19,20], for example, the kinematic hardening model established by introducing a back stress tensor into the isotropic hardening model. The kinematic hardening model was first proposed by Prager to capture the Bauschinger effect, wherein the yield surface translates with the deformation but its shape remains the same [21]. Subsequently, some kinematic hardening models with a nonlinear relationship between the increments of back stress and plastic strain were proposed to more accurately capture the flow stress under reverse loading or cyclical loading conditions [22,23]. In fact, the typical hardening behavior of metals is often a mixture of isotropic hardening and kinematic hardening, i.e., the yield surface changes both in size and position. Hence, the mixed hardening mode was proposed [24,25].

As an alternative to the kinematic hardening model, Barlat et al. [26] proposed the homogeneous anisotropic hardening (HAH) model, which is a distortional hardening model without a back stress tensor. In the HAH model, a fluctuating component is added into the traditional yield function to change the shape of yield surface, and a microstructure deviator related to the loading history is introduced to reflect the state of the microstructure. In order to further describe the latent hardening, more variables were introduced into the model [27,28]. More importantly, the HAH model can be regarded as a theoretical framework for the anisotropic hardening model, in which any homogeneous yield function and hardening law can be used, and the yield surface can be distorted depending on the specific loading path [17]. Some researchers have successfully implemented the HAH model in numerical simulations, and the prediction accuracy has improved clearly [29–31].

More recently, in order to capture complex and varied anisotropic hardening behavior and improve prediction accuracy, more parameters were introduced, for example, the HAH-20 model (using, at most, 23 coefficients [17]) and the HEXAH model (using, at most, 15 coefficients [32]). On the other hand, in order to reduce the computational cost in numerical simulation, some efficient algorithms were implemented, including a multi-step return-mapping algorithm for the HAH model [33], a fully implicit numerical algorithm that can solve a complete set of residuals [34], and a fast and robust stress-update algorithm based on the general cutting-plane method [35].

It should be pointed out that the HAH model and these improvements were proposed based on the second-order stress deviator with six independent components, corresponding to the fully 3D (three dimensional) stress state. However, in the plastic forming of thin-walled components, the normal stress is usually much smaller than the in-plane principal stresses, which means that the stress condition can be simplified to the plane stress state with only three components. Therefore, shell elements and material constitutive models for plane stress states are widely used in numerical simulations of sheet metal forming. When the HAH model is selected to analyze the deformation of sheet metals under nonlinear loading paths, it is required to transform the in-plane stress components to a stress deviator, then calculate the strain increments and new stress deviator after a deformation increment using an iterative method, and lastly, transform the new stress deviation tensor into in-plane stress components for output. Obviously, the plane stress problem is transformed into a more complex 3D stress problem, which makes the calculation complicated and costly. Therefore, it is of great significance to establish an anisotropic hardening model in plane stress state to analyze the deformation of sheet metals subjected to complex loading.

In this paper, an anisotropic hardening model for plane stress problems such as thin-walled metal forming will be proposed based on the framework of the HAH model. For convenience, the proposed model is named the HAH-2d model in this work. In the meantime, the performance of the HAH-2d model in predicting the evolution of yield loci and the stress–strain curve will be analyzed under typical nonlinear loading paths. Moreover, the new model will be verified by experimental results from two-step uniaxial tension tests.

2. Fundamentals of the HAH Model

The original HAH model was proposed by Barlat et al. in the second-order stress deviator space [26]. The yield function, as well as the plastic potential, is as follows:

$$\Phi(\mathbf{s}) = \left[\phi^q + \phi_h^q\right]^{\frac{1}{q}} = \left[\phi^q(\mathbf{s}) + f_1^q\left|\hat{\mathbf{h}}:\mathbf{s} - \left|\hat{\mathbf{h}}:\mathbf{s}\right|\right|^q + f_2^q\left|\hat{\mathbf{h}}:\mathbf{s} + \left|\hat{\mathbf{h}}:\mathbf{s}\right|\right|^q\right]^{\frac{1}{q}} = \bar{\sigma}\left(\bar{\varepsilon}\right) \quad (1)$$

where, $\mathbf{s}$ and $\hat{\mathbf{h}}$ are the stress deviator and microstructure deviator, respectively, and ":" denotes the double dot product.

The yield function $\Phi(\mathbf{s})$ consists of a stable component ϕ and a fluctuating component ϕ_h. Any yield function may be used as the stable component after being reduced to a homogeneous function of degree 1 with the form of $\phi(\mathbf{s}) = \bar{\sigma}$. $\bar{\sigma}\left(\bar{\varepsilon}\right)$ is a hardening law, where $\bar{\sigma}$ and $\bar{\varepsilon}$ are the equivalent stress and equivalent plastic strain, respectively. f_1 and f_2 are state variables related to deformation history leading to the distortion of the yield surface, and q is a constant exponent. If $f_1 = f_2 = 0$, Equation (1) reduces to a conventional yield function, $\Phi(\mathbf{s}) = \phi(\mathbf{s}) = \bar{\sigma}\left(\bar{\varepsilon}\right)$.

The microstructure deviator $\hat{\mathbf{h}}$, a normalized tensorial state variable, is proposed to capture the deformation history and reflect the microstructure evolution of the material. It defines an axis, not a direction; namely, $\hat{\mathbf{h}}$ and $-\hat{\mathbf{h}}$ represent the same microstructural state. Its initial value $\hat{\mathbf{h}}^0$ is defined as the normalized stress deviator $\hat{\mathbf{s}}^0$ corresponding to the initial yield, as in the following equation, where the factor 8/3 is used for convenience [26,27].

$$\hat{\mathbf{h}}^0 = \hat{\mathbf{s}}^0 = \frac{\hat{\mathbf{s}}^0}{\sqrt{\frac{8}{3}s_{ij}^0 s_{ij}^0}} \quad (2)$$

If the material is reloaded in a different stress state, $\hat{\mathbf{h}}$ will rotate gradually towards the new stress deviator $\mathbf{s}$ when $\cos\chi \geq 0$, or towards $-\mathbf{s}$ when $\cos\chi < 0$. Additionally, $\cos\chi$ determined by Equation (3) is the cosine of the angle between $\mathbf{s}$ and $\hat{\mathbf{h}}$, representing the variation of the loading path [27,28]. Monotonic, reverse and cross-loading sequences are, respectively, represented by $\cos\chi$ = 1, −1 and 0.

$$\cos\chi = \frac{\mathbf{s}:\hat{\mathbf{h}}}{\|\mathbf{s}\| \cdot \|\hat{\mathbf{h}}\|} \quad (3)$$

where, $\|\mathbf{s}\| = \sqrt{\mathbf{s}:\mathbf{s}} = \sqrt{s_{ij}s_{ij}}$ denotes the norm of the second-order tensor $\mathbf{s}$.

3. HAH-2d Model in Plane Stress State

3.1. Stress Vector and Microstructure Vector

During the forming process of sheet metal, the material is mainly in plane stress state; the three in-plane stress components can be simply expressed as a stress vector $\boldsymbol{\sigma} = (\sigma_{11}, \sigma_{22}, \sigma_{12})^{\mathrm{T}}$, while the other stress components such as σ_{33}, σ_{13} and σ_{23} are equal to 0. In order to describe the microstructure evolution of the sheet metal under plane stress conditions, a normalized microstructure vector $\hat{\mathbf{H}} = (\hat{H}_{11}, \hat{H}_{22}, \hat{H}_{12})^{\mathrm{T}}$ is defined in terms of stress vector in this paper. Its initial value $\hat{\mathbf{H}}^{0}$ also corresponds to the normalized stress vector $\hat{\boldsymbol{\sigma}}^{0}$ leading to the first increment of plastic deformation.

$$\hat{\mathbf{H}}^{0} = \hat{\boldsymbol{\sigma}}^{0} = \frac{\boldsymbol{\sigma}^{0}}{\sqrt{\boldsymbol{\sigma}^{0} : \boldsymbol{\sigma}^{0}}} \tag{4}$$

Considering that the plastic deformation of metals is driven by the stress deviator $\mathbf{s}$, it is still necessary to discuss the variation of loading path based on stress deviator when analyzing plastic deformation processes under plane stress conditions. The second-order deviatoric tensors, $\mathbf{s}$ and $\mathbf{h}$, corresponding to the stress vector $\boldsymbol{\sigma}$ and the microstructure vector $\hat{\mathbf{H}}$, can be obtained through the following two equations, respectively.

$$\mathbf{s} = \begin{pmatrix} \sigma_{11} & \sigma_{12} & 0 \\ \sigma_{12} & \sigma_{22} & 0 \\ 0 & 0 & 0 \end{pmatrix} - \frac{\sigma_{11} + \sigma_{22}}{3} \begin{pmatrix} 1 & 0 & 0 \\ 0 & 1 & 0 \\ 0 & 0 & 1 \end{pmatrix} \tag{5}$$

$$\mathbf{h} = \begin{pmatrix} \hat{H}_{11} & \hat{H}_{12} & 0 \\ \hat{H}_{12} & \hat{H}_{22} & 0 \\ 0 & 0 & 0 \end{pmatrix} - \frac{\hat{H}_{11} + \hat{H}_{22}}{3} \begin{pmatrix} 1 & 0 & 0 \\ 0 & 1 & 0 \\ 0 & 0 & 1 \end{pmatrix} \tag{6}$$

Then, the double dot product $\mathbf{s} : \mathbf{h}$ can be expressed as:

$$\mathbf{s} : \mathbf{h} = \frac{1}{3}\left(2\sigma_{11}\hat{H}_{11} + 2\sigma_{22}\hat{H}_{22} - \sigma_{11}\hat{H}_{22} - \sigma_{22}\hat{H}_{11} + 6\sigma_{12}\hat{H}_{12}\right) = \boldsymbol{\sigma}^{\mathrm{T}}\mathbf{D}\hat{\mathbf{H}} \tag{7}$$

where the matrix $\mathbf{D} = \frac{1}{3}\begin{pmatrix} 2 & -1 & 0 \\ -1 & 2 & 0 \\ 0 & 0 & 6 \end{pmatrix}$.

Therefore, the value of $\cos\chi$ can be calculated through vectors $\boldsymbol{\sigma}$ and $\hat{\mathbf{H}}$, as follows:

$$\cos\chi = \frac{\boldsymbol{\sigma}^{\mathrm{T}}\mathbf{D}\hat{\mathbf{H}}}{\sqrt{\boldsymbol{\sigma}^{\mathrm{T}}\mathbf{D}\boldsymbol{\sigma}}\sqrt{\hat{\mathbf{H}}^{\mathrm{T}}\mathbf{D}\hat{\mathbf{H}}}} \tag{8}$$

Moreover, $\hat{\mathbf{H}}$ defines an axis in the vector space of plane stress. When the sheet metal undergoes loading-path changes, $\hat{\mathbf{H}}$ will remain the same if $\cos\chi = \pm 1$, or rotate gradually towards the new stress vector $\boldsymbol{\sigma}$ if $0 \le \cos\chi < 1$, or rotate towards $-\boldsymbol{\sigma}$ if $-1 \le \cos\chi < 0$. In this work, a possible evolution for $\hat{\mathbf{H}}$ is given as follows:

$$\begin{cases} d\hat{\mathbf{H}} = k\left(\hat{\boldsymbol{\sigma}} - \hat{\mathbf{H}}\right)d\bar{\varepsilon}, \ \cos\chi \ge 0 \\ d\hat{\mathbf{H}} = k\left(-\hat{\boldsymbol{\sigma}} - \hat{\mathbf{H}}\right)d\bar{\varepsilon}, \ \cos\chi < 0 \end{cases} \tag{9}$$

$$\mathbf{H}^{i+1} = \overset{\wedge}{\mathbf{H}}^{i} + d\overset{\wedge}{\mathbf{H}} \tag{10}$$

$$\overset{\wedge}{\mathbf{H}}^{i+1} = \frac{\mathbf{H}^{i+1}}{\sqrt{\left(H_{11}^{i+1}\right)^2 + \left(H_{22}^{i+1}\right)^2 + \left(H_{12}^{i+1}\right)^2}} \tag{11}$$

in which $\overset{\wedge}{\boldsymbol{\sigma}}$ is the normalized stress vector of $\boldsymbol{\sigma}$, k is a constant that controls the rate of rotation, and $d\bar{\varepsilon}$ is the increment of equivalent strain $\bar{\varepsilon}$, which can be calculated according to the plastic work increment dW^p:

$$\bar{\sigma} d\bar{\varepsilon} = \boldsymbol{\sigma} \cdot d\boldsymbol{\varepsilon} = dW^p \tag{12}$$

$$d\boldsymbol{\varepsilon} = (d\varepsilon_{11}, d\varepsilon_{22}, d\gamma_{12})^{\mathrm{T}} \tag{13}$$

where $d\boldsymbol{\varepsilon}$ is the plastic strain increment vector, $d\varepsilon_{11}$ and $d\varepsilon_{22}$ are strain increments along directions 1 and 2, and $d\gamma_{12}$ is the shear strain increment.

The evolution of $\overset{\wedge}{\mathbf{H}}$ is shown schematically in Figure 1. $\overset{\wedge}{\mathbf{H}}$ rotates in the plane determined by $\overset{\wedge}{\mathbf{H}}^{i}$ and $\overset{\wedge}{\boldsymbol{\sigma}}$, and finally, $\overset{\wedge}{\mathbf{H}}$ takes the direction of $\overset{\wedge}{\boldsymbol{\sigma}}$ or $-\overset{\wedge}{\boldsymbol{\sigma}}$.

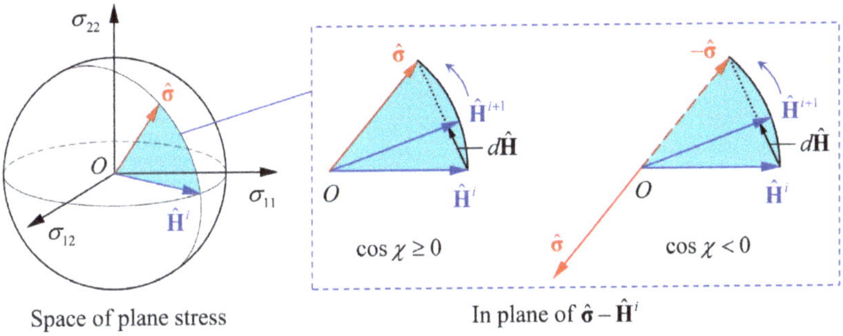

Figure 1. Schematic diagram of the evolution of microstructure vector $\overset{\wedge}{\mathbf{H}}$.

3.2. Formulation of the HAH-2d Model

3.2.1. Distortional Yield Function

The distortional yield function of the HAH model in the plane stress space (HAH-2d model) is as follows:

$$\Phi(\boldsymbol{\sigma}) = \frac{1}{F_L}[\phi^q + \phi_h{}^q]^{\frac{1}{q}} = \frac{1}{F_L}\left[\phi^q(\boldsymbol{\sigma}) + f_1^q|\sigma_p - |\sigma_p||^q + f_2^q|\sigma_p + |\sigma_p||^q\right]^{\frac{1}{q}} = \bar{\sigma}\left(\bar{\varepsilon}\right) \tag{14}$$

$$\sigma_p = ||\mathbf{s}|| \cos\chi = \frac{\boldsymbol{\sigma}^{\mathrm{T}}\mathbf{D}\overset{\wedge}{\mathbf{H}}}{\sqrt{\overset{\wedge}{\mathbf{H}}^{\mathrm{T}}\mathbf{D}\overset{\wedge}{\mathbf{H}}}} \tag{15}$$

The stable component ϕ can be any homogeneous yield function of degree 1 for plane stress. The fluctuating component ϕ_h is adopted to cover the Bauschinger and permanent softening effects, and the multiplier component $1/F_L$ is used to capture the latent hardening effect. Correspondingly, $\Phi(\boldsymbol{\sigma})$ is also a homogeneous function of degree 1, i.e., $\Phi(a\boldsymbol{\sigma}) = a\Phi(\boldsymbol{\sigma})$ holds for any real number a. In addition, σ_p is the projection of the stress deviator $\mathbf{s}$ in the direction of the microstructure deviator $\mathbf{h}$.

It is important to note that the associated flow rule is applied in this paper, the yield function also serves as the plastic potential. In other words, the plastic strain increment vector $d\boldsymbol{\varepsilon}$ is defined as

$$d\boldsymbol{\varepsilon} = \frac{\partial \Phi(\boldsymbol{\sigma})}{\partial \boldsymbol{\sigma}} d\lambda \tag{16}$$

where $d\lambda \geq 0$ is the plastic multiplier.

3.2.2. Bauschinger Effect and Permanent Softening Effect

A schematic diagram of yield loci in the $\sigma_{11} - \sigma_{22}$ plane with $f_1 > 0$, $f_2 > 0$ and $F_L = 1$ is shown in Figure 2. $\boldsymbol{\sigma}_f$ and $\boldsymbol{\sigma}_r$ are yield stress vectors on the current distortional yield locus along the positive and negative directions of $\hat{\mathbf{H}}$, respectively, and $\boldsymbol{\sigma}_{\text{iso}}$ is the yield stress vector on the isotropic yield locus. State variables g_1 and g_2 are defined as Equation (17) to represent the relationship between $\boldsymbol{\sigma}_f$, $\boldsymbol{\sigma}_r$ and $\boldsymbol{\sigma}_{\text{iso}}$, and the parameters f_1 and f_2 in Equation (14) can be expressed as Equation (18).

$$\begin{cases} g_1 = -\boldsymbol{\sigma}_r / \boldsymbol{\sigma}_{\text{iso}} \leq 1 \\ g_2 = \boldsymbol{\sigma}_f / \boldsymbol{\sigma}_{\text{iso}} \leq 1 \end{cases} \tag{17}$$

$$f_i = \frac{\sqrt{6}}{4} \left(g_i^{-q} - 1 \right)^{\frac{1}{q}},\ i = 1,\ 2 \tag{18}$$

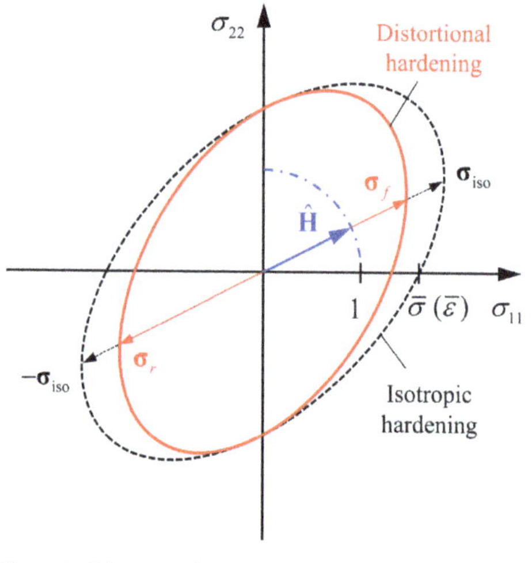

Figure 2. Schematic diagram of yield loci in $\sigma_{11} - \sigma_{22}$ plane with $f_1 > 0$, $f_2 > 0$ and $F_L = 1$.

For a sheet metal exhibiting the Bauschinger effect, the normalized yield surface in the opposite direction of loading tends to contract during forward loading, and the value of f_1 should increase from 0 towards a finite value. As a result, the yield stress of reverse loading is reduced compared with the final forward-loading stress, which is the Bauschinger effect. During reverse loading, the flat yield surface at the side of reverse loading tends to recover the yield surface determined by the isotropic hardening function $\phi(\boldsymbol{\sigma}) = \bar{\sigma}$. At the same time, the yield surface opposite to the reverse loading contracts, and f_2 starts to increase in the same way as f_1. If the yield surface at the side of reverse loading cannot recover to the level of isotropic hardening, the permanent softening effect will occur during reverse loading correspondingly.

A possible evolution for g_1 and g_2 is given in Equation (19) which has been successfully applied in existing HAH models [26–28]. State variables g_3 and g_4 represent the maximum saturation values of g_2 and g_1, respectively. Material coefficients k_1 and k_2 control the evolution rate of g_1 and g_2, and k_3 controls their lower bound; k_4 controls the minimum value of g_3 and g_4, and k_5 controls their evaluation rate. Meanwhile, the value of k_4 is usually slightly less than 1, according to the permanent softening effect. When the initial values of g_1 and g_2 are both 1 and $k_1 = k_2 = 0$, the ability to describe the Bauschinger effect will be suppressed, and when $g_3 = g_4 = 1$ and $k_4 = 1$ (or $k_5 = 0$), the ability to describe the permanent softening effect will be suppressed.

$$\begin{array}{ll} \text{If } \cos\chi \geq 0 & \text{If } \cos\chi < 0 \\ \frac{dg_1}{d\bar{\varepsilon}} = k_2\left(k_3\frac{\bar{\sigma}(0)}{\bar{\sigma}\left(\bar{\varepsilon}\right)} - g_1\right) & \frac{dg_1}{d\bar{\varepsilon}} = k_1\frac{g_4-g_1}{g_1} \\ \frac{dg_2}{d\bar{\varepsilon}} = k_1\frac{g_3-g_2}{g_2} & \frac{dg_2}{d\bar{\varepsilon}} = k_2\left(k_3\frac{\bar{\sigma}(0)}{\bar{\sigma}\left(\bar{\varepsilon}\right)} - g_2\right) \\ \frac{dg_4}{d\bar{\varepsilon}} = k_5(k_4 - g_4) & \frac{dg_3}{d\bar{\varepsilon}} = k_5(k_4 - g_3) \end{array} \tag{19}$$

3.2.3. Latent Hardening Effect

For some materials, the flow stress during reloading may overshoot the stress during monotonic loading under the condition of the same equivalent strain. The phenomenon of overshooting was explained as the latent hardening effect. Additionally, the degree of overshooting is usually maximum for cross loading, in which the angle between the two corresponding normalized stress deviators $\hat{\mathbf{s}}_1$ and $\hat{\mathbf{s}}_2$ is 90° ($\hat{\mathbf{s}}_1 : \hat{\mathbf{s}}_2 = 0$). As the degree of plastic deformation increases in the second step, $\hat{\mathbf{h}}$ rotates gradually towards $\hat{\mathbf{s}}_2$ or $-\hat{\mathbf{s}}_2$.

Suppose $\mathbf{s}_v$ and $\mathbf{s}'_v$ are the stress deviators vertical to $\hat{\mathbf{h}}$ on the distortional yield surface and the isotropic one, respectively. Additionally, the corresponding plane stress vectors are $\boldsymbol{\sigma}_v$ and $\boldsymbol{\sigma}'_v$. Then, a state variable g_L can be defined as Equation (20), which represents that the distortional yield surface at $\cos\chi = 0$ expands to g_L times the yield surface under monotonic loading, and its minimum value is 1.

$$g_L = \frac{\phi(\boldsymbol{\sigma}_v)}{\phi(\boldsymbol{\sigma}'_v)} = \frac{\mathbf{s}_v}{\mathbf{s}'_v} \tag{20}$$

Because the degree of the latent hardening effect is different under different strain path changes, it is important to determine the yield surface expansion for arbitrary χ. The stress deviator $\mathbf{s}$ can be decomposed into $\mathbf{s}_{\|}$ parallel to $\hat{\mathbf{h}}$ and $\mathbf{s}_{\perp}$ perpendicular to $\hat{\mathbf{h}}$, as shown in Figure 3 [28]. Considering that the latent hardening effect is mainly caused by the vertical component $\mathbf{s}_{\perp}$, a new deviator $\mathbf{s}'$ is composed as

$$\mathbf{s}' = \mathbf{s}_{\perp}/g_L + \mathbf{s}_{\|} \tag{21}$$

Then, the expansion ratio of the yield surface for arbitrary χ can be represented by

$$F_L = \frac{\|\mathbf{s}\|}{\|\mathbf{s}'\|} = \frac{g_L}{\sqrt{(g_L{}^2 - 1)\cos^2\chi + 1}} \tag{22}$$

Thus, the surface of $\phi(\boldsymbol{\sigma})/F_L = \bar{\sigma}\left(\bar{\varepsilon}\right)$ expands faster than that of $\phi(\boldsymbol{\sigma}) = \bar{\sigma}\left(\bar{\varepsilon}\right)$ under any strain path change except $\cos\chi = \pm 1$. Furthermore, the evolution equation of g_L is as follows [27]:

$$\frac{dg_L}{d\bar{\varepsilon}} = k_L\left[\left(\frac{\bar{\sigma}\left(\bar{\varepsilon}\right) - \bar{\sigma}(0)}{\bar{\sigma}\left(\bar{\varepsilon}\right)}\right)\left(\sqrt{L(1-\cos^2\chi) + \cos^2\chi} - 1\right) + 1 - g_L\right] \tag{23}$$

where L and k_L are constants controlling the upper bound and evolution rate of g_L, respectively. When the initial value of g_L is 1 and $L = 1$, F_L will be identical to 1, and the latent hardening effect is suppressed.

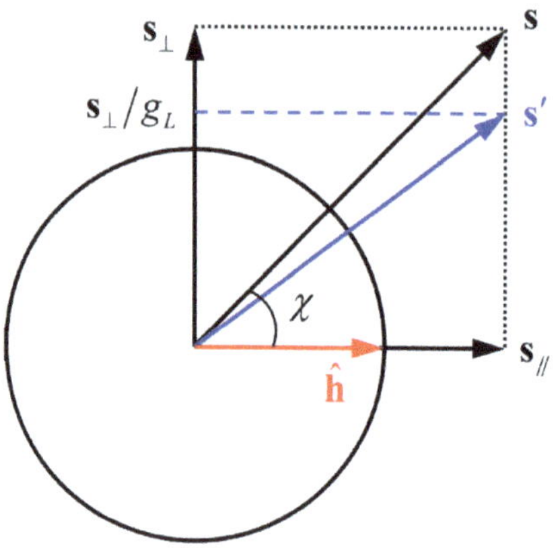

Figure 3. Schematic of the decomposition and composition of stress deviator [28]. Redraw with reference to International Journal of Plasticity, Vol 58, Barlat, F.; Vincze, G.; Grácio, J.J.; Lee, M.G.; Rauch, E.F.; Tomé, C.N., Enhancements of homogenous anisotropic hardening model and application to mild and dual-phase steels, 201–218, Copyright Elsevier (2014), with permission from Elsevier.

3.3. Coefficient Identification

It is recommended to identify the coefficients in the HAH-2d model in the following order: yield function $\phi(\boldsymbol{\sigma})$, isotropic hardening law $\bar{\sigma}\left(\bar{\varepsilon}\right)$, and distortional hardening [27]. The coefficients of $\phi(\boldsymbol{\sigma})$ and $\bar{\sigma}\left(\bar{\varepsilon}\right)$ can be determined independently in a conventional way; for example, the coefficients of $\phi(\boldsymbol{\sigma})$ can be calculated based on the yield stresses and anisotropy coefficients along different directions of the sheet, and the coefficients of $\bar{\sigma}\left(\bar{\varepsilon}\right)$ can be obtained by fitting a monotonic stress–strain curve. Additionally, the nine coefficients related to the distortion, namely, q, k, k_1, k_2, k_3, k_4, k_5, k_L and L, can be identified using an optimization method in which the input experimental data should come from reverse-loading and cross-loading tests.

When the stress–strain data of forward-reverse loading are available, coefficients k_1, k_2, k_3, k_4 and k_5 can be determined independently. Otherwise, the five coefficients have to be evaluated together with k, k_L, L and q. In particular, the Bauschinger, permanent softening and latent hardening effects can be expressed simultaneously in the stress–strain curve from a two-step tensile test with $-1 < \cos\chi < 0$, for example, re-tension in the direction orthogonal to the fist tensile direction ($\cos\chi = -0.5$). Given that the orthogonal two-step tensile test is easy to carry out, all the coefficients associated with the distortion could be identified based on the stress–strain curve.

4. Hardening Behavior Predicted under Typical Nonlinear Loading Paths

In this section, the stress–strain curves and yield locus evolutions of a generic sheet material under several typical nonlinear loading paths were predicted by the HAH-2d model. The typical nonlinear loading paths in the range of $-1 \le \cos\chi < 1$ include reverse loading ($\cos\chi = -1$), cross loading ($\cos\chi = 0$) and two-step tensions conducted at 45° ($\cos\chi = 0.25$) and 90° ($\cos\chi = -0.5$) from the first loading direction.

The initial yield condition of the generic sheet material is assumed to satisfy the Mises isotropic yield criterion, see Equation (24), and the hardening satisfies the Swift hardening law [36], see Equation (25). The material has a Bauschinger effect, permanent softening effect and latent hardening effect simultaneously. The coefficients of the swift hardening law and coefficients associated with the distortion are listed in Table 1.

$$\phi_{\text{Mises}}(\boldsymbol{\sigma}) = \sqrt{\sigma_{11}^2 + \sigma_{22}^2 - \sigma_{11}\sigma_{22} + 3\sigma_{12}^2} = \bar{\sigma} \quad (24)$$

$$\bar{\sigma} = K\left(\bar{\varepsilon} + \varepsilon_0\right)^n \quad (25)$$

where K, ε_0 and n are material coefficients.

Table 1. Coefficients of the generic material.

Coefficients	Swift Hardening Law			Coefficients Associated with the Distortion								
	K (MPa)	ε_0	n	q	k	k_1	k_2	k_3	k_4	k_5	L	k_L
Values	500	0.01	0.25	2	20	100	50	0.5	0.9	20	2.0	300

4.1. Reverse Loading

The evolution of the normalized yield locus of the generic material, predicted by the HAH-2d model under a reverse loading, namely, uniaxial compression (UC) to true strain of 0.10, followed by uniaxial tension (UT) in the reverse direction, is shown in Figure 4. The microstructure vector $\hat{\mathbf{H}}$ maintains $(-1, 0, 0)$ during the whole deformation. The normalized yield loci in $\sigma_{11} - \sigma_{22}$ plane are divided into left and right sides by the line $\sigma_{22}/\sigma_{11} = 2.0$. This line is chosen because the two stress deviators corresponding to stress vectors $(1, 2, 0)^T$ and $(1, 0, 0)^T$ are orthogonal. During the compression step, the yield locus on the left side remains unchanged, while the yield locus on the right side contracts inward, which will result in a reduction in re-loading yield stress when tension occurs in the opposite direction. As the tensile strain increases in the second step, the yield locus on the left side contracts inward instead. The yield locus on the right side recovers towards the isotropic hardening yield, but not completely, because of the permanent softening effect.

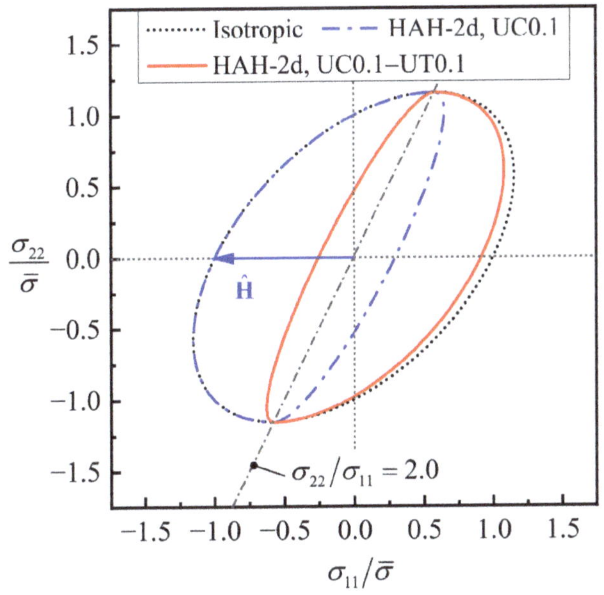

Figure 4. Evolution of the normalized yield locus of the generic material predicted by the HAH-2d model in a uniaxial compression-tension loading path.

The predicted stress–strain curve during reverse loading is given in Figure 5. The reloading yield stress after compression pre-deformation is significantly lower than the monotonic flow stress under the same deformation level, which is the Bauschinger effect. When $k_4 = 0.9$, the stress will not reach the monotonic uniaxial tensile curve, which is the permanent softening effect. Additionally, the permanent softening phenomenon disappears when $k_4 = 1.0$. Therefore, the HAH-2d model can describe the Bauschinger effect and permanent softening effect during reverse loading.

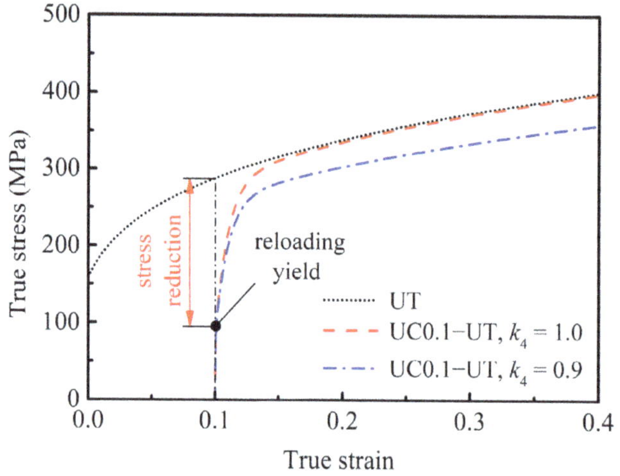

Figure 5. True stress–strain curves of the generic material predicted by the HAH-2d model in a uniaxial compression-tension loading path.

The sensitivity of the HAH-2d model to parameters k_1, k_2, and k_3 should be considered when predicting the Bauschinger effect during reverse loading. The state of the yield surface associated with the Bauschinger effect is reflected by state variables g_1 and g_2, and their evolutions follow the same rule. Therefore, the effects of k_1, k_2, and k_3 on the predicted responses, such as the evaluations of g_1 and the stress–strain curves, were analyzed as shown in Figure 6. It can be found from Figure 6a,b that k_1 affects the evaluation rate of g_1 during reloading. The greater k_1 is, the greater the increase rate of g_1, and the faster the flow stress reaches the saturation state. Figure 6c shows that k_2 affects the evaluation rate of g_1 during preloading. However, the corresponding stress–strain curves of reloading under the condition of $30 \leq k_2 \leq 50$, $k_1 = 100$ and $k_3 = 0.5$ are almost coincident, as shown in Figure 6d. The reason is that there is little difference in g_1 after a pre-strain of 0.1, and the difference is quickly covered by the evolution during reverse loading. For the same reason, the difference among the curves in Figure 6f under different k_3 is not obvious too. Even so, it should be noted that the value of g_1 during preloading and the reloading yield stress are significantly affected by k_3; the smaller the value of k_3, the smaller the value of g_1 and the reloading yield stress.

Figure 7 shows the effects of k_4 and k_5 on the predicted responses of the permanent softening effect during reverse loading, where the state variable g_4 represents the maximum saturation value of the reloading stress. It can be seen that the permanent softening effect is mainly affected by k_4, and the smaller the value of k_4, the lower the value of g_4 and the saturation value of reloading stress. In the meantime, when using a value of k_4 less than 1.0, the larger the value of k_5, the faster the value of g_4 approaches the saturation value, and the lower the reloading stress–strain curve will be after a certain pre-strain. Of course, if the pre-strain is large enough, the influence of k_5 on the reloading stress–strain curve will be significantly reduced or may even disappear.

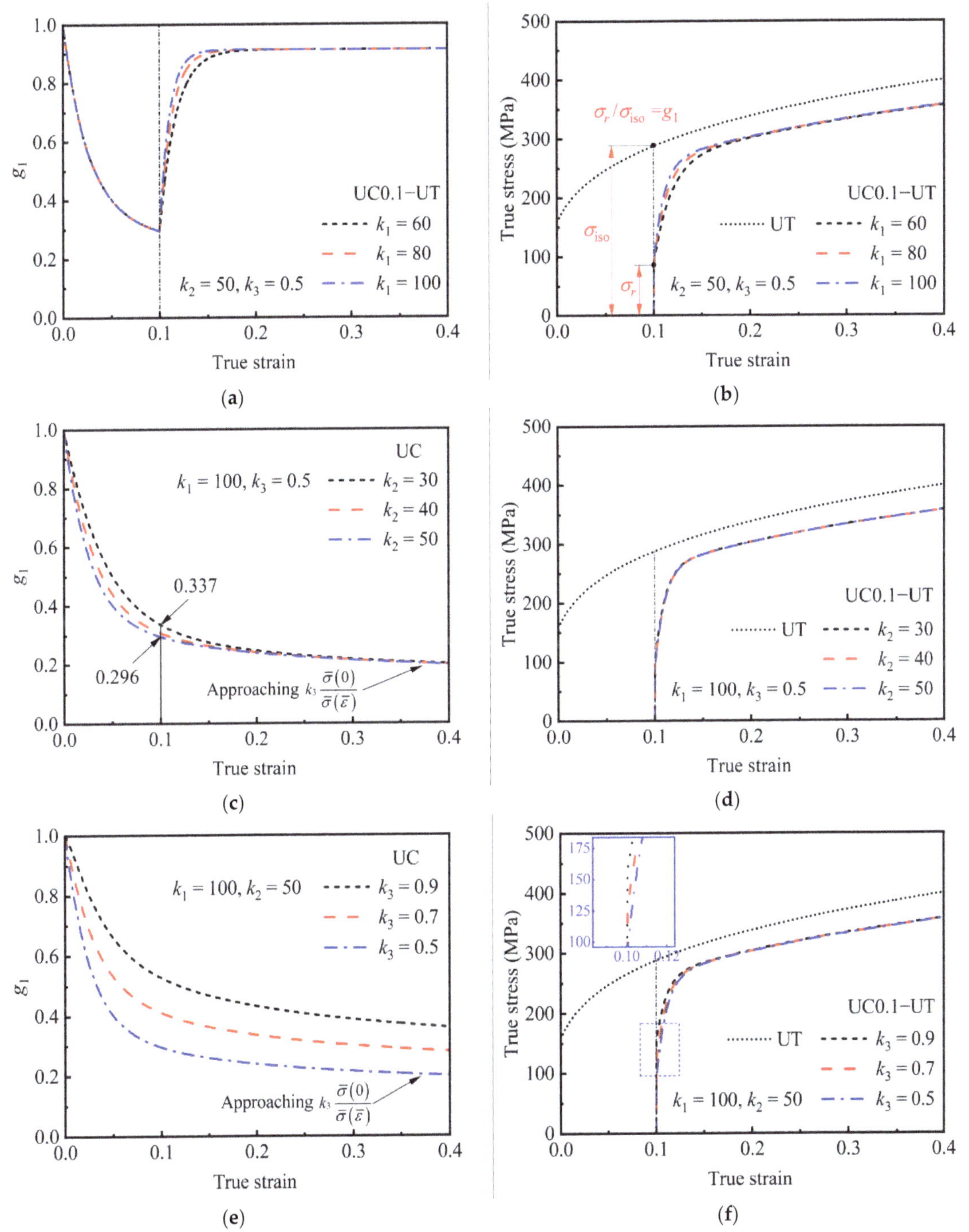

Figure 6. Predicted responses of the Bauschinger effect during reverse loading; evaluation of g_1 during (**a**) UC0.1−UT loading under different k_1; (**c**) UC loading under different k_2; and (**e**) UC loading under different k_3; and stress–strain curves during UC0.1−UT loading under (**b**) different k_1; (**d**) different k_2 and (**f**) different k_3.

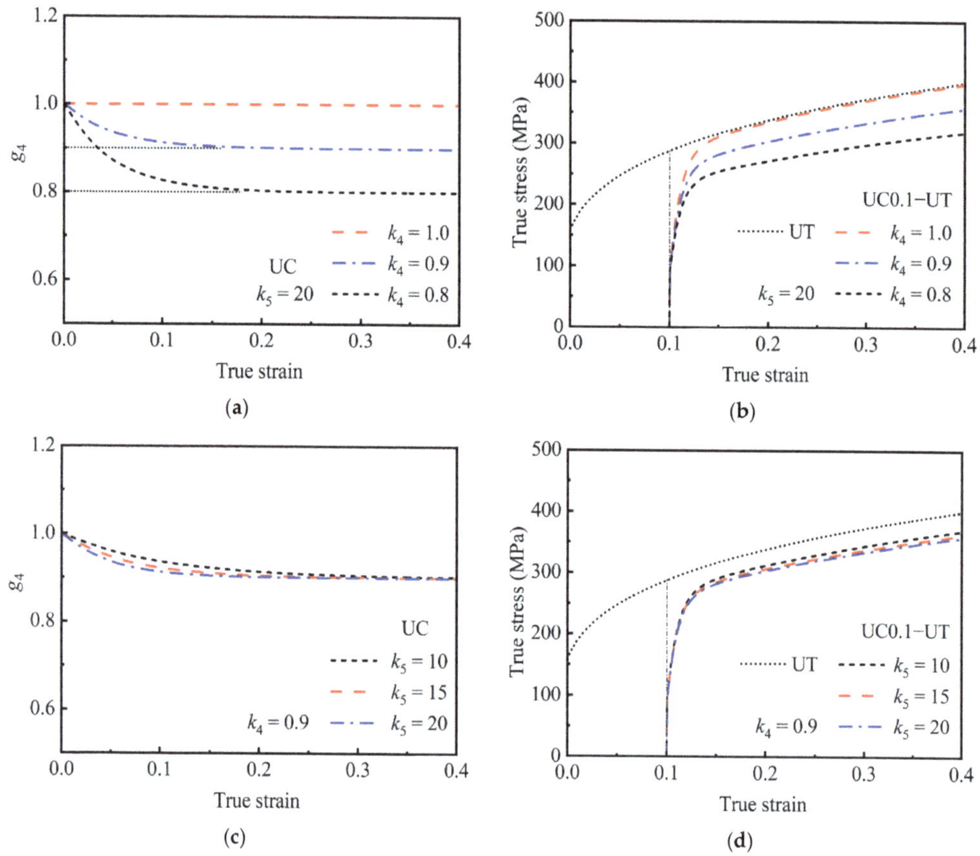

Figure 7. Predicted responses of permanent softening effect during reverse loading; evaluation of g_4 during UC loading under (**a**) different k_4 and (**c**) different k_5; and stress–strain curves during UC0.1−UT loading under (**b**) different k_4 and (**d**) different k_5.

4.2. Cross Loading

It is known that the two stress deviators corresponding to $\hat{\boldsymbol{\sigma}}_{PT} = (1/\sqrt{5},\ 2/\sqrt{5},\ 0)^T$ and $\hat{\boldsymbol{\sigma}}_{UT} = (1,\ 0,\ 0)^T$ are orthogonal to each other, and the corresponding deformation types are plane strain tension and uniaxial tension, respectively. Therefore, the loading sequence of a plane strain tension (PT) followed by a uniaxial tension at an angle of 90° from the first tension direction is a typical cross loading path, and $\cos\chi = 0$ at the start of the second loading.

The evolution of the normalized yield locus of the generic material, predicted by the HAH-2d model during uniaxial tension after plane strain tension to equivalent strain of 0.10, is shown in Figure 8. In the plane strain tension process, there is $\hat{\mathbf{H}} = \hat{\boldsymbol{\sigma}}_{PT}$, the yield locus above the line $\sigma_{22} = 0$ remains unchanged, and the yield locus below the line $\sigma_{22} = 0$ contracts inward because of the Bauschinger effect. In the second loading step, $\hat{\mathbf{H}}$ rotates gradually towards $\hat{\boldsymbol{\sigma}}_{UT}$ with the increase in tensile strain, and the yield locus near $\hat{\boldsymbol{\sigma}}_{UT}$ expands outward first and then returns to the isotropic hardening yield locus.

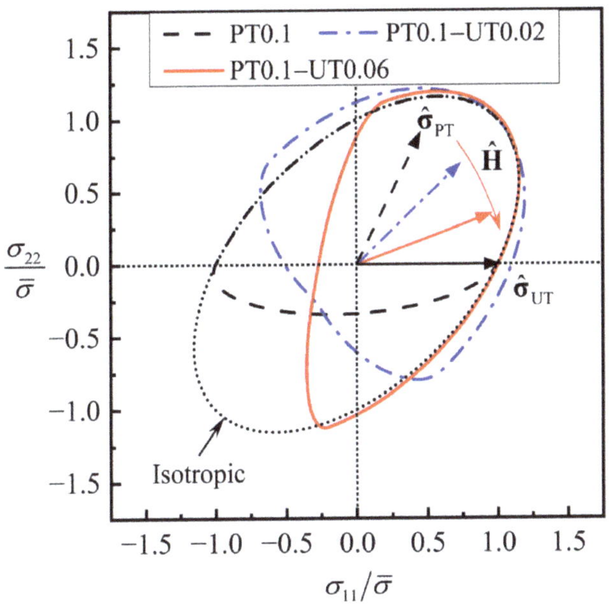

Figure 8. Evolution of the normalized yield locus of the generic material predicted by HAH-2d model in plane strain tension–uniaxial tension.

The uniaxial tensile stress–strain curves in the second loading step after different degrees of plane strain pre-tension are shown in Figure 9. Obvious stress overshooting is observed, and the higher the pre-deformation the greater the overshooting. When the tensile strain reaches about 0.06 in the second step, the tensile stress returns to the level of single uniaxial tension. It indicates that the HAH-2d model can capture the latent hardening in cross loading.

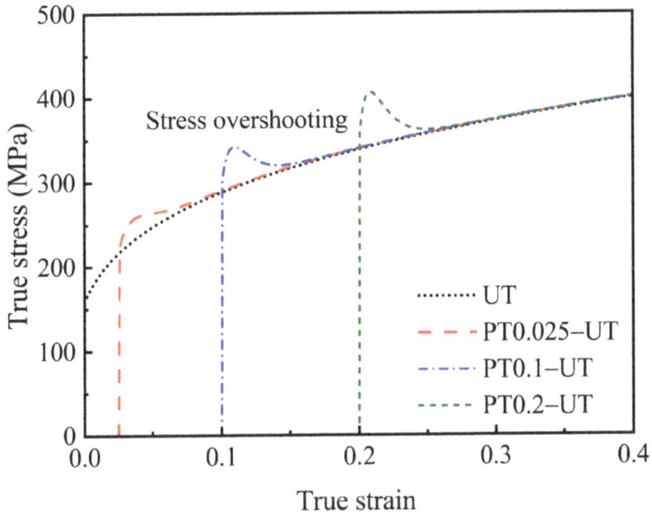

Figure 9. Stress–strain curves predicted by the HAH-2d model in plane strain tension–uniaxial tension with different pre-strain levels.

The latent hardening effect predicted by the HAH-2d model during cross loading is affected mainly by parameters k, L, and k_L. Therefore, the sensitivity of the model to these three parameters was analyzed. Figure 10 shows the predicted responses of latent hardening effect under different values of k during a typical cross loading, PT0.1−UT. With the increase in the reloading strain, the value of χ gradually decreases from 90° to 0°, and the value of g_L rapidly increases to a certain value and then gradually decreases to 1.0. As a result, the phenomenon of stress overshooting is observed. Meanwhile, the larger the value of k, the faster the value of χ and g_4 decrease, and the faster the overshoot stress recovers.

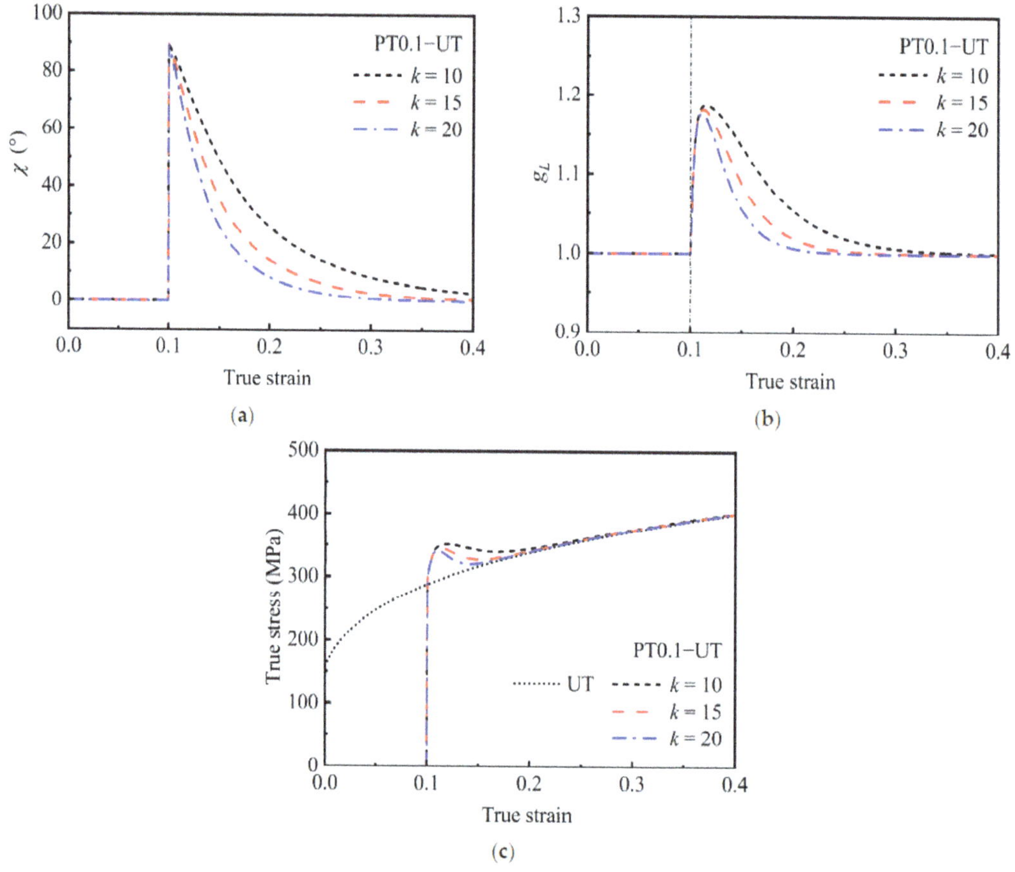

Figure 10. Predicted responses of latent hardening effect during cross loading of PT0.1−UT under different k: (**a**) evaluation of χ, (**b**) evaluation of g_L and (**c**) stress–strain curves.

Figure 11 shows the effects of L and k_L on the predicted responses of latent hardening effect during cross loading. It can be seen from Figure 11a,b that the larger the value L, the larger g_L can be obtained and the more significant the stress overshooting. On the other hand, the latent hardening effect is also influenced by the value of k_L, as shown in Figure 11c,d. Additionally, the larger the value of k_L, the faster the stress value reaches its peak, and the larger the maximum stress value is. However, the sensitivity of the model to k_L is much lower than that to L. In addition, neither L nor k_L affects the strain span of the phenomenon of stress overshooting.

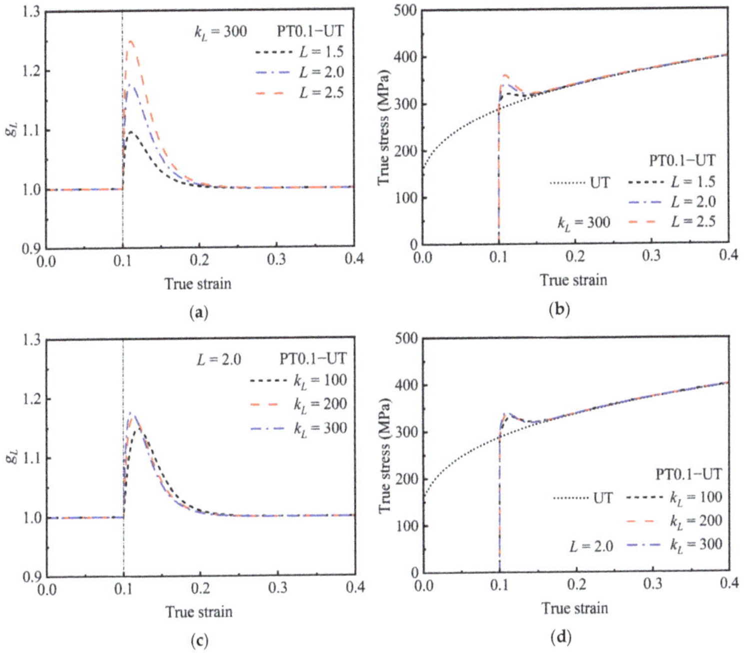

Figure 11. Predicted responses of latent hardening effect during cross loading of PT0.1−UT; evaluation of g_L under (**a**) different L and (**c**) different k_L; and stress–strain curves under (**b**) different L and (**d**) different k_L.

4.3. Two-Step Uniaxial Tension

Stress–strain curves in two-step uniaxial tensions with tensile axes at 45° and 90° from each other were predicted by the HAH-2d model and are shown in Figures 12 and 13, respectively. The curves with 45° express only latent hardening, while the curves with 90° express the Bauschinger effect, permanent softening and latent hardening at the same time. Therefore, it is better to optimize the coefficients of the HAH-2d model by using the stress–strain data of a two-step uniaxial tension with 90°, or the orthogonal two-step tension mentioned in Section 3.3.

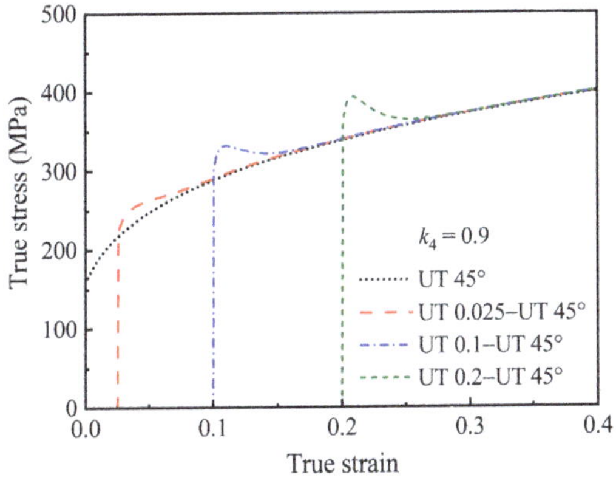

Figure 12. Stress–strain curves predicted by the HAH-2d model in two-step uniaxial tensions with tensile axes at 45° from each other.

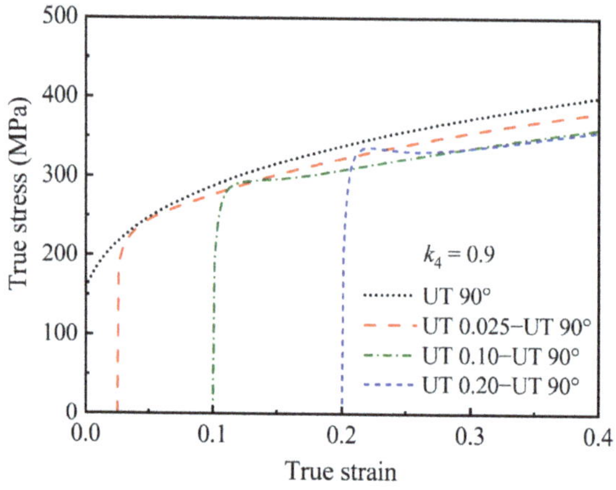

Figure 13. Stress–strain curves predicted by the HAH-2d model in two-step uniaxial tensions with tensile axes at 90° from each other.

5. Model Validation

5.1. Determination of Material Coefficients

In this paper, the HAH-2d model will be applied to describe the hardening behavior of DP780 and EDDQ steel sheets with 1.2 mm thickness that were tested by Ha et al. using two-step uniaxial tension tests [6]. As shown in Figure 14, the sub-specimens were cut along different directions at the center of the pre-deformed big-specimen [6]. Additionally, the experimental results from tests consisted first of tension in the rolling direction (RD) and second of tension at 90° (TD, the transverse direction), 60° and 45° from RD were used in this work. Coefficients of the Yld2000-2d anisotropic yield function (see Appendix A) and the Swift hardening law (see Equation (25)) for the two materials are listed in Table 2, where $m = 6$ was adopted as recommended for BCC metals [6].

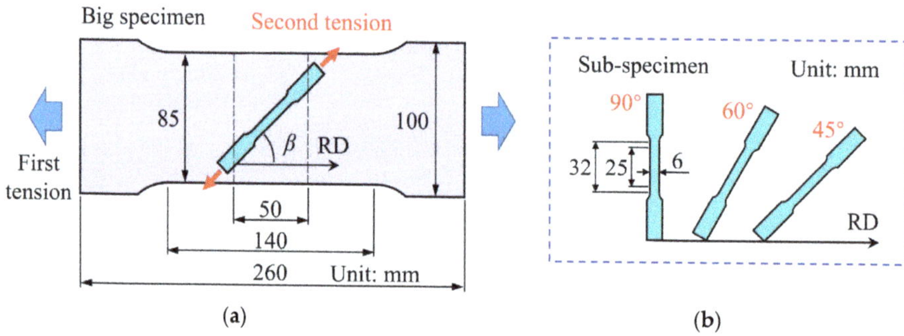

Figure 14. Schematic illustration of the specimens used in the two-step uniaxial tension tests: (**a**) the sub-specimen at the center of the big-specimen and (**b**) directions of the second tension [6].

Table 2. Coefficients of yield function and hardening law for DP780 and EDDQ (Data from [6]).

Material	Yld2000-2d Yield Function									Swift Hardening Law		
	α_1	α_2	α_3	α_4	α_5	α_6	α_7	α_8	m	K (MPa)	ε_0	n
DP780	0.946	1.022	1.015	1.000	1.011	0.968	1.010	1.006	6	1295	0.0008	0.142
EDDQ	1.014	1.118	0.931	0.892	0.904	0.811	1.029	0.918	6	538	0.0075	0.267

To quantitatively evaluate the error between the predicted and experimental data, the root mean square error (RMSE) and the average absolute relative error (AARE) were used in this paper. The definition of RMSE and AARE can be expressed by Equations (26) and (27).

$$\text{RMSE} = \sqrt{\frac{1}{N}\sum_{1}^{N}(\sigma_P - \sigma_E)^2} \tag{26}$$

$$\text{AARE }(\%) = \frac{1}{N}\sum_{1}^{N}\left|\frac{\sigma_P - \sigma_E}{\sigma_E}\right| \times 100\% \tag{27}$$

where σ_P is the predicted stress, σ_E is the experimental stress, and N is the number of data.

The coefficients associated with the distortion of the HAH-2d model for each material were identified with an optimization method using the stress–strain data from an orthogonal two-step tensile test with 10% pre-strain in RD, as shown in Figure 15. In the optimization method, the objective is to minimize the RMSE. Considering that the DP780 steel expresses no permanent softening and latent hardening, k_4 = 1.0, k_5 = 0, k_L = 0 and L = 1.0 were set directly. Similarly, as the EDDQ steel expresses no permanent softening, k_4 = 1.0 and k_5 = 0 were set. In the meantime, the value of q was set to 2.0 for convenience. The obtained coefficients of the HAH-2d models for DP780 and EDDQ are listed in Table 3. Additionally, the predicted stress–strain curves are plotted in Figure 15a,b, both of which agree well with the experimental data. Additionally, the values of RMSE for DP780 and EDDQ are 9.81 and 1.85, respectively. Incidentally, the reason for using the experimental data from reference [27] in Figure 15 is that the experimental data used in reference [27] are also from reference [6], but more points are given for this testing condition.

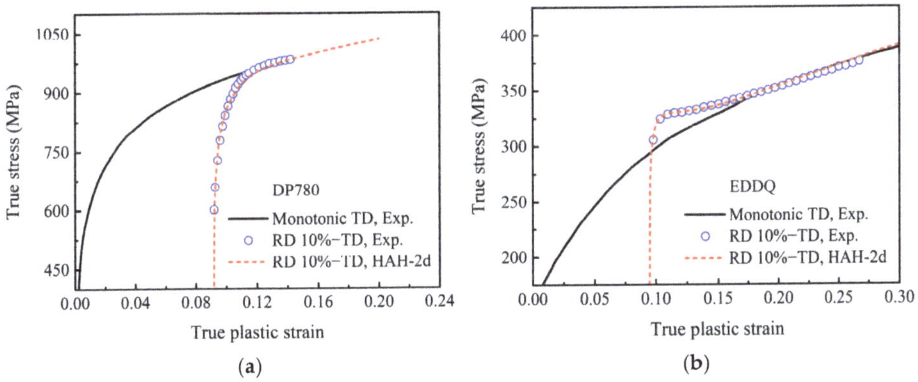

Figure 15. Experimental and predicted stress–strain curves of re-tension in TD with 10% uniaxial pre-strain in RD for: (**a**) DP780 steel and (**b**) EDDQ steel. The experimental data are derived from reference [27]: International Journal of Plasticity, Vol 46, Barlat, F.; Ha, J.; Grácio, J.J.; Lee, M.; Rauch, E.F.; Vincze, G., Extension of homogeneous anisotropic hardening model to cross-loading with latent effects, 130–142, Copyright Elsevier (2013), with permission from Elsevier.

Table 3. Coefficients of HAH-2d models for DP780 and EDDQ.

Material	k	k_1	k_2	k_3	k_4	k_5	k_L	L	q
DP780	45.0	135.2	39.6	0.475	1.0	0	0	1.0	2.0
EDDQ	12.80	773.9	295.9	0.551	1.0	0	205.9	2.226	2.0

5.2. Comparison of Prediction and Experimental Results

In order to validate the HAH-2d model, the hardening curves in two-step uniaxial tensions of DP780 and EDDQ sheets, which consisted of 4% or 10% pre-strain in RD and re-tension at 90° (TD), 60° or 45° from RD, were predicted and compared with the experimental results. Incidentally, the two-step tension with tensile axes at 60° from each other can be called "pseudo cross loading", because it is close to the ideal cross-loading condition, about 55° between the first and second tensile directions.

The predicted stress–strain curves of the HAH-2d model for DP780 under different two-step uniaxial tensions are compared with the experimental curves in Figure 16, and the corresponding prediction errors are listed in Table 4. It can be seen from Figure 16a that the HAH-2d model captures the transient hardening behavior of DP780 fairly well for re-tension in TD after a 4% pre-strain in RD, and the AARE value is only 1.28%. However, the prediction at 60° shown in Figure 16b slightly overestimates the re-loading yield stress and leads to a lower hardening rate. As a result, the values of RMSE and AARE increase to 37.85 and 3.44% under the condition of 10% pre-strain. On the other hand, as shown in Figure 16c, the phenomenon of cross-loading softening was observed in re-loading stress–strain curves at 45°, but the predicted re-loading yield stresses at 45° reach the monotonic curve immediately. Additionally, the high RMSE values of 24.49 and 47.13 are mainly due to the significant difference between the predicted stress and the experimental stress at the beginning of the second tension.

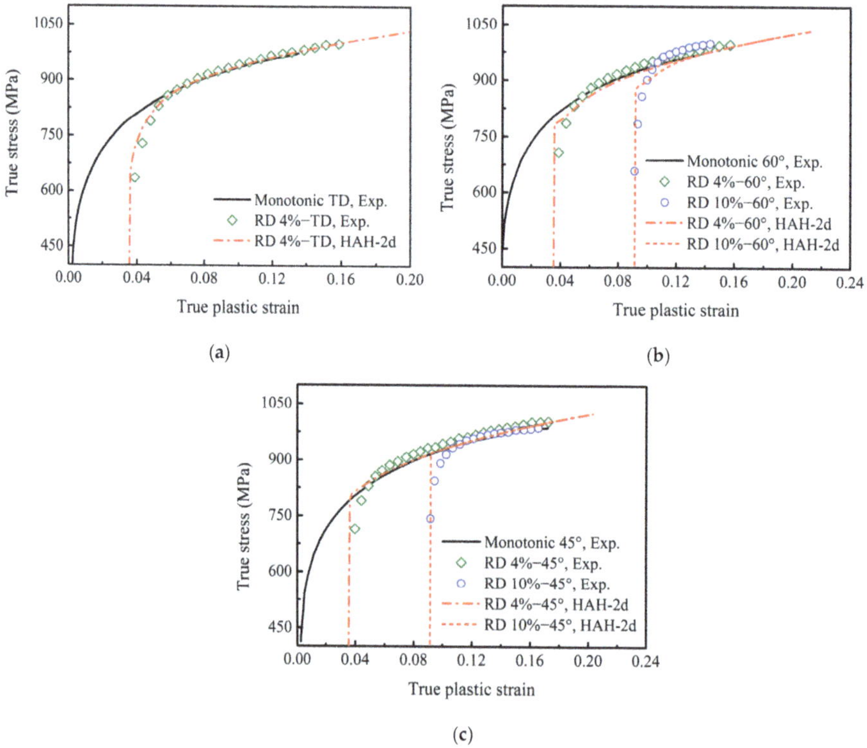

Figure 16. Experimental and predicted stress–strain curves of re-tension at (**a**) 90°, (**b**) 60° and (**c**) 45° from RD with 4% or 10% uniaxial pre-strain in RD for DP780. The experimental data are derived from reference [6]: Mechanics of Materials, Vol 64, Ha, J.; Lee, M.; Barlat, F., Strain hardening response and modeling of EDDQ and DP780 steel sheet under non-linear strain path, 11–26, Copyright Elsevier (2013), with permission from Elsevier.

Table 4. Prediction errors for DP780 under different two-step uniaxial tensions.

Conditions	RD 4%−TD	RD 4%–60°	RD 10%–60°	RD 4%–45°	RD 10%–45°
RMSE	20.11	24.07	37.85	24.49	47.13
AARE (%)	1.28	2.14	3.44	1.94	2.58

Figure 17a–c show the predicted reloading flow stress curves for EDDQ in different two-step uniaxial tensions with pre-tension in RD and re-tension at 90° (TD), 60° and 45° from RD. The predicted curves capture all the experimental results well, even though there is a little overestimation of the flow stress in the re-tension at 60° from RD. The corresponding prediction errors are listed in Table 5. The maximum values of RMSE and AARE, 9.86 and 2.00%, both occur in the condition of re-tension at 60° after a 10% pre-tension. This indicates that the HAH-2d model can capture anisotropic hardening responses such as flow stress overshooting and strain-hardening stagnation.

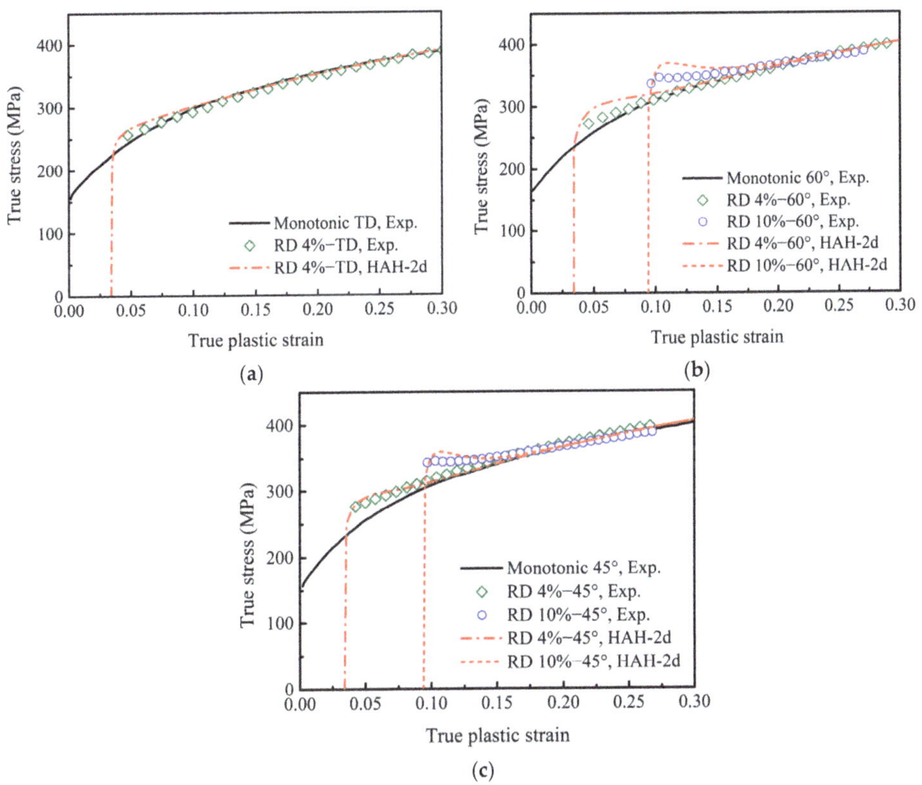

Figure 17. Experimental and predicted stress–strain curves of re-tension at (**a**) 90°, (**b**) 60° and (**c**) 45° from RD with 4% or 10% uniaxial pre-strain in RD for EDDQ. The experimental data are derived from reference [6]: Mechanics of Materials, Vol 64, Ha, J.; Lee, M.; Barlat, F., Strain hardening response and modeling of EDDQ and DP780 steel sheet under non-linear strain path, 11–26, Copyright Elsevier (2013), with permission from Elsevier.

Table 5. Prediction errors for EDDQ under different two-step uniaxial tensions.

Conditions	RD 4%−TD	RD 4%–60°	RD 10%–60°	RD 4%–45°	RD 10%–45°
RMSE	5.59	8.65	9.86	4.74	5.40
AARE (%)	1.53	1.78	2.00	1.33	1.13

6. Conclusions

In this work, an HAH model in plane stress state was proposed, and the evolution of the yield loci and the stress–strain curve of a generic material under some typical nonlinear loading paths was analyzed using this model. Furthermore, the model was validated by experimental results from two-step uniaxial tension tests. The conclusions are as follows:

(1) The HAH-2d model was proposed in plane stress state. Both the loading stress vector and microstructure vector contain only three in-plane components. Compared with the HAH model in fully 3D space, the transformation between the plane stress vector and the second-order stress deviator and the complicated calculation in the stress deviator space can be avoided in the HAH-2d model. The computational cost in finite element applications is therefore reduced.

(2) The Bauschinger effect and permanent softening effect in reverse loading and the latent hardening effect in cross loading were investigated in this paper. The evolution of the yield surface and the stress–strain curve of the sheet metal with these effects can be predicted by the HAH-2d model.

(3) The HAH-2d model can reasonably predict experimental phenomena such as the reduction of re-loading yield stress in two-step loading with $\cos\chi < 0$ and stress overshooting in cross loading. However, the phenomenon of cross-loading softening observed in two-step loading with $0 \leq \cos\chi < 1$ cannot be captured by the model.

Author Contributions: Conceptualization, H.Z. and Z.H..; methodology, H.Z. and Y.L.; validation, Y.L. and K.C.; formal analysis, Y.L. and K.C.; investigation, H.Z., Y.L. and Z.H.; resources, Z.H. and S.Y.; data curation, H.Z. and Y.L.; writing—original draft preparation, H.Z., Y.L. and K.C.; writing—review and editing, H.Z., Y.L., K.C. and Z.H.; supervision, Z.H. and S.Y.; project administration, S.Y.; funding acquisition, Z.H. All authors have read and agreed to the published version of the manuscript.

Funding: This research was funded by the National Natural Science Foundation of China, grant numbers: 52075075, 51575131; and the National Key R&D Program of China, grant numbers: 2017YFB0304404.

Institutional Review Board Statement: Not applicable.

Informed Consent Statement: Not applicable.

Data Availability Statement: Not applicable.

Acknowledgments: The present work was financed by the National Natural Science Foundation of China (52075075, 51575131) and the National Key R&D Program of China (2017YFB0304404). The authors would like express their sincere appreciation to the funds.

Conflicts of Interest: The authors declare no conflict of interest.

Appendix A

Yld2000-2d anisotropic yield criterion [37] is defined as:

$$\phi = |X'_1 - X'_2|^m + |X''_1 + 2X''_2|^m + |2X''_1 + X''_2|^m = 2\bar{\sigma}^m \quad \text{(A1)}$$

where $\bar{\sigma}$ is the equivalent stress, X'_i and X''_i (i = 1, 2) are the principal values of tensors $\mathbf{X}' = \mathbf{L}' \cdot \boldsymbol{\sigma}$ and $\mathbf{X}'' = \mathbf{L}'' \cdot \boldsymbol{\sigma}$, respectively. $\boldsymbol{\sigma}$ is Cauchy stress tensor. $\mathbf{L}'$ and $\mathbf{L}''$ can be written in the following form

$$[\mathbf{L}'] = \begin{bmatrix} L'_{11} & L'_{12} & 0 \\ L'_{21} & L'_{22} & 0 \\ 0 & 0 & L'_{66} \end{bmatrix} = \frac{1}{3}\begin{bmatrix} 2\alpha_1 & -\alpha_1 & 0 \\ -\alpha_2 & 2\alpha_2 & 0 \\ 0 & 0 & \alpha_7 \end{bmatrix} \tag{A2}$$

$$[\mathbf{L}''] = \begin{bmatrix} L''_{11} & L''_{12} & 0 \\ L''_{21} & L''_{22} & 0 \\ 0 & 0 & L''_{66} \end{bmatrix} = \frac{1}{9}\begin{bmatrix} 8\alpha_5 - 2\alpha_3 - 2\alpha_6 + 2\alpha_4 & 4\alpha_6 - 4\alpha_4 - 4\alpha_5 + \alpha_3 & 0 \\ 4\alpha_3 - 4\alpha_5 - 4\alpha_4 + \alpha_6 & 8\alpha_4 - 2\alpha_6 - 2\alpha_3 + 2\alpha_5 & 0 \\ 0 & 0 & 9\alpha_8 \end{bmatrix} \tag{A3}$$

where α_i ($i = 1, 2 \ldots 8$) are pending coefficients.

The principal values of **X** are

$$\begin{cases} X_1 = \frac{1}{2}\left(X_{11} + X_{22} + \sqrt{(X_{11} - X_{22})^2 + 4X_{12}^2} \right) \\ X_2 = \frac{1}{2}\left(X_{11} + X_{22} - \sqrt{(X_{11} - X_{22})^2 + 4X_{12}^2} \right) \end{cases} \tag{A4}$$

In the Yld2000-2d criterion, only the plane stress components, σ_{xx}, σ_{yy} and σ_{xy} are relevant, where x and y indicate the RD and TD, respectively. For the exponent m in Equation (A1), it is recommended to use 6 for BCC material and 8 for FCC material.

References

1. Yuan, S.; Fan, X. Developments and perspectives on the precision forming processes for ultra-large size integrated components. *Int. J. Extrem. Manuf.* **2019**, *1*, 022002. [CrossRef]
2. Zheng, K.; Politis, D.J.; Wang, L.; Lin, J. A review on forming techniques for manufacturing lightweight complex—Shaped aluminium panel components. *Int. J. Lightweight Mater. Manuf.* **2018**, *1*, 55–80. [CrossRef]
3. Bouvier, S.; Gardey, B.; Haddadi, H.; Teodosiu, C. Characterization of the strain-induced plastic anisotropy of rolled sheets by using sequences of simple shear and uniaxial tensile tests. *J. Mater. Process. Tech.* **2006**, *174*, 115–126. [CrossRef]
4. Qin, J.; Holmedal, B.; Hopperstad, O.S. A combined isotropic, kinematic and distortional hardening model for aluminum and steels under complex strain-path changes. *Int. J. Plast.* **2018**, *101*, 156–169. [CrossRef]
5. He, J.; Gu, B.; Li, Y.; Li, S. Forming limits under stretch-bending through distortionless and distortional anisotropic hardening. *J. Manuf. Sci. Eng.: Trans. ASME* **2018**, *140*, 121013. [CrossRef]
6. Ha, J.; Lee, M.; Barlat, F. Strain hardening response and modeling of EDDQ and DP780 steel sheet under non-linear strain path. *Mech. Mater.* **2013**, *64*, 11–26. [CrossRef]
7. Mánik, T.; Holmedal, B.; Hopperstad, O.S. Strain-path change induced transients in flow stress, work hardening and *r*-values in aluminum. *Int. J. Plast.* **2015**, *69*, 1–20. [CrossRef]
8. Bauschinger, J. Changes of the elastic limit and the modulus of elasticity on various metals. *Ziivilingenieur* **1881**, *27*, 289–348.
9. Hasegawa, T.; Yakou, T.; Karashima, S. Deformation behaviour and dislocation structures upon stress reversal in polycrystalline aluminium. *Mater. Sci. Eng.* **1975**, *20*, 267–276. [CrossRef]
10. Hasegawa, T.; Yakou, T. The effect of strain reversal and thermal recovery on stress vs strain behavior in aluminum. *Scr. Metall.* **1980**, *14*, 1083–1087. [CrossRef]
11. Schmitt, J.H.; Fernandes, J.V.; Gracio, J.J.; Viera, M.F. Plastic behaviour of copper sheets during sequential tension tests. *Mater. Sci. Eng. A* **1991**, *147*, 143–154. [CrossRef]
12. Barlat, F.; Ferreira Duarte, J.M.; Gracio, J.J.; Lopes, A.B.; Rauch, E.F. Plastic flow for non-monotonic loading conditions of an aluminum alloy sheet sample. *Int. J. Plast.* **2003**, *19*, 1215–1244. [CrossRef]
13. Ben Bettaieb, M.; Abed-Meraim, F. Localized necking in elastomer-supported metal layers: Impact of kinematic hardening. *J. Manuf. Sci. Eng. Trans. ASME* **2017**, *139*, 61008. [CrossRef]
14. Min, J.; Guo, N.; Hou, Y.; Jiang, K.; Chen, X.; Carsley, J.E.; Lin, J. Effect of tension-compression testing strategy on kinematic model calibration and springback simulation of advanced high strength steels. *Int. J. Mater. Form.* **2020**, *14*, 435–448. [CrossRef]
15. Yetna N'Jock, M.; Badreddine, H.; Labergere, C.; Yue, Z.; Saanouni, K.; Dang, V.T. An application of fully coupled ductile damage model considering induced anisotropies on springback prediction of advanced high strength steel materials. *Int. J. Mater. Form.* **2021**, *14*, 739–752. [CrossRef]
16. Clausmeyer, T.; Güner, A.; Tekkaya, A.E.; Levkovitch, V.; Svendsen, B. Modeling and finite element simulation of loading-path-dependent hardening in sheet metals during forming. *Int. J. Plast.* **2014**, *63*, 64–93. [CrossRef]
17. Barlat, F.; Yoon, S.; Lee, S.; Wi, M.; Kim, J. Distortional plasticity framework with application to advanced high strength steel. *Int. J. Solids Struct.* **2020**, *202*, 947–962. [CrossRef]
18. Cao, J.; Banu, M. Opportunities and challenges in metal forming for lightweighting: Review and future work. *J. Manuf. Sci. Eng. Trans. ASME* **2020**, *142*, 110813. [CrossRef]

19. Lee, M.G.; Kim, C.; Pavlina, E.J.; Barlat, F. Advances in sheet forming—Materials modeling, numerical simulation, and press technologies. *J. Manuf. Sci. Eng. Trans. ASME* **2011**, *133*, 61001. [CrossRef]
20. Banabic, D.; Barlat, F.; Cazacu, O.; Kuwabara, T. Advances in anisotropy of plastic behaviour and formability of sheet metals. *Int. J. Mater. Form.* **2020**, *13*, 749–787. [CrossRef]
21. Prager, W. A new method of analyzing stresses and strains in work-hardening plastic solids. *J. Appl. Mech.* **1956**, *23*, 493–496. [CrossRef]
22. Frederick, C.; Armstrong, P. A mathematical representation of the multiaxial Bauschinger effect. *Mater. High Temp.* **2007**, *24*, 1–26. [CrossRef]
23. Chaboche, J.L. Time-independent constitutive theories for cyclic plasticity. *Int. J. Plast.* **1986**, *2*, 149–188. [CrossRef]
24. Bouhamed, A.; Jrad, H.; Said, L.B.; Wali, M.; Dammak, F. A non-associated anisotropic plasticity model with mixed isotropic-kinematic hardening for finite element simulation of incremental sheet metal forming process. *Int. J. Adv. Manuf. Technol.* **2018**, *100*, 929–940. [CrossRef]
25. Grilo, T.J.; Valente, R.A.F.; Alves De Sousa, R.J. Modelling non-quadratic anisotropic yield criteria and mixed isotropic-nonlinear kinematic hardening: Analysis of forward- and backward-Euler formulations. *Int. J. Mater. Form.* **2014**, *8*, 533–547. [CrossRef]
26. Barlat, F.; Gracio, J.J.; Lee, M.; Rauch, E.F.; Vincze, G. An alternative to kinematic hardening in classical plasticity. *Int. J. Plast.* **2011**, *27*, 1309–1327. [CrossRef]
27. Barlat, F.; Ha, J.; Grácio, J.J.; Lee, M.; Rauch, E.F.; Vincze, G. Extension of homogeneous anisotropic hardening model to cross-loading with latent effects. *Int. J. Plast.* **2013**, *46*, 130–142. [CrossRef]
28. Barlat, F.; Vincze, G.; Grácio, J.J.; Lee, M.G.; Rauch, E.F.; Tomé, C.N. Enhancements of homogenous anisotropic hardening model and application to mild and dual-phase steels. *Int. J. Plast.* **2014**, *58*, 201–218. [CrossRef]
29. Choi, J.; Lee, J.; Bae, G.; Barlat, F.; Lee, M. Evaluation of springback for DP980 S rail using anisotropic hardening models. *JOM* **2016**, *68*, 1850–1857. [CrossRef]
30. Chen, S.; Liao, J.; Xiang, H.; Xue, X.; Pereira, A.B. Pre-strain effect on twist springback of a 3D P-channel in deep drawing. *J. Mater. Process. Technol.* **2021**, *287*, 116224. [CrossRef]
31. Choi, H.; Choi, S.; Kang, S.; Kim, C.; Lee, M. Prediction of anisotropic strengths of steel plate after prior bending-reverse bending deformation: Application of distortional hardening model. *Int. J. Mech. Sci.* **2021**, *204*, 106512. [CrossRef]
32. Reyne, B.; Barlat, F. A new concept for continuum distortional plasticity. *Int. J. Plast.* **2022**, *155*, 103303. [CrossRef]
33. Hc, A.; Jwya, B. Stress integration-based on finite difference method and its application for anisotropic plasticity and distortional hardening under associated and non-associated flow rules. *Comput. Methods Appl. Mech. Eng.* **2019**, *345*, 123–160. [CrossRef]
34. Choi, H.; Choi, S.; Kang, S.C.; Lee, M.G. Fully Implicit stress update algorithm for distortion-based anisotropic hardening with cross-loading effect: Comparative algorithmic study and application to large-size forming problem. *Appl. Sci.* **2021**, *11*, 5509. [CrossRef]
35. Lee, J.; Bong, H.J.; Ha, J. Efficient and robust stress integration algorithm for anisotropic distortional hardening law under cross-loading with latent hardening. *Eur. J. Mech.-A/Solids* **2022**, *96*, 104775. [CrossRef]
36. Swift, H. Plastic instability under plane stress. *J. Mech. Phys. Solids* **1952**, *1*, 1–18. [CrossRef]
37. Barlat, F.; Brem, J.C.; Yoon, J.W.; Chung, K.; Dick, R.E.; Lege, D.J.; Pourboghrat, F.; Choi, S.H.; Chu, E. Plane stress yield function for aluminum alloy sheets—Part 1: Theory. *Int. J. Plast.* **2003**, *19*, 1297–1319. [CrossRef]

materials

Article

Optimization of Processing Parameter and Mechanical Response Analysis of Advanced Heterogeneous Laminated Composites Using Ni/Al Foils by In Situ Reaction Synthesis

Ying Sun [1] and Shijian Yuan [2,*]

[1] Institute of Precision Forming for High Performance, Dalian University of Technology, Dalian 116024, China
[2] National Key Lab for Precision Hot Processing of Metals, Harbin Institute of Technology, Harbin 150001, China
* Correspondence: syuan@hit.edu.cn

Abstract: The advanced heterogeneous laminated composites were successfully fabricated by vacuum hot pressing using Ni and Al foils by in situ solid-state reaction synthesis. The effects of holding time and temperature on the microstructure and phase distribution were analyzed using scanning electron microscopy. Based on the optimized processing parameters, the microstructure and phase transformation, and the relationship between the microstructure and the corresponding mechanical properties were discussed in detail. To clarify the mechanical response of the laminated structure, the deformation microstructure and fracture characteristics were studied by scanning electron microscopy and electron backscatter diffraction. The results indicated that the evolution of the interfacial phases in the laminated composite occurred via the sequence: $NiAl_3$, Ni_2Al_3, NiAl, and Ni_3Al. An interface between the Ni and Ni_3Al layers without cracks and voids formed due to the uniform pressure applied during hot pressing. The laminated composites hot pressed under 620 °C/5 MPa/1 h + 1150 °C/10 MPa/2 h exhibited the best ultimate tensile strength of 965 MPa and an elongation of 22.6% at room temperature. Extending the holding time during the second stage of the reaction synthesis decreased the thickness of the Ni_3Al layer. This decreased the tensile strength of the laminated composite at 1000 °C but improved the tensile strength at room temperature. Moreover, the layer–thickness relationship of the laminated structure and the matching pattern were important factors affecting the strength and elongation of the laminated composites. The reinforcement form of the materials was not limited to a lamellar structure but could be combined with different forms of reinforcement to achieve continuous reinforcement over a wide range of temperatures.

Keywords: heterogeneous laminated composites; parameter optimization; intermetallics; microstructure; mechanical properties

1. Introduction

Ni_3Al has attracted great attention for high-temperature structures in aerospace applications due to its excellent mechanical properties at elevated temperatures, including its high melting point, low density, and oxidation resistance [1]. However, the intrinsic brittleness of Ni_3Al alloy leads to its poor ductility and fracture toughness. Metal–intermetallic composites offer an attractive combination of physical and mechanical properties, such as a high strength, high toughness, and low density [2]. The mechanical properties of laminated composites can be optimized by group element materials, group element layer thickness, interface thickness, and interface configuration. This ensures that metal–intermetallic composites can maintain the high strength and stiffness of intermetallic compounds while retaining the high toughness and ductility of metals. The metal–intermetallic composites are attractive in a wide range of engineering applications. Because the tailored mechanical properties can be obtained by controlling processing parameters, the composites are suitable for a service environment at high temperatures and also meet the requirements of the assembly conditions at ambient temperatures. The applications include compressor vanes and cylinder heads of automotive engines, as well as high-performance aerospace applications [2–4]. Therefore, the design and manufacturing of laminated metal–intermetallic compound composites by introducing microstructural heterogeneity can realize useful intermetallic compounds.

Rawers et al. [5] and Alman et al. [6] demonstrated a near-net shaping method that can be used to prepare metal–intermetallic composites via in situ reaction synthesis between constituent Ni/Al foil elements. Because the starting material was metal foils, this approach was a reasonably inexpensive process. Compared with pre-alloyed intermetallic foils, this method can control the distribution of phases and protect the interfaces between layers. Compared to Ni and Al powder sintering or $NiAl_3$ spray deposition, the solid-state reactive synthesis method by the hot pressing of Ni and Al foils can approach large-scale industrial production for the fabricating of thin sheets and thin-walled components at low cost. Moreover, it is possible to achieve the metal–intermetallic laminated composite with low oxygen contents by the synthetic approach compared to the sintering of standard aluminum and nickel powders. Reaction synthesis is an essential bridge between the starting foils and the Ni-Al metal–intermetallic laminated composites, which involves the self-propagating high-temperature synthesis mode and the thermal explosion mode [7]. Many researchers [8–11] have studied the phase transition and the reaction products at different reaction stages by using Ni/Al foil diffusion couples. Kim et al. [7] investigated the reaction synthesis of a NiAl/Ni microlaminated composite fabricated by alternatively stacked Ni and Al foils. The reaction products were converted to NiAl and Ni_3Al at the temperature of 900–950 °C and the applied pressure of 50–100 MPa, with the initial thickness ratio (Ni:Al) of 1:1. Seyring and Rettenmayr [12] studied the formation of the Ni_3Al phase at the Ni/NiAl interface and established a specific crystallographic relationship between the Ni_3Al phase and the neighboring Ni/NiAl interface, which was evaluated using EBSD and STEM-EDX. As reported by Wu et al. [13], the microstructure and comprehensive properties of Ni_3Al-based alloys were greatly affected by heat treatments. As investigated by Wu et al. [14], the microstructure evolution and mechanical properties affect, at room temperature or high temperature, the multiphase Ni_3Al-based alloy during the laser remelting treatment. Ogneva et al. [15] investigated the microstructure and phase transitions of Ni-Al multilayer composites obtained by spark plasma sintering with Ni and Al foils. During the sintering process, the Ni_2Al_3, NiAl, and Ni_3Al layers were formed in the composites under the condition of 900–1100 °C and the pressure of 10–30 MPa. However, the reaction products were studied based on certain processing parameters. Despite the similarity of the phase constituents of the composites, the processes of phase formation and disappearance of the lamellar structure and the role of process parameters in controlling the thickness of each layer have not been investigated sufficiently.

Konieczny [16] studied the tensile properties of the laminated $Ni/Ni_2Al_3/NiAl_3$ and $Ni/Ni_3Al/NiAl$ composites fabricated by reaction synthesis using 0.4 mm Ni foils and 0.15 mm Al foils under the conditions of 620 °C/2 h/1 MPa and 1150 °C/4 h/1 MPa. Mizuuchi et al. [17] studied the microstructure and mechanical properties of Ni-aluminide-reinforced Ni matrix composites fabricated by 0.05 mm Ni/0.012 mm Al foils and 0.05 mm Ni/0.024 mm Al foils by pulsed current hot pressing equipment. Kim et al. [18] investigated the tensile and fracture properties of NiAl/Ni microlaminated composites prepared by reaction synthesis. Ogneva et al. [15] studied the effect of sintering temperature and pressure on the microstructure and mechanical properties of Ni-Al metal–intermetallic laminated composites. However, optimization of processing parameters was more important for providing desirable mechanical properties. In addition, the relevance of the high ductility of the Ni matrix in providing desirable mechanical properties for Ni-Al metal–intermetallic laminated composites has not been fully investigated. Wang et al. [19] studied the effect of Ni foil thickness on the microstructure and tensile properties of reaction synthesized Ni/Ni_3Al multilayer composites. Xi [20] studied the microstructure and mechanical properties of reaction synthesized Ni-Al intermetallic sheets by different thickness ratios of Ni and Al foils under the conditions of 640 °C/20 MPa/1 h + 1200 °C/20 MPa/1 h. However, the process parameters have an influence on the mechanical properties of the laminated materials, which are controlled essentially by adjusting the phase distribution and phase components to control the mechanical properties. It is difficult to obtain the desired mechanical properties by changing the thickness ratios of Ni and Al foils under a certain process parameter. Overall, the influences of the hot pressing temperature and holding time on the microstructures and phase composition have not been sufficiently researched. It is essential to clarify the microstructure and mechanical properties of the advanced laminated composites under the optimized processing parameters to provide guidance for practical industrial applications.

In this study, the advanced heterogeneous laminated composites were prepared via hot pressing at various temperatures and holding times to investigate the effect on the microstructure and mechanical properties. The microstructures and phase distribution of the laminated composites were characterized in detail. Moreover, the mechanical properties were measured to establish structure–processing property relationships to provide a new strategy for manufacturing Ni/Ni_3Al thin-walled components.

2. Experimental Procedure

2.1. Materials and Fabrication Processes

Commercial pure Ni foils (99.93% wt.% pure) and Al foils (99.99% wt.% pure) were used as the raw materials for fabricating the laminated composites. Pure Ni foils and pure Al foils were cut into 100 mm × 100 mm squares and treated in an ultrasonic acetone bath for 30 min. The Ni foils (19 pieces) with a thickness of 0.06 mm and Al foils (18 pieces) with a thickness of 0.02 mm were alternately stacked to prepare sandwich structure with a sequence of Ni/Al/Ni/Al ... ... Ni/Al/Ni. The Ni foils were placed in the topmost and lowest positions of the Ni/Al composited structure. Figure 1 shows the schematics of the preparation procedure for the laminated composites. The stacking laminates Ni/Al foils of Ni and Al were prepared and then hot pressed in a graphite mold in a vacuum hot pressing furnace. The stacked laminates Ni/Al foils were first preheated at 620 °C for 1 h and then heated to 1100–1200 °C with a holding time of 0.5–2 h under a pressure of 10 MPa and then cooled to room temperature in the furnace vacuum chamber. The working vacuum degree of the furnace was 5×10^{-3} Pa, and the pressure control accuracy of the hydropress was ±5%. Table 1 summarizes the hot pressing parameters and their corresponding marks for the convenience of expression and presentation.

Table 1. The hot pressing processing parameters for the laminated Ni/Al foils and their corresponding marks.

Marks	Parameters
N1	620 °C/5 MPa/1 h + 1100 °C/10 MPa/2 h
N2	620 °C/5 MPa/1 h + 1150 °C/10 MPa/2 h
N3	620 °C/5 MPa/1 h + 1200 °C/10 MPa/2 h
N4	620 °C/5 MPa/1 h + 1150 °C/10 MPa/0.5 h
N5	620 °C/5 MPa/1 h + 1150 °C/10 MPa/1 h

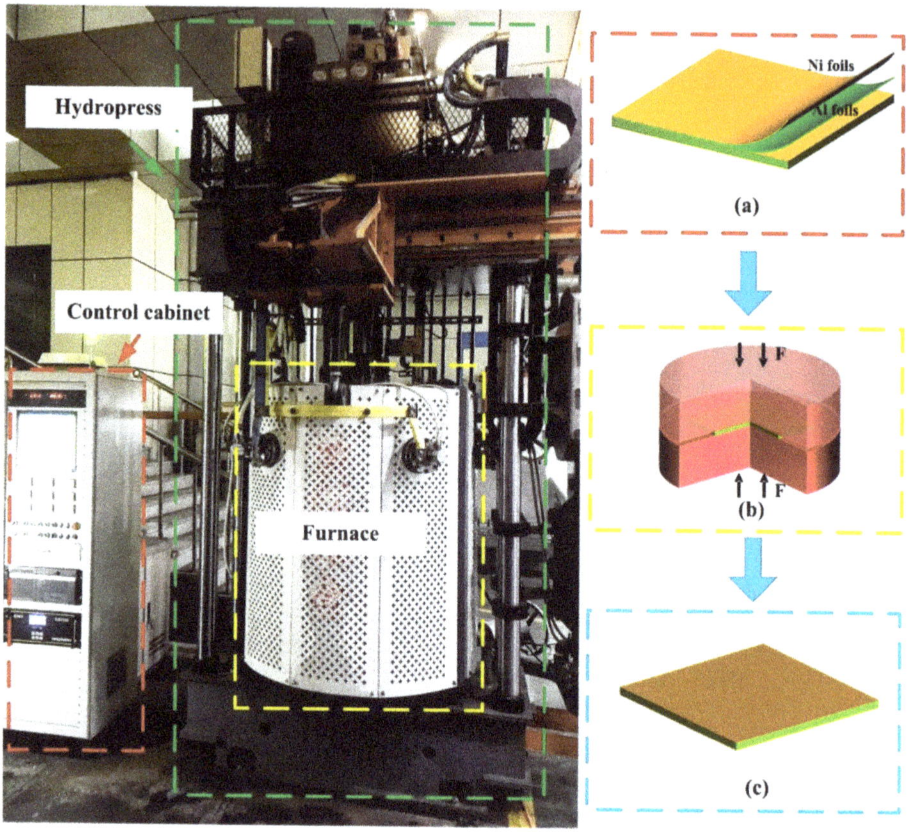

Figure 1. Schematic of the preparation procedure for the laminated composites. (**a**) Stacked alternately Ni and Al foils; (**b**) hot pressing to obtain the laminated composites; (**c**) the final laminated composites.

2.2. Materials Characterization

A scanning electron microscope (SEM, ZEISS SUPRA 55) equipped with energy-dispersive spectroscopy (EDS) was used for microstructure observation, chemical composition measurement, and fracture surface analysis. The specimens for BSE and EBSD were ground and then subjected to electro-chemical polishing with a solution of 6% perchloric acid, 35% butarol, and 59% carbinol (vol%) at −30 °C. The polishing voltage and current density were 25 V and 0.6 A, respectively. The geometrically necessary dislocation (GNDs) maps were calculated from the EBSD data by Channel 5 analysis software (Version 5.0.9.0). The EBSD data were collected with a step size of 0.5 μm. Uniaxial tensile tests were performed using a Shimadzu AG-X Plus instrument with an initial strain rate of 1×10^{-3} s^{-1} at room temperature. High-temperature tensile tests were conducted on an electronic universal material testing machine (Shimadzu AG-X Plus, Kyoto, Japan) with a strain rate of 1×10^{-3} s^{-1} at 700–1000 °C. The detailed specimen geometry is shown in Figure 2. The tensile specimens were cut by electrical discharge machining. Before testing, the tensile specimens were mechanically ground and subsequently polished with a diamond suspension containing 0.5 μm particles. Strain gauges were attached to the specimens to measure tensile strain, and the tensile tests were repeated three times to guarantee repeatability. Each sample was maintained at the designated temperature for 5 min to obtain a uniform deformation temperature. After deformation, the samples for high-temperature tensile tests were quickly removed from the resistance furnace and quenched in water.

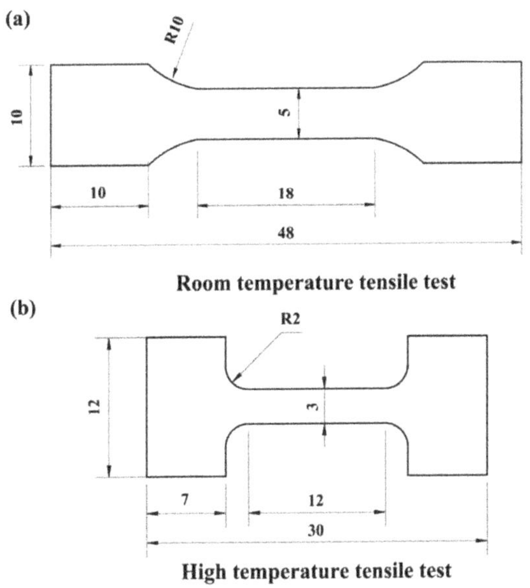

Figure 2. Dimensions of the specimens used for tensile testing. (**a**) Room-temperature tensile test; (**b**) high-temperature tensile test (all dimensions are in mm).

3. Results

3.1. Microstructural Analysis of the Laminated Composites

Figure 3 shows the cross-sectional microstructure of the laminated composites after hot pressing under 620 °C/5 MPa/1 h + 1150 °C/10 MPa/2 h, in which a uniform multilayered structure with straight bonding can be clearly observed. The element distributions are displayed in the high-magnification SEM image in Figure 3b. The EDS point scanning results are listed in Table 2. As indicated in Figure 3, the microstructure mainly consisted of Ni_3Al layers (dark gray region) and Ni layers (dark white region), as detected by SEM-EDS. The original Al layers were not detected, indicating that the Al foils were alloyed entirely, leading to the formation of a reaction synthesis region, which was visible as a dark gray region. Sound interfacial bonding was obtained by the interdiffusion of Ni and Al foils in the normal direction. The uniform thickness of the original foils contributed to the formation of a straight bond between each layer. Region 1 was detected as the Ni_3Al layer. Regions 2 and 3 were detected as Ni_3Al precipitates and the Ni matrix, respectively. These three regions constituted a dual-phase laminated structure. The structure was composed of Ni matrix layers and Ni_3Al layers that were parallel to each other. There were almost no microvoids or cracks observed in the middle of the Ni and Ni_3Al layers or their interface due to the uniform pressure applied during hot pressing bonding. This indicated that a well-bonded laminated composite was achieved.

Table 2. Compositions of reacted Ni-Al system products determined by EDS in Figure 3.

Position	Average Chemical Composition/at. %		Phase Identity
	Ni	Al	
1	77.3	22.8	Ni_3Al
2	77.6	22.4	Ni_3Al
3	83.0	17.0	Ni + Ni_3Al (Precipitates)

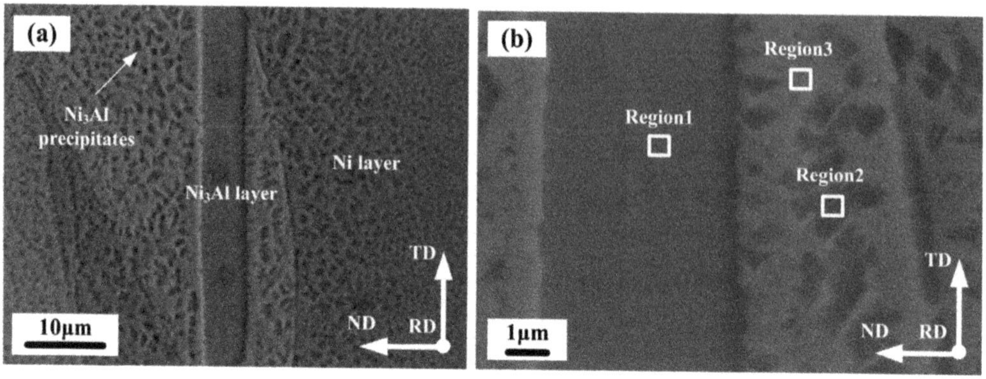

Figure 3. Low-magnification SEM image (**a**) and high-magnification SEM image (**b**) of the laminated composites after hot pressing under 620 °C/5 MPa/1 h + 1150 °C/10 MPa/2 h. RD, ND, and TD are the rolling, normal, and transverse directions, respectively.

3.2. Structural Design and Fabrication Processing Parameter Optimization

A laminated metal–intermetallic composite was designed based on the Ni-Al material system. The fabrication process of the laminated composite strongly depended on the interdiffusion between Ni and Al atoms. Therefore, the microstructure and mechanical properties were highly dependent on the temperature and holding time during hot pressing. Figure 4 depicts the microstructure of samples N1, N2, and N3 fabricated under different processing parameters. As shown in Figure 4a, the microstructure mainly consisted of three phases: Ni, Ni_3Al, and Ni_5Al_3. Thin Ni_5Al_3 layers were observed between Ni_3Al layers, where the Ni_5Al_3 layers presented a discontinuous pattern in the Ni_3Al layers. The Ni_5Al_3 phase can nucleate above 600 °C, and the formation of the $Ni/Ni_3Al/Ni_5Al_3$ lamellar structure in Figure 4a was a result of the Ni_5Al_3 phase not completely changing into the equilibrium mixture of the Ni_3Al + NiAl phases above 700 °C [21]. Microvoids were not observed in the middle of the Ni_5Al_3 layers. The average thicknesses of the Ni_3Al and Ni_5Al_3 layers were 21 μm and 2 μm, respectively. Fine Ni_3Al precipitates were observed in the Ni layers and showed a gradient distribution. As the hot pressing temperature rose to 1150 °C (Figure 4b), the average thickness of the Ni_3Al layers gradually decreased to 6 μm, and the Ni_5Al_3 layers were undetected. The average size of the Ni_3Al precipitates in Ni layers gradually increased with the temperature. When the hot pressing temperature rose to 1200 °C, the laminated structure disappeared (Figure 4c). The results suggested that the coarse Ni_3Al precipitates were homogeneously distributed in the Ni layers. The average size of the Ni_3Al precipitates was about 200 μm.

Figure 5 shows the cross-section BSE image of multilayer composites fabricated under various holding times. It can be seen from Figure 5a that the microstructure was mainly composed of Ni, NiAl, and Ni_3Al phases based on our EDS results. The thickness of the Ni layer was the largest, and a NiAl layer was observed in the middle of the Ni_3Al layer with a continuous distribution. The NiAl and Ni_3Al intermetallics were also observed after reaction under heat treatment at high temperature by Kim et al. [7]. The average thicknesses of the Ni_3Al and NiAl layers were 21 μm and 4.2 μm, respectively. The Ni_3Al layer was adjacent to the Ni layer, followed by the NiAl layer. This sequential phase distribution was consistent with the experimental results of our previous study in the high-temperature reaction synthesis stage, suggesting a certain dependence between the Ni and Ni_3Al layers during the reaction. The Ni_3Al phase formed at the interface between the Ni and NiAl phases, and the Ni_3Al precipitates formed in the Ni phase. The Ni_3Al phase formed in the Ni phase in the following orientation: $\{100\}_{Ni3Al}|\{100\}_{Ni}$ and $<001>_{Ni3Al}|<001>_{Ni}$. This resulted in a flat bond between the Ni and Ni_3Al layers [12]. The Ni_3Al matrix was larger closer to the Ni_3Al layer, which contributed to grain boundary diffusion [12]. As the holding time increased to 1 h, the NiAl layer disappeared, and the microstructure only consisted of Ni and Ni_3Al layers. The average thickness of the Ni_3Al layer decreased to 18.3 μm, as shown in Figure 5b. When the holding time increased to 2 h, the average thickness of the Ni_3Al layer decreased to 6 μm. The closer to the Ni_3Al layer, the larger the size and volume fraction of Ni_3Al precipitates, as shown in Figure 5c.

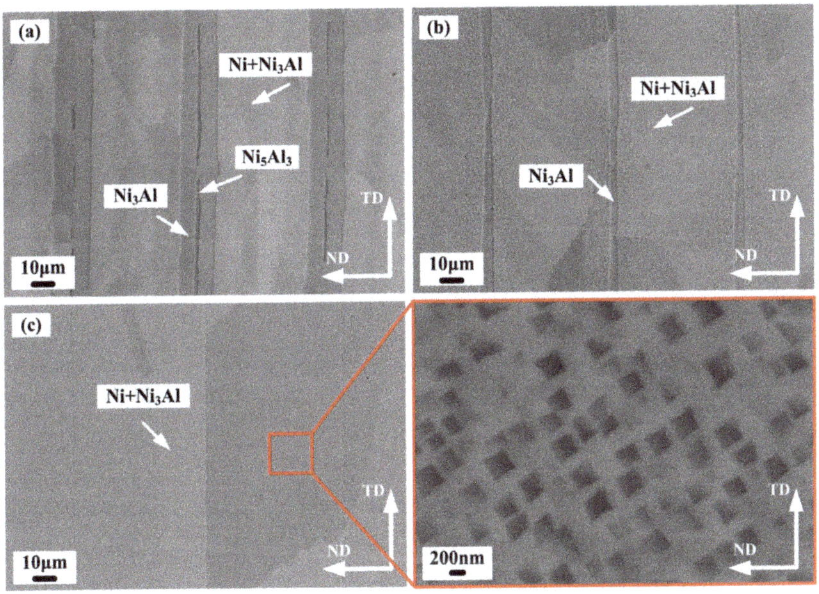

Figure 4. BSE images of the laminated composites. (**a**) N1; (**b**) N2; (**c**) N3.

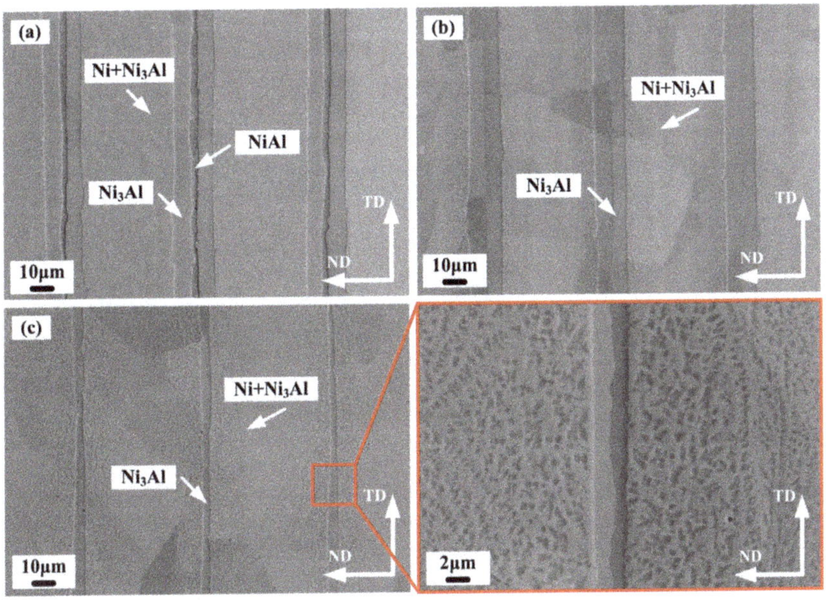

Figure 5. BSE images of the laminated composites: (**a**) N4; (**b**) N5; (**c**) N2.

3.3. *Mechanical Properties of the Laminated Composites*

The typical tensile stress–strain curves of the laminated composites under different hot pressing conditions at room temperature with a strain rate of 1×10^{-3} s^{-1} are shown in Figure 6. The yield strength, ultimate tensile strength, and elongation are listed in Table 3. The yield strength and ultimate tensile strength of the Ni foils with a thickness of 0.06 mm at room temperature were 119.8 MPa and 345.9 MPa, respectively [22]. Additionally, the yield strength and ultimate tensile strength of the Al foils with a thickness of 0.1 mm were 35.6 MPa and 66.8 MPa, respectively [22]. It can be seen that the tensile strength values of the laminated composites were significantly improved compared with the Ni foils and Al foils.

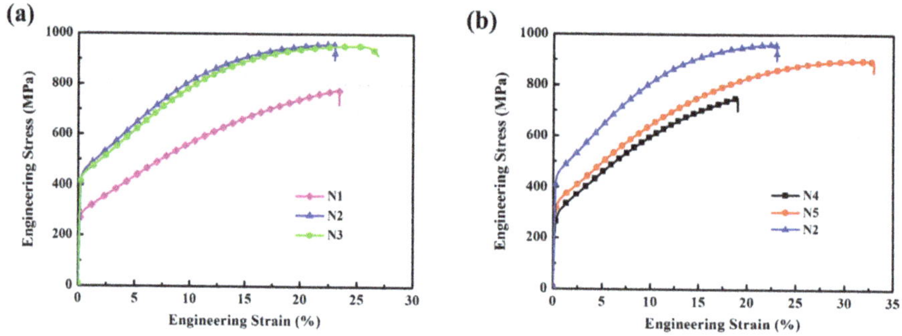

Figure 6. Tensile stress–strain curves of the laminated composites at room temperature for: (**a**) N1, N2, and N3; (**b**) N4, N5, and N2.

Table 3. Tensile properties of the laminated composites at room temperature.

Sample	Yield Strength/MPa	Ultimate Tensile Strength/MPa	Elongation/%
N1	290	778	22.7
N2	443	965	22.6
N3	450	960	27.2
N4	281	748	18.5
N5	319	887	36.5

Figure 6a displays the tensile stress–strain curves of the laminated composites (marked N1, N2, and N3) fabricated under 620 °C/5 MPa/1 h + 1100–1200 °C/10 MPa/2 h. As the hot pressing temperature increased from 1100 °C to 1150 °C, the yield strength increased from 290 MPa to 443 MPa, and the ultimate tensile strength increased from 778 MPa to 965 MPa. However, the yield strength and ultimate tensile strength did not change significantly when the hot pressing temperature further increased from 1150 °C to 1200 °C. Figure 6b displays the tensile strength of the laminated composites (marked N4, N5, and N2) processed under 620 °C/5 MPa/1 h + 1150 °C/10 MPa/0.5–2 h. When the holding time increased from 0.5 h to 1 h, the yield strength increased from 281 MPa to 319 MPa, and the ultimate tensile strength increased from 748 MPa to 887 MPa. When the holding time was increased to 2 h, the yield strength increased to 443 MPa and the tensile strength increased to 965 MPa. It can be seen that the Ni_3Al layers enhanced the room-temperature mechanical properties of the laminated materials. However, improvement in the room-temperature strength of the laminated composites was limited by the continued addition of discontinuously distributed Ni_5Al_3 layers and continuously distributed NiAl layers in the Ni_3Al layers. In addition, Ni_3Al existed as a strengthening phase, mainly in the form of dispersed particles and layers. The effects of these two forms on the strength of the laminated composites were similar when the total content of the Ni_3Al reinforcing phase was constant. However, the effect on elongation enhancement was greater when Ni_3Al

was present in the form of a lamellar reinforcing phase. The greater thickness of the Ni_3Al layers contributed to the strength of the laminated composites but reduced the elongation.

Based on the previous room-temperature tensile tests, the laminated composite sheets exhibited an excellent combination of strength and toughness. The laminated composites prepared under different hot-pressing parameters exhibited different work-hardening abilities during deformation. It was essentially the variability of the microstructure that affected its strength and plasticity [23]. The work-hardening curves are an essential method for statistically evaluating the work-hardening behavior. Figure 7a depicts the plotted work hardening rate vs. true strain for materials fabricated under different processes. The work hardening rate was calculated by $\Theta = (\partial\sigma/\partial\varepsilon)_{\dot{\varepsilon}}$. In the linear hardening stage (Stage I), the strain hardening rate dropped linearly upon increasing the strain for all samples, and the strain hardening rate curves overlapped. Then, the curve of sample N1 dropped linearly during the homogenous hardening stage (Stage II). In this stage, the peaks of the curves for samples N2 and N3 appeared at a strain of 0.05, followed by a rapid decline to necking instability. In contrast to samples N2 and N3, samples N4 and N5 exhibited comparable tendencies, with a strain peak at 0.06, followed by a subsequent decline. The strain hardening rate of samples N2 and N3 was higher than that of the samples N4 and N5 (Figure 7b), indicating the presence of extra strain hardening. The interaction and mutual entanglement of dislocations led to a rapid proliferation of dislocations, which was manifested by a small increase in the strain hardening rate. The increase in samples N2 and N3 was significantly greater than that of N4 and N5. It can be seen that the morphology, size, and distribution of the intermetallic compound reinforcement phases significantly impacted the strain hardening rate of the layered materials. The contribution of the layered NiAl phase was lower than that of the layered Ni_3Al phase. Both the laminated and precipitated Ni_3Al phases increased the work hardening rates of the laminated composites, resulting in an enhanced capacity to form, drag, and pin dislocations. The higher strength and ductility were thought to be the result of heterogeneous deformation-induced strengthening and strain hardening.

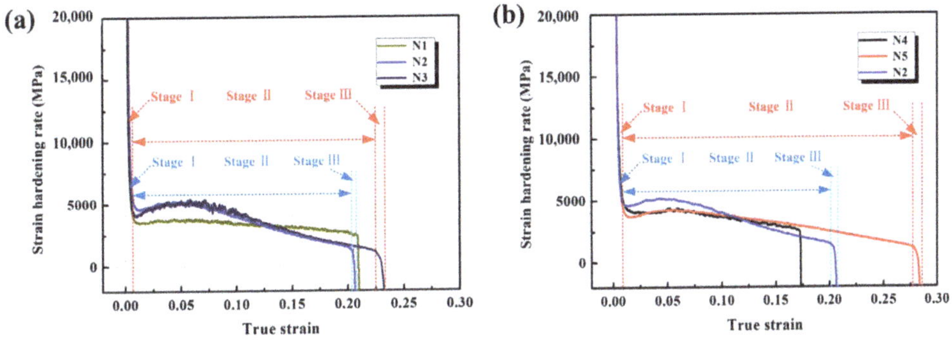

Figure 7. Work hardening rate curves of the sheet at room temperature for (**a**) N1, N2, and N3; (**b**) N4, N5, and N2.

Figure 8 shows the tensile stress–strain curves of the laminated composites obtained under different hot pressing parameters tested at elevated temperatures. The yield strength, ultimate tensile strength, and elongation of the laminated composites at temperatures ranging from 800 °C to 1000 °C and strain rate of 1×10^{-3} s^{-1} are listed in Table 4. The thickness of the NiAl and Ni_3Al layers significantly affected the high-temperature tensile strength of the laminated composites. The thickness of the intermetallic compound layer was positively correlated with the strength of the laminated composites at 1000 °C. The effect of the Ni_3Al layer on improving the high-temperature strength of the laminated composites was greater than that of the NiAl layer, but the Ni_3Al layer did not improve the elongation. The results showed that the dispersed Ni_3Al phase had a greater effect on the strength of the laminated composites than when in the form of a layered structure. When the Ni_3Al phase was distributed in a laminar structure, NiAl and Ni_3Al were thinner, which resulted in a higher strength and greater elongation of the laminated composites.

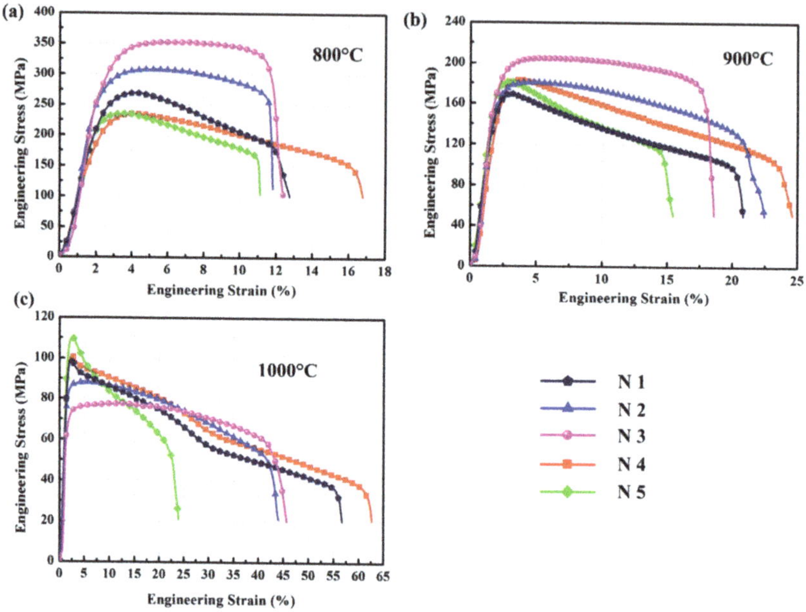

Figure 8. Tensile stress–strain curves of the laminated composites at elevated temperatures: (**a**) 800 °C; (**b**) 900 °C; (**c**) 1000 °C.

Table 4. Tensile properties of the laminated composites at elevated temperatures.

Sample	Temperature/°C	Yield Strength/MPa	Ultimate Tensile Strength/MPa	Elongation/%
N1	800	216	269	12.8
	900	145	169	20.8
	1000	94	99	56.6
N2	800	248	307	11.8
	900	142	181	22.4
	1000	76	89	43.8
N3	800	263	352	12.4
	900	156	204	18.6
	1000	64	77	45.6
N4	800	192	234	16.8
	900	137	183	24.6
	1000	92	101	62.6
N5	800	203	235	11.1
	900	158	181	15.4
	1000	99	110	23.9

3.4. Fracture Characteristics of the Laminated Composites

Figure 9 depicts the fracture morphologies after uniaxial tensile tests at different temperatures. The sample was fabricated under 620 °C/5 MPa/1 h + 1100 °C/10 MPa/2 h. As shown in Figure 9a, plastic and intermetallic compound layers were clearly distributed at the fracture sites of the specimens at room temperature. The fracture surface was relatively rough and revealed noticeable grooves, and the specimens failed in a jagged fracture mode along the loading direction. Many dimples were present at the fracture site of the Ni layers, and the height of the failed Ni layers was greater than that in the intermetallic compound layers, which indicated that final local ductile failure occurred in the Ni layer. In addition, clear delamination was observed between the Ni layer and the intermetallic compound layer, indicating that the dislocation density at the interface exceeded the bonding strength of the interface. The release of local stress was achieved through debonding. Figure 9b–d show that the intermetallic layers cracked parallel to the direction of the applied force under tensile stress at higher temperatures. The intermetallic layers experienced delamination only near the fracture surface, and nickel layers underwent ductile fracture. The morphology of the high-temperature tensile specimen fractured at 700–800 °C was significantly different from that of the room-temperature fracture. The depth of the grooves on the fracture surface was significantly lower than that of the room-temperature fracture. Upon increasing the temperature, the interface between the Ni plastic layer and the intermetallic compound layer at the fracture surface was gradually blurred, and delamination tended to be weaker. This indicated that the plastic deformation capacity of both the Ni layer and the intermetallic compound layer was improved at high temperatures. The improved synergistic deformation ability was accompanied by the mitigation of local stress concentration at the interface.

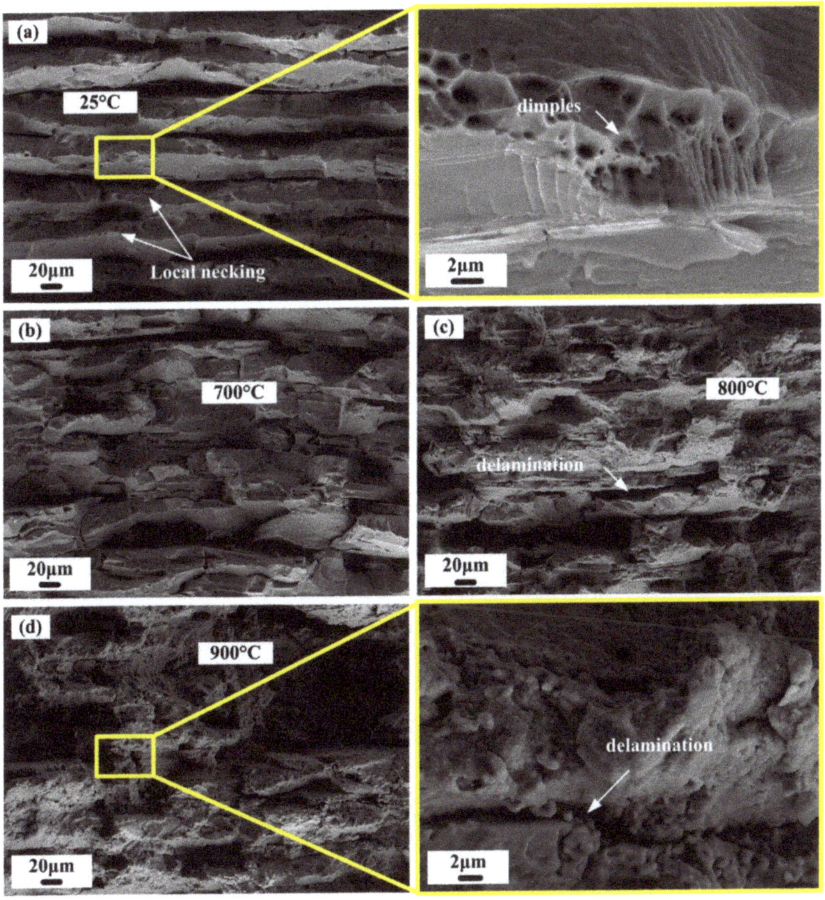

Figure 9. Fracture morphology of the tensile specimens at (**a**) 25 °C; (**b**) 700 °C; (**c**) 800 °C; (**d**) 900 °C.

The fracture behavior of the laminated composites can be explained using the classical cantilever beam mechanics model [24], as follows:

$$\sigma = \frac{10}{11} \cdot \frac{\delta \cdot E \cdot t_{IMCs}^{3}}{L^{4}} \tag{1}$$

where σ is out-of-plane stress perpendicular to the thickness direction between Ni and the intermetallic layers (IMCs), δ is the deflection of the intermetallic layers, t_{IMCs} is the thickness of the intermetallic layers, L is the length of the cantilever beam, and E is the Young's modulus of the intermetallic layers.

The interlayer normal stress was positively correlated with the deflection at the fracture of the IMCs for the same strain at fracture. The deformation temperature was negatively correlated with the deflection at the fracture of the IMCs. Therefore, decreasing the temperature increased the normal stress between the layers; the increase in temperature, in turn, decreased the normal stress between the layers. δ/σ can be used to represent the confinement effect between the interface of the Ni layer and the IMCs. Thus, it can be concluded that δ/σ was positively correlated with the strain at fracture, i.e., the larger the strain, the stronger the interfacial confinement effect on the material. The interlayer confinement was enhanced at low strain rates or high temperatures. Conversely, the interlayer confinement effect was weakened.

3.5. *Interfacial Evolution during Hot Tensile Deformation*

During hot tensile deformation, the interface and each group element shared the load transfer, strain distribution, and control during plastic deformation. The interface played an important role in coordinating the deformation among the group members. The inhomogeneous strain distribution was often related to the accumulation and stacking of geometrically necessary dislocations (GNDs) during plastic deformation. Therefore, GNDs in Ni and Ni_3Al layers and their evolution trends can be obtained by quantitative calculations of the EBSD data. The kernel average misorientation (KAM) method can be used to calculate the local misorientation from the EBSD data to indirectly obtain the deformation laws under the limiting effect of the laminate materials. Misorientations greater than 3° were not taken into account during the calculation of local misorientation to exclude subgrain boundaries. The local misorientation is defined as the average misorientation between a central point (0.1 × 0.1 μm²) and its eight surrounding points:

$$\theta_{local} = \sum_{i=1}^{8} \theta_i \cdot I_{(\theta_i<\alpha)} / \sum_{i=1}^{8} I_{(\theta_i<\alpha)} \tag{2}$$

where θ_{local} is the calculated local misorientation for the corresponding point, and θ_i is the misorientation between point i and its surrounding points; $I_{(\theta_i<\alpha)}$ is an indicator function, and α is the misorientation threshold (3 here). The GND density of a pixel (ρ_{GND}) is presented by the Kubin strain gradient theory [25]:

$$\rho_{GND} = \frac{2\theta}{ub} \tag{3}$$

where ρ_{GND} is the GND density, θ is the local misorientation, u is the unit length (100 nm here), and b is the Burger vector (0.253 nm for face-centered cubic, and 0.248 nm for body-centered cubic). Figure 10 shows the GND density maps of the laminated composites at different tensile strains hot pressed under 620 °C/5 MPa/1 h + 1100 °C/10 MPa/2 h. Overall, the density of the GND increased significantly upon increasing the tensile strain. This was a typical deformation phenomenon and was the same as the results reported in a previous study [25]. The GND density of the Ni layers increased from 4.19×10^{14} m^{-2} to 5.74×10^{14} m^{-2} after deformation. The GND density of the intermetallic layers increased from the original 4.81×10^{14} m^{-2} to 6.48×10^{14} m^{-2}, with higher values than that of the Ni layer. However, the GND density distribution was non-uniform and showed a regular gradient distribution. The GND density in the Ni layer was significantly lower than that in the intermetallic layer. The non-homogeneity specific to the material organization inside, such as grain boundaries, phase interfaces, and crystal orientation, resulted in an inhomogeneous GND density during plastic deformation. The GND density was closely related to the strain gradient, indicating that the intermetallic layers needed to adapt to the overall deformation by more GND in the form of dislocation stacking when the laminate material was deformed in concert.

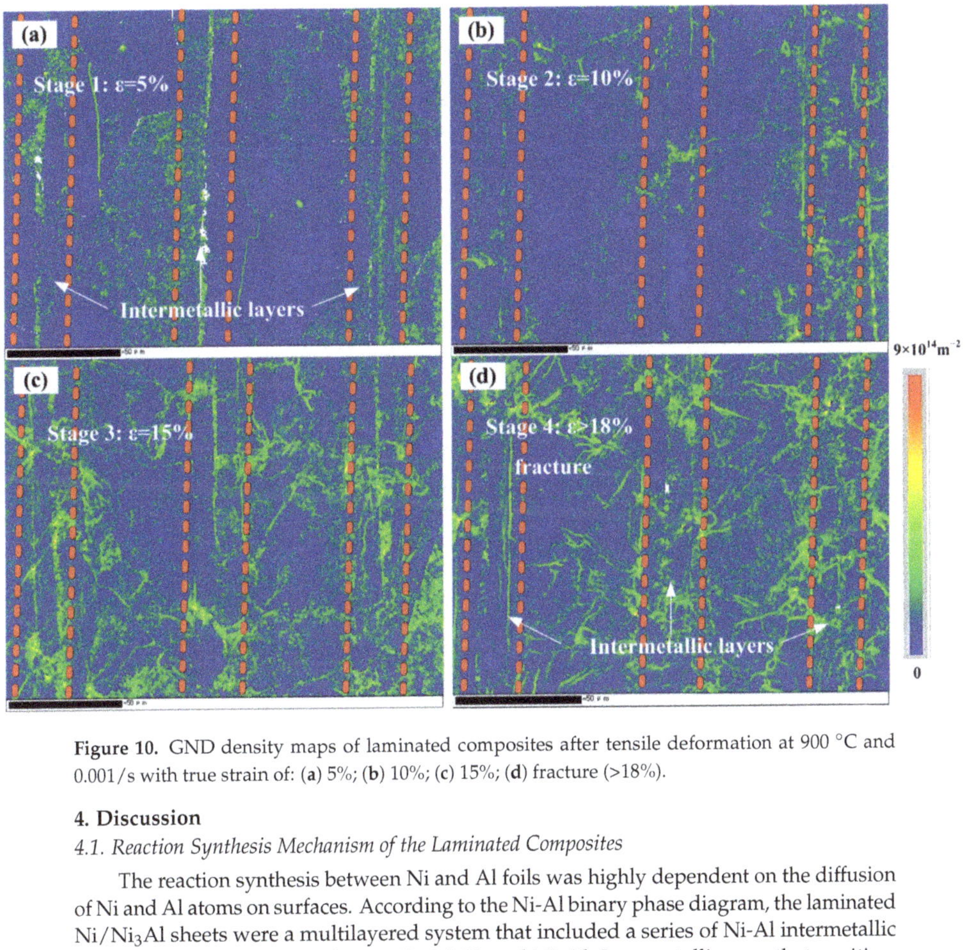

Figure 10. GND density maps of laminated composites after tensile deformation at 900 °C and 0.001/s with true strain of: (**a**) 5%; (**b**) 10%; (**c**) 15%; (**d**) fracture (>18%).

4. Discussion

4.1. Reaction Synthesis Mechanism of the Laminated Composites

The reaction synthesis between Ni and Al foils was highly dependent on the diffusion of Ni and Al atoms on surfaces. According to the Ni-Al binary phase diagram, the laminated Ni/Ni_3Al sheets were a multilayered system that included a series of Ni-Al intermetallic compounds, including $NiAl_3$, Ni_2Al_3, NiAl, and Ni_3Al. Intermetallic growth, transition, and stabilization were the three vital stages of intermetallic layer generation. According to previous reports [26–28] and our studies [22,29], the synthesis from step 1 to step 4 is illustrated below:

$$\text{Stage 1}: \ Ni + Al \rightarrow NiAl_3 + \Delta H_1 \quad (4)$$

$$\text{Stage 2}: \ NiAl_3 + Ni \rightarrow Ni_2Al_3 + \Delta H_2 \quad (5)$$

$$\text{Stage 3}: \ Ni_2Al_3 + Ni \rightarrow NiAl + \Delta H_3 \quad (6)$$

$$\text{Stage 4}: \ NiAl + Ni \rightarrow Ni_3Al + \Delta H_4 \quad (7)$$

Figure 11 displays a schematic illustration of the reaction synthesis mechanism of the laminated composites. The temperature-induced reaction synthesis can be divided into four steps, ranging from room temperature to 1150 °C. In the first stage, the $NiAl_3$ layer emerged at the Ni/Al interface and then developed continuously along the thickness direction. In this stage, the diffusion reaction occurred rapidly, followed by the second stage. $NiAl_3$ was an intermediate product in this process, but the $NiAl_3$ and Ni_2Al_3 phases usually coexisted. Ni/Al underwent a long-term solid-state reaction synthesis at a constant temperature of 620 °C. In this stage, the Al layer provided sufficient Al atoms for the diffusion reaction of the $NiAl_3$ and Ni_2Al_3 phases, resulting in a continual increase in the thickness of the $NiAl_3$ and Ni_2Al_3 layers. However, as the thickness of the Al layer gradually diminished, the supply of Al atoms also gradually diminished, resulting in the development and expansion of Ni_2Al_3, which depended even more on the $NiAl_3$ layer. When the reaction synthesis temperature reached 854 °C, a small amount of a liquid–solid mixed phase formed at the phase boundary between $NiAl_3$ and Ni_2Al_3, leading to an intensification of the reaction. The $NiAl_3$ layer subsequently disappeared and developed into a $Ni/Ni_2Al_3/Ni$ lamellar structure. Upon increasing the temperature, the NiAl phase gradually generated and occupied the position of Ni_2Al_3 phase. When the temperature increased to 1133 °C, a solid–liquid mixture appeared in the reaction zone. At this time, the decomposition efficiency and reaction rate of the remaining Ni_2Al_3 phase reached an extreme value. In the third and fourth stages, as the NiAl phase was generated, the Ni_3Al phase formed at the Ni/NiAl phase boundary, which was related to the controlled induction of grain boundary diffusion. As the diffusion time increased, the thickness of the NiAl phase gradually decreased until it finally disappeared, and a $Ni/Ni_3Al/Ni$ lamellar structure formed. The Ni_3Al phase existed mainly as a lamellar structure or in the form of precipitates in the Ni matrix. As the diffusion time continued to increase, the thickness of the Ni_3Al layer gradually decreased until it finally disappeared. However, the volume fraction of Ni_3Al precipitates in the Ni matrix showed a gradient distribution in the thickness direction. When the diffusion time was sufficient, the volume fraction of Ni_3Al precipitates in the Ni matrix was gradually homogenized.

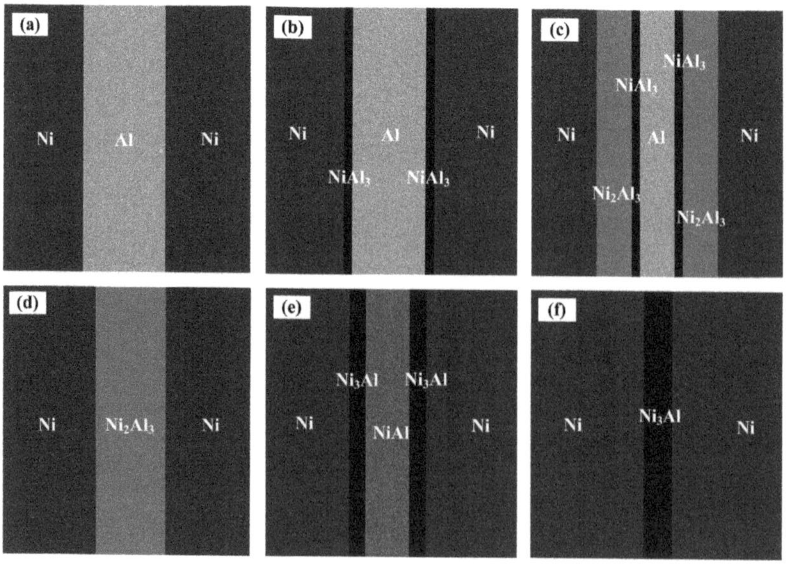

Figure 11. Schematic illustration of the reaction synthesis mechanism of the laminated composites. (**a**) The hot-pressed Ni/Al/Ni laminated composites; (**b**) $NiAl_3$ layers are formed between Ni and Al layers; (**c**) Ni_2Al_3 layers are formed between $NiAl_3$ and Ni layers; (**d**) $NiAl_3$ disappears and forms the $Ni/Ni_2Al_3/Ni$ lamellar structure; (**e**) Ni_3Al layers are formed between NiAl and Ni layers; (**f**) NiAl disappears and forms the $Ni/Ni_3Al/Ni$ lamellar structure.

4.2. Mechanical Response of the Laminated Structure

The layer –thickness relationship of the laminated structure and its matching pattern were important factors affecting the strength and elongation of the laminated composites. The strength and strain-hardening capacity of the Ni and Ni_3Al layers were very different. As a result, the laminated composites displayed significant heterogeneity during overall deformation, in contrast to the homogeneous material. During the deformation process, the Ni layer first yielded, and the Ni_3Al layer underwent elastic deformation. The deformation of the Ni layer was restricted by the Ni_3Al layer. When sufficient interfacial bonding was maintained between the Ni and Ni_3Al layers, the laminated material underwent simultaneous plastic deformation, and a large strain gradient appeared near the interface between the Ni and Ni_3Al layers. This was fully verified by the experimental results shown in Figure 10. At the initial stage of deformation, significant dislocation density accumulation was observed near the intermetallic compound layer and the interface region, demonstrating the storage of GNDs in the heterogeneous interface region. Upon increasing the plastic strain, the inhomogeneous distribution of GNDs at the layer interface was obvious, and the heterogeneous structure led to a complex stress state transmitted by the additional stresses near the interface. This contributed to dislocation propagation and entanglement, and the stresses began to gradually expand toward the Ni layer along the thickness direction. The GNDs in the Ni layer originated from the generation of their own plastic deformation and accumulation when accommodating the strain gradient, according to the strain gradient plasticity theory [25].

The average thicknesses of the Ni_3Al layers in samples N2 and N5 were about 6 μm and 18.3 μm, respectively. The average thicknesses of the Ni layers in samples N2 and N5 were about 65.4 μm and 55.2 μm, respectively. The stress of the laminates was influenced by the thickness of each layers according to the rule of mixtures [30,31].

$$\sigma_m = V_i\sigma_i + (1 - V_i)\sigma_{\mathrm{Ni}} \tag{8}$$

$$V_i = \frac{t_i}{t_i + t_{\mathrm{Ni}}} \tag{9}$$

where σ_m is the fracture strength of the material, σ_{Ni} is the fracture strength of the Ni layer, σ_i is the fracture strength of the Ni_3Al intermetallic layer, V_i is the volume fraction of the Ni_3Al intermetallic, t_i and t_{Ni} refer to the thicknesses of the Ni_3Al intermetallic layer and Ni layer, respectively, measured by SEM. Therefore, it is very important to control the microstructure and thickness of the obtained Ni_3Al layers. In samples N2 and N5, the thickness of the Ni layer increased, and the thickness of the Ni_3Al layer decreased in the non-homogeneous laminated material. GNDs were accommodated more fully in the Ni layer, producing a significant hetero-deformation-induced strengthening and hardening effect, which in turn increased the yield strength at room temperature. This enhancement continued until near 900 °C. At higher temperatures (1000 °C), the hetero-deformation-induced strengthening and hardening effect of GNDs in the Ni layer decreased, while the high-strength effect of Ni_3Al was fully manifested. Thus, the desired strengthening effect was not achieved. This conclusion was confirmed by experiments (combined with samples N2 and N4). In other words, the material's strength could not be enhanced by simply adding intermetallic compound phase layers, which required the design of matching models for the Ni_3Al, NiAl, and Ni layers. The structural form of the Ni_3Al phase had a significant effect on the strength and plasticity of the stacked material, as illustrated by combining samples N2 and N3. The laminated Ni_3Al phase had a significant high-temperature strengthening effect compared with the uniform distribution in the Ni matrix, but it could not achieve a high strength effect at room temperature or a low temperature. Figure 12 shows the microstructure schematic of different phase structures. The insight gained from this experimental phenomenon was that the reinforcement form of the material should not be limited to a lamellar structure but should be enhanced by adding different reinforcement effects to achieve overall strengthening at all temperature intervals. This, in turn, provided a reference for the preparation of high-performance metal–intermetallic compound materials.

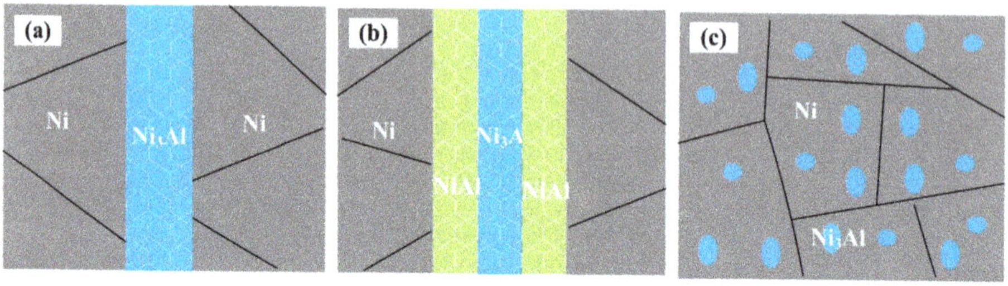

Figure 12. Microstructure schematics of different phase structure forms. (**a**) Dual-phase with Ni matrix layer phases that contains Ni_3Al layers; (**b**) ternary-phase with Ni matrix layers phase that contains Ni_3Al and NiAl layers; (**c**) dual-phase with Ni matrix phase that contains Ni_3Al particles.

5. Conclusions

The Ni/Ni_3Al laminated composites were successfully fabricated by hot pressing with 0.6 mm Ni and 0.02 mm Al foils through an in situ solid-state reaction synthesis. The main conclusions are summarized as follows:

(1) The evolution of the interface phases in the Ni/Ni_3Al laminated composite was shown to be $NiAl_3$, Ni_2Al_3, NiAl, and Ni_3Al in sequence. An excellent interface between the Ni and Ni_3Al layers without cracks and voids was observed due to the uniform pressure applied during hot pressing.

(2) The laminated composites hot pressed under 620 °C/5 MPa/1 h + 1150 °C/10 MPa/2 h exhibited a better ultimate tensile strength of 965 MPa and an elongation of 22.6% at room temperature, accompanied with the high-temperature tensile properties with an ultimate tensile strength of 89 MPa and an elongation of 43.8% at 1000 °C.

(3) The layer–thickness relationship of the laminated structure and its matching pattern were important factors affecting the strength and elongation of the laminated composites over a wide range of temperatures, i.e., the alternating Ni_3Al and the Ni matrix layers, as well as the dispersed Ni_3Al precipitate phases.

Future work will elucidate the mechanical response mechanism of different phase structures with quantitative index and establish the correspondence relationship between Ni/Al foil thickness ratios and mechanical properties.

Author Contributions: Y.S.: methodology, formal analysis, investigation, and writing—original draft. S.Y.: conceptualization, and supervision. All authors have read and agreed to the published version of the manuscript.

Funding: This research was supported by the National Natural Science Foundation of China (No. 52105321) and the China Postdoctoral Science Foundation (No. 2021M700659 and No. 2022T150082).

Data Availability Statement: The data presented in this study are available upon request from the corresponding author.

Acknowledgments: The authors would like to thank Jicai Feng and Lixia Zhang of Harbin Institute of Technology.

Conflicts of Interest: The authors declare no conflict of interest.

References

1. Czeppe, T.; Wierzbinski, S. Structure and mechanical properties of NiAl and Ni3Al-based alloys. *Int. J. Mech. Sci.* **2000**, *42*, 1499–1518. [CrossRef]
2. Vecchio, K.S. Synthetic multifunctional metallic-intermetallic laminate composites. *Jom* **2005**, *3*, 25–31. [CrossRef]
3. Wang, Y.; Wang, H.; Liu, X.; Vecchio, K.S. Microstructure evolution in pure Ni and Invar-based metallic-intermetallic laminate (MIL) composites. *Mater. Sci. Eng. A* **2017**, *682*, 454–465. [CrossRef]

4. Konieczny, M.; Mola, R.; Thomas, P.; Kopcial, M. Processing, microstructure and properties of laminated Ni-intermetallic composites synthesized using Ni sheets and Al foils. *Arch. Metall. Mater.* **2011**, *56*, 693–702. [CrossRef]
5. Rawers, J.C.; Hansen, J.S.; Alman, D.E.; Hawk, J.A. Formation of sheet metal-intermetallic composites by self-propagating high-temperature reactions. *J. Mater. Sci. Lett.* **1994**, *13*, 1357–1360. [CrossRef]
6. Alman, D.E.; Dogan, C.P.; Hawk, J.A.; Rawers, J.C. Processing, structure and properties of metal-intermetallic composites. *Mater. Sci. Eng. A* **1995**, *192*, 624–632. [CrossRef]
7. Kim, H.Y.; Chung, D.S.; Hong, S.H. Reaction synthesis and microstructures of NiAl/Ni micro-laminated composites. *Mater. Sci. Eng. A* **2005**, *396*, 376–384. [CrossRef]
8. Qiu, X.; Liu, R.; Guo, S.; Graeter, J.H.; Kecskes, L.; Wang, J. Combustion Synthesis Reactions in Cold-Roll ed Ni/Al and Ti/Al Multilayers. *Metall. Mater. Trans. A* **2009**, *40*, 1541–1546. [CrossRef]
9. Kim, H.; Chung, D.; Hong, S. Intermixing criteria for reaction synthesis of Ni/Al multilayered microfoils. *Scr. Mater.* **2006**, *54*, 1715–1719. [CrossRef]
10. Qiu, X.; Wang, J. Experimental evidence of two-stage formation of Al3Ni in reactive Ni/Al multilayer foils. *Scr. Mater.* **2007**, *56*, 1055–1058. [CrossRef]
11. Mumtaz, K.; Echigoya, J.; Nakata, C.; Husain, S.W. Effect of cold rolling and subsequent annealing on hot pressed Ni/Al laminates. *J. Mater. Sci.* **2001**, *36*, 3981–3987. [CrossRef]
12. Seyring, M.; Rettenmayr, M. Impact of crystallography at Ni/NiAl interfaces on the nucleation of Ni3Al. *Acta Mater.* **2021**, *208*, 116713. [CrossRef]
13. Wu, Y.; Li, C.; Li, Y.; Wu, J.; Xia, X.; Liu, Y. Effects of heat treatment on the microstructure and mechanical properties of Ni3Al-based superalloys: A review. *Int. J. Miner. Metall. Mater.* **2021**, *28*, 553–566. [CrossRef]
14. Wu, J.; Yu, J.; Wang, W.; Gao, S.; Luo, J.; Li, C.; Ma, Z.; Liu, Y. Influences of laser remelt-ing treatments on microstructure and tensile properties of multiphase Ni3Al-based in-termetallic alloy. *Mater. Sci. Eng. A* **2020**, *857*, 144087. [CrossRef]
15. Ogneva, T.S.; Bataev, I.A.; Mali, V.I.; Anisimov, A.G.; Lazurenko, D.V.; Popelyukh, A.I.; Emurlaeva, Y.Y.; Bateaev, A.A.; Tanaka, S.; Yegoshin, K.D. Effect of sintering pressure and temperature on structure and properties of Ni single bond Al metal-intermetallic composites produced by SPS. *Mater. Charact.* **2021**, *180*, 111415. [CrossRef]
16. Konieczny, M. Microstructural characterisation and mechanical response of laminated Ni-intermetallic composites synthesised using Ni sheets and Al foils. *Mater. Charact.* **2012**, *70*, 117–124. [CrossRef]
17. Mizuuchi, K.; Inoue, K.; Sugioka, M.; Itami, M.; Lee, J.; Kawahara, M. Properties of Ni-aluminides-reinforced Ni-matrix laminates synthesized by pulsed-current hot pressing (PCHP). *Mater. Sci. Eng. A* **2006**, *428*, 169–174. [CrossRef]
18. Kim, H.Y.; Chung, D.S.; Enoki, M.; Hong, S.H. Tensile and fracture properties of NiAl/Ni micro-laminated composites prepared by reaction synthesis. *J. Mater. Res.* **2006**, *21*, 1141–1149. [CrossRef]
19. Wang, H.B.; Han, J.C.; Du, S.Y.; Northwood, D.O. Effects of Ni foil thickness on the microstructure and tensile properties of reaction synthesized multilayer composites. *Mater. Sci. Eng. A* **2007**, *445–446*, 517–525. [CrossRef]
20. Xi, X.; Lin, P.; Meng, L.; Xue, S.; Lin, F.; Zhang, Z.; Wang, T. The microstructure and mechanical properties of reaction synthesized Ni-Al intermetallic sheets by pure Ni/Al foils with different thickness ratios. *J. Alloy Compd.* **2022**, *902*, 163750. [CrossRef]
21. Lasek, J.; Chraska, T.; Krecek, P.; Bartuska, P. Phase transformations in nickel rich NiAl using electrical resistivity measurements. *Scr. Mater.* **1997**, *37*, 897–902. [CrossRef]
22. Sun, Y.; Lin, P.; Yuan, S.J. A novel process for integrated forming and reaction synthesis of NiAl alloy curved shells. *J. Mater. Process. Technol.* **2020**, *285*, 116798. [CrossRef]
23. Chen, M.; Fan, G.; Tan, Z.; Yuan, C.; Guo, Q.; Xiong, D.; Chen, M.L.; Zheng, Q.; Li, Z.Q.; Zhang, D. Heat treatment behavior and strengthening mechanisms of CNT/6061Al composites fabricated by flake powder metallurgy. *Mater. Charact.* **2019**, *153*, 261–270. [CrossRef]
24. Liang, F.; Tan, H.; Zhang, B.; Zhang, G. Maximizing necking-delayed fracture of sandwich-structured Ni/Cu/Ni composites. *Scr. Mater.* **2017**, *134*, 28–32. [CrossRef]
25. Kubin, L.P.; Mortensen, A. Geometrically necessary dislocations and straingradient plasticity: A few critical issues. *Scripta. Mater.* **2003**, *48*, 119–125. [CrossRef]
26. Xia, Z.; Liu, J.; Zhu, S.; Zhu, S.; Zhao, Y. Fabrication of laminated metal–intermetallic composites by interlayer in-situ reaction. *J. Mater. Sci.* **1999**, *34*, 3731–3735. [CrossRef]
27. Zhu, P.; Li, J.; Liu, C.T. Reaction mechanism of combustion synthesis of NiAl. *Mater. Sci. Eng. A* **2002**, *329*, 57–68. [CrossRef]
28. Zhu, H.X.; Abbaschian, R. Reactive processing of nickel-aluminide intermetallic compounds. *J. Mater. Sci.* **2003**, *38*, 3861–3870. [CrossRef]
29. Sun, Y.; Lin, P.; Yuan, S.J. A novel method for fabricating NiAl alloy sheet components using laminated Ni/Al foils. *Mater. Sci. Eng. A* **2019**, *754*, 428–436. [CrossRef]
30. Cohades, A.; Cetin, A.; Mortensen, A. Designing laminated metal composites for tensile ductility. *Mater. Des.* **2015**, *66*, 412–420. [CrossRef]
31. Lyu, S.; Sun, Y.; Li, G.; Xiao, W.; Ma, C. Effect of layer sequence on the mechanical properties of Ti/TiAl laminates. *Mater. Des.* **2018**, *143*, 160–168. [CrossRef]

materials

MDPI

Article

Surface Roughening Behavior of the 6063-T4 Aluminum Alloy during Quasi-in Situ Uniaxial Stretching

Yang Cai [1,2], Xiaosong Wang [1,*] and Yan Du [2]

[1] School of Materials Science and Engineering, Harbin Institute of Technology, Harbin 150090, China
[2] School of Materials Science and Engineering, Heilongjiang University of Science and Technology, Harbin 150022, China
* Correspondence: hitxswang@hit.edu.cn; Tel.: +86-0451-86415754

Abstract: Owing to orange-peel defects, the industrial application of light alloy structural members is significantly restricted. In this study, a quasi-in situ axial tensile experiment was conducted on a 6063-T4 aluminum alloy sample. The surface morphology and microstructure evolution of the tagged area were scanned simultaneously using laser scanning confocal microscopy and electron backscattered diffraction, and the surface roughening behavior of the polycrystal aluminum alloy surface, caused by deformation, was quantitatively analyzed. As the concave–convex features at the surface appear in pairs with increasing global strain, the width of the concave features increases, whereas that of the convex features decreases gradually, resulting in the initially increasing surface roughness, which subsequently remains unchanged. During the stretching process, the small-sized grains in the 37~102 μm range show weak strain localization and the highest coordination of deformation. The deformation mode of medium-sized grains in the 114–270 μm range tends to grain deflection, and others tend to slip.

Keywords: surface roughening; orange peel; uniaxial stretching; in situ-EBSD; polycrystal aluminum alloy; deformation coordination

1. Introduction

The production of lightweight components has become a point of discussion in advanced manufacturing. They have been developed for the application in new energy vehicles that help reduce energy consumption and realize resource economization and efficiency [1,2]. Orange-peel defects, which are surface defects, are often observed on the surfaces of lightweight components, such as aluminum, magnesium, and titanium [3,4], as shown in Figure 1. Distinct deteriorations in surface roughness [5], surface performance [6], weldability, fatigue life [7–9], and even forming limits [10,11] arise from roughened surfaces. Thus, the application of light alloy components in the industry is considerably restricted.

Surface roughening is a commonly occurring surface phenomenon with a series of concave–convex features and is caused by plastic deformation. It is known as a surface orange-peel defect when the surface roughening sufficiently grows to a macroscopically visible defect [12]. The surface roughening of polycrystalline alloys [13] is affected by two aspects: plastic deformation (including stress and strain state [14], stress–strain path [15], strain rate, and work-hardening ability) and microstructure (grain size, grain orientation, texture, microdefects [16], and other factors).

Experiments have demonstrated that the surface roughness increases with increasing plastic strain [17], or first increases then gradually decreases at a maximum roughness [18] for pure copper and pure aluminum. The evolution trend of surface roughness is determined by the coordination deformation ability that predominantly depends on grain size. Generally, the strong matrix texture and grain refinement are both advantageous for deformation coordination [19,20].

Fundamentally, the formation of each bump or pit on a rough surface is closely related to the slip system, precipitations, and inclusions of the grains. The peak appears near the grain boundary, where it is difficult for slippages occurring [21]. The presence of inclusions in a subsurface layer also affects the surface features, and bumps and pits form above the hard and soft inclusions, respectively [22,23]. The height of the surface undulations is affected by the inclusion-to-surface distance [24]. Soft-oriented grains are more prone to plastic deformation than hard-oriented grains [25].

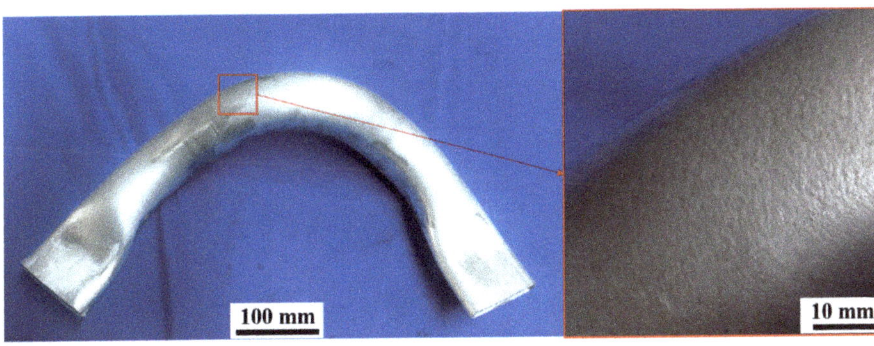

Figure 1. Aluminum alloy structural member and "orange-peel" defects.

In addition, the coordination deformation capability is also related to the strain state, such as uniaxial stretching, biaxial stretching [26], and bending deformation [27], and results in different strain localization. The degree of surface roughness produced in the biaxial stress state is significantly less than that in the uniaxial stress state [28–30]. The strain localization is a primary factor that induces surface roughening [31,32].

However, few studies have analyzed the quantitative relationship between macroscopic surface roughness and strain, dynamic evolution between surface roughening behavior, and grain-size effect. In this study, the surface morphology and microstructure of an aluminum alloy sample were observed via quasi-in situ tensile experiments. The surface morphology evolution process and corresponding microstructure were both quantificationally investigated. The dimensional variations of bumps and pits were qualitatively discussed, and the grain-orientation variations and strain localization were quantitatively analyzed in every size range in this study, thereby laying a theoretical foundation for avoiding orange-peel defects on the surfaces of light alloy components.

2. Experiments

2.1. Quasi-in Situ Tension Testing

To study the formation mechanism of surface roughening behavior and the phenomenon of concave–convex features caused by the plastic deformation of polycrystal aluminum alloy, a quasi-in situ tensile experiment was conducted. The experimental material was a 6063-T4 aluminum alloy extruded tube. The axial direction of the tube is denoted as AD, the tangential direction as TD, and the normal direction as ND. The in situ tensile specimens were prepared using electrical discharge machining wire cutting along the AD of the tube, and the thickness along the ND was 1 mm. The size and sampling direction are shown in Figure 2a, and the available tensile area of the specimen is 6 × 4 mm. The tensile test was performed using an electronic universal material testing machine (Instron 5569R, Norwood, MA, USA) at a stretching velocity of 1 mm/min at 25 °C. The strain rate negligibly influenced the microstructure at a temperature in the range of −20 °C to 20 °C [33]. The brief unloading was performed at tensile deformations of 2.14%, 6.32%, 9.62%, and 14.12% for SEM and EBSD testing. Photographs of the in situ tensile specimens before and after stretching are shown in Figure 2b, showing that the specimen surface resembles a mirror before stretching, and then becomes a rough surface after stretching.

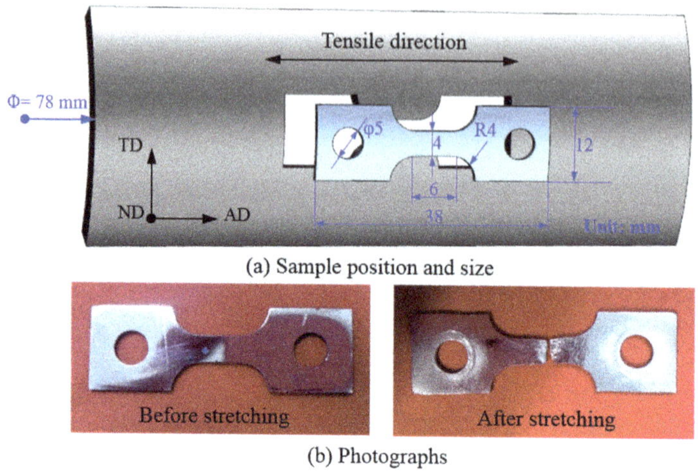

Figure 2. (**a**) Sample position and size. (**b**) Photographs of quasi-in situ sample.

2.2. Surface Roughness Characterization

The surface roughness (R_a) was characterized within an area of 2560 × 2560 μm to reasonably represent surface roughening and was tested using laser scanning confocal microscopy (OLS 3000, Tokyo, Japan). The magnification of the microscope ranges from 120× to 14,400×, and it was used to characterize the surface morphology and surface roughness at different strain stages. The surface roughness, R_a, of the sample is the arithmetic average of the roughness profile [34]:

$$R_a = \frac{1}{l}\int_0^l |f(x)|dx \tag{1}$$

where l is the sampling length and $f(x)$ is the offset distance from the profile within the sampling length. The surface roughness of the polished surface was measured before stretching and was observed to be 0.25 μm. It is considered that the surface roughness of a mirror surface does not exceed 0.4 μm; therefore, it can be assumed that the surface of the sample in this experiment was a mirror before stretching [35].

2.3. Microstructure Characterization

Electron backscattered diffraction (EBSD) was performed using the field emission environmental scanning electron microscopy (FE-SEM, Quanta 200FEG, FEI, Hillsboro, OR, USA) with a step size of 5.5 μm. Before EBSD analysis, the tensile specimens were electropolished. The electrolyte ratio was 10% $HClO_4$ to 90% C_2H_5OH, the temperature was approximately −20 °C, the voltage was 15 V, and the polishing time was approximately 50 s. However, the SEM and EBSD areas were not selected as the same area, and only the changes in surface morphology and microstructure were observed.

3. Results and Discussion

3.1. Surface Morphology

A significant advantage of in situ-SEM is the ability to show the real-time corresponding relations between the concave–convex features, slip lines, and global strain. The line profiles can be obtained by 2D images using laser scanning confocal microscopy. Figure 3a–d shows that the concave and convex features simultaneously and continuously appear in the surface morphology. The corresponding positions are predominantly formed with shallow concave features at the initial deformation stage and are indicated using the green dashed line, and adjacent concaves are indicated using red lines. As stretching progresses, the concave and convex features appear more distinctly. Numerous slip traces (indicated by blue "X") are observed at a global strain of 2.14%, as shown in Figure 3a. There are few or no slip traces in certain areas, but there are marked or even cross slips in other areas. This signifies that the distribution of plastic deformation is non-uniform in the initial stage of plastic deformation [36].

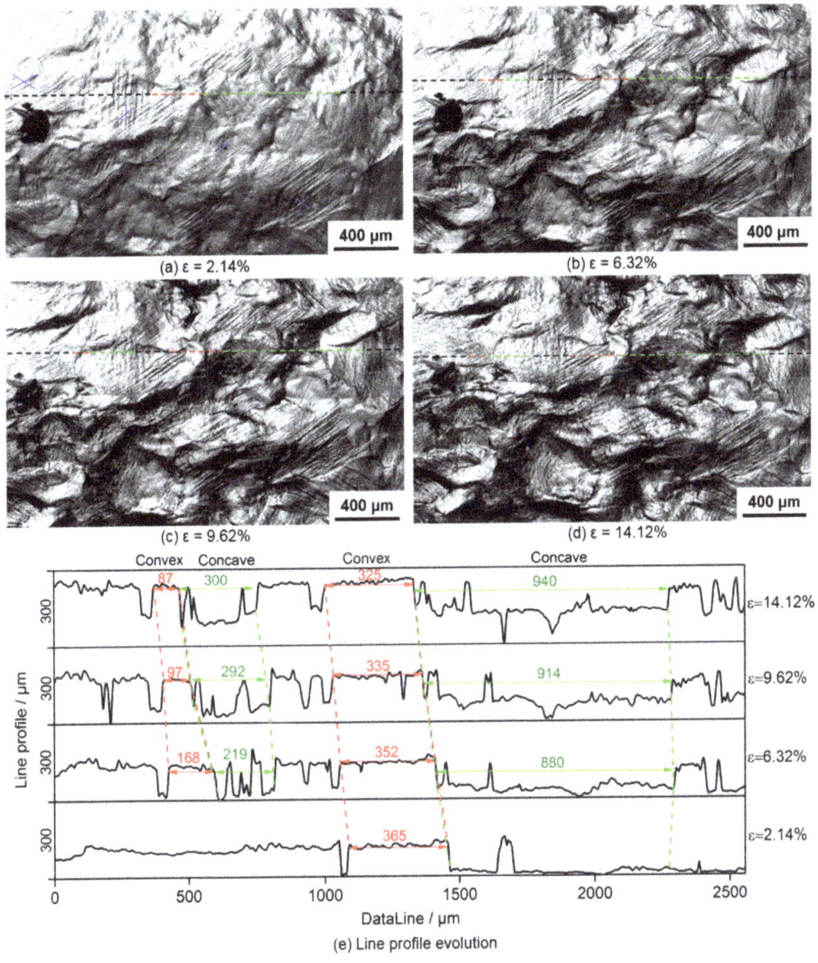

Figure 3. Surface morphologies of the marked area when the sample was stretched to global strains of (**a**) 2.14%; (**b**) 6.32%; (**c**) 9.62%; (**d**) 14.12%; and (**e**) line profile evolution.

Figure 3e shows the line profile evolution of the sample at different strain stages. The line is scanned along the tensile direction and the length is 2560 μm. The line profile appears to have undergone profile changes, with five turning points that divide the concave and convex features, and the maximum height of the undulation is approximately 150 μm when the global strain is 2.14%. The new concave–convex features appear, and the maximum height of the undulation increases to approximately 170 μm when the global strain reaches 6.32%. Then, a few new concave–convex features appear when the deformation exceed 9.62%. Line scanning evolution shows that the widths of the concave and adjacent convex features increase and decrease, respectively. The maximum height of the undulation increases at strains of 6.32% and 9.62% and then remains virtually constant at a strain of 14.12%.

3.2. Surface Roughness

The resulting surface roughness is a macroscopic manifestation of the microscopic concave and convex features. The surface shifted from a mirror finish to a rough surface owing to deformation. To measure the surface roughness of the deformation-induced rough surface, the surface roughness, R_a, of the line profile was calculated using Equation (1) at different tensile strains, as shown in the experimental points of Figure 4a. It is true that surface roughness variation must be continuous due to the continuity of deformation. From the view of experimental points, the surface roughness, R_a, first increases and then remains unchanged with increasing tensile strain. It can be observed that the gradient of the R_a shows an obvious transformation. To research the variation of R_a deeply and quantificationally, the derivatives of R_a were calculated and shown in Figure 4b. It is obvious that the derivatives of R_a were exponentially dependent on the tensile strain. The fitting type is the exponential fit by Origin software, refer to the literature [28], and the R-squared value is 0.99 showing a higher confidence coefficient. It shows that the derivative drops rapidly from a high value of 900 to almost 0. That means the surface roughness increases rapidly at the beginning and then stays at the same level. Secondly, the mathematical expression for surface roughness, R_a, is obtained by integrating the derivative, as shown in Figure 4a. The obtained integral curve fits well with the experimental points. There are deviations between the fitting curve and the experimental points, where the maximum deviation is 0.4 μm when the strain is 6.32%, and the other deviations are all within 0.3 μm. In conclusion, with increasing tensile deformation, the surface roughness, R_a, first increased rapidly and then remained at the same level.

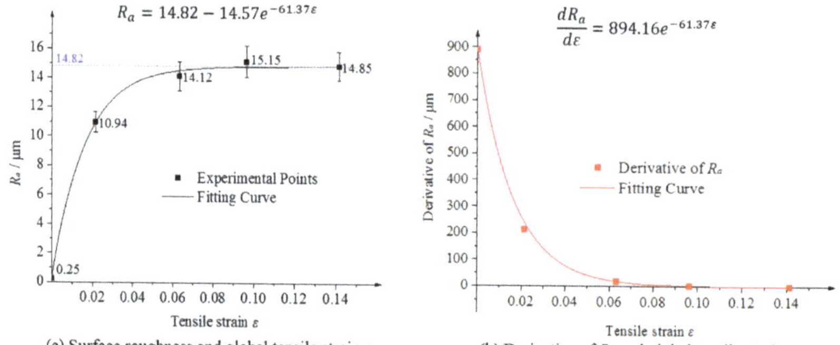

(a) Surface roughness and global tensile strain ε (b) Derivative of R_a and global tensile strain ε

Figure 4. Surface roughness of the sample at different strain stages.

3.3. Microstructure

To study the microscopic evolution process of surface roughening on polycrystal aluminum alloy during stretching, an in situ-EBSD test was performed, and the dynamic microstructure evolution process at the sample surfaces was obtained. The microstructure evolution process was studied with respect to the aspects of grain-orientation variation and strain localization.

3.3.1. Grain-Orientation Variation

There were 33 grains with grain sizes between 37 and 444 μm in the observable field, except for some grains, such as the grains with incomplete boundaries or those that are considerably small. We considered the grain sizes of 282–444 μm as large grains, 114–270 μm as medium grains, and 37–102 μm as small grains, and their grain distribution and grain-size distribution are shown in Figure 5. Surface roughening is initiated by the deflection of the surface grains relative to the original surface during plastic deformation.

Therefore, to characterize the correlation between surface roughening behavior and grain size, the deflection degrees of the large, medium, and small grains were separately analyzed.

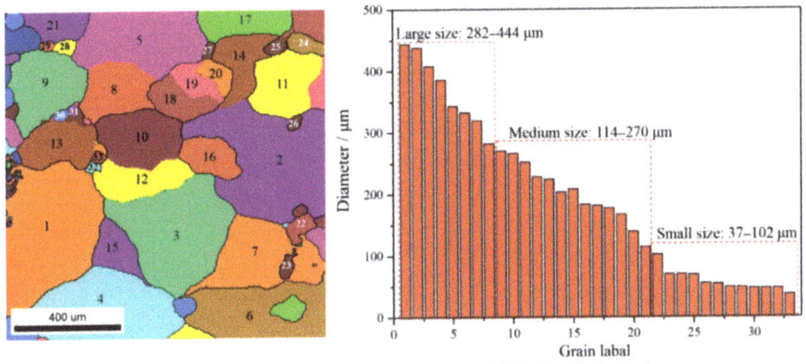

(a) Grain distribution (b) Grain size distribution

Figure 5. Grain and their grain-size distribution observed by EBSD.

The grain orientations of the surface grains, with random Schmidt factors, were randomly distributed in the initial state and then inevitably deflected with stretching. The Schmidt factor was used to characterize the grain orientation, and the deflection of the surface grains was reflected via the change in the Schmidt factor. The Schmidt factor distribution of the observable field is shown in Figure 6. The Schmidt factors of these 33 grains were randomly distributed in the 0.43–0.50 range before stretching; after stretching, the Schimidt factors of the grains changed. The change in the Schmidt factor increased with increasing tensile deformation. Even the deformation of some small-sized grains was too severe, where it could not be recognized when the deformation was 14.12%.

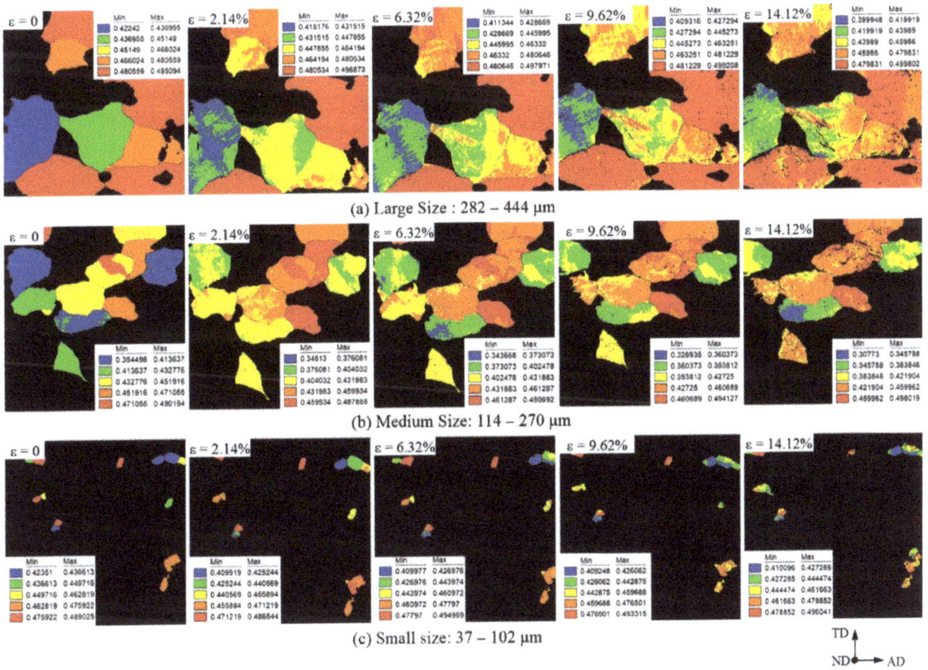

Figure 6. Schmidt factor distribution of the observable field observed by EBSD.

To quantitatively characterize the degree of deflection of surface grains of different sizes, the average value of the Schmidt factor for each grain was calculated and shown in Figure 7. The deflection degree of the large-sized grains is the smallest, that of the small-sized grains is medium, and that of the medium-sized grains is the largest. Medium-sized grains, i.e., a grain size of 114–270 μm, were prone to be deflected during stretching. The Schmidt factor changes toward smaller values until 0.36.

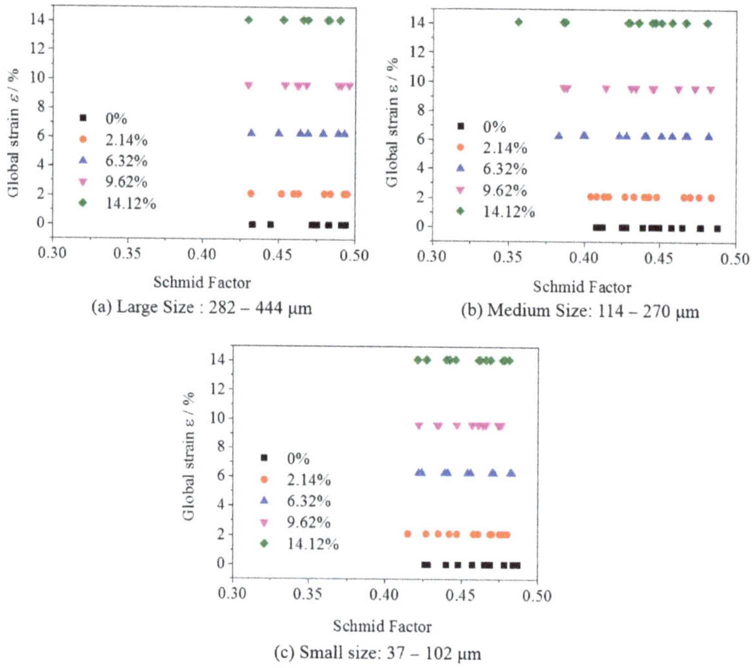

Figure 7. Statistics of Schmidt factors of the observable field observed by EBSD.

3.3.2. Strain Localization

The kernel average misorientation (KAM) at different strains is shown in Figure 8; the strain localization degrees were indicated using different colors. Red represents the most severe plastic deformation, that is, the highest degree of strain localization. Orange represents the second-most severe, yellow represents the general degree, green represents less plastic deformation, and blue indicates no plastic deformation. Geometrically necessary dislocations (GNDs) are dislocations generated to maintain the continuity of crystal deformation and can be represented by KAM [37].

As observed in the KAM distribution, a global strain of 2.14% already induces strain localization. The deformation predominantly occurs at the grain boundaries and then spreads from the grain boundary into the grain with increasing global strain. The quantitative statistics of the KAM are shown in Figure 9a and show a strong regularity. The geometrically necessary dislocations, which are expressed using KAM, increase with increasing strain. The KAM increases the fastest in large-sized grains, indicating that the GND in large-sized grains increases most distinctly, and the deformation coordination is moderately weak during the axial stretching process. Accordingly, the deformation coordination in the small-sized grains is the highest.

In general, numerous microcosmic defects in the polycrystalline microstructure can be caused by plastic deformation, and these defects can be represented by the generation of lower-angle grain boundaries (LAGBs, 2–15°). The quantitative statistics of the content of LAGBs are shown in Figure 9b, which shows its correlation with grain size. No strong regularity is observed in the variation of the density of LAGBs with global strain. In small-sized and large-sized grains, the LAGB content first increased and then decreased, whereas, in medium-sized grains, it first decreased and then increased. This is likely because when the tensile deformation is approximately 6%, the medium-sized grains mainly coordinate the continuity of plastic deformation by deflection, whereas the small-sized and large-sized grains mainly coordinate the continuity of deformation by slipping.

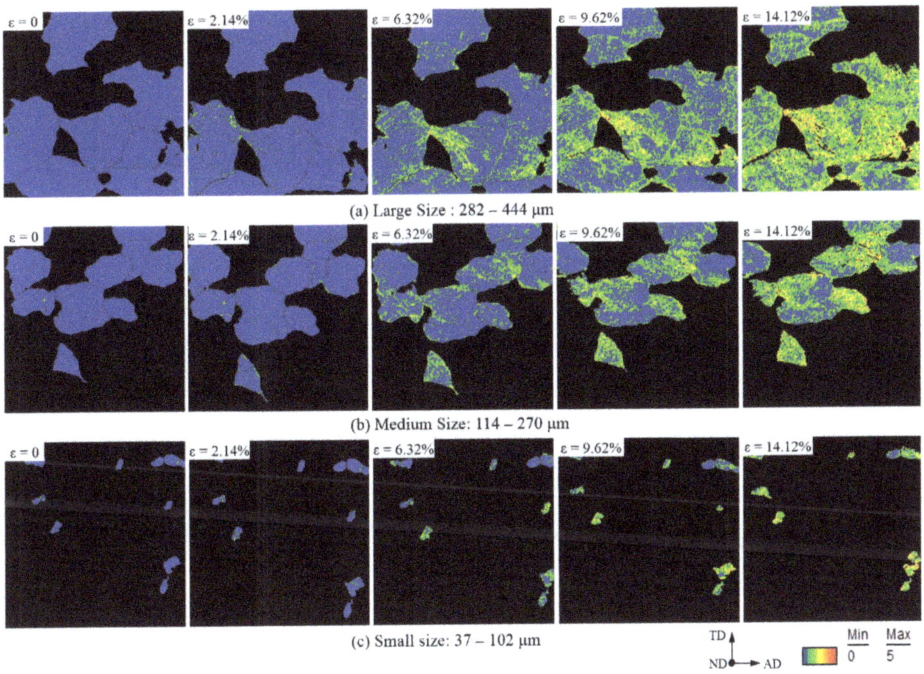

Figure 8. KAM distribution of the observable field observed by EBSD.

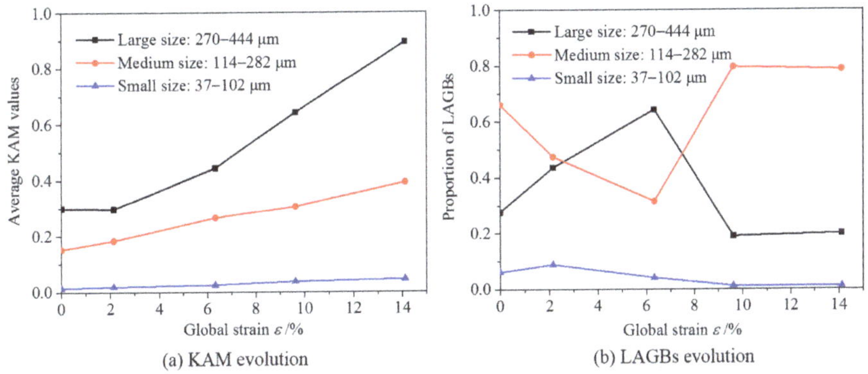

Figure 9. KAM and LAGBs evolutions of the observable field observed by EBSD.

4. Conclusions

(1) The deformation-induced concave and convex features simultaneously and continuously appear in the surface morphology and have shown a significant inhomogeneity of

deformation. The widths of the concave features gradually expand, whereas the widths of the convex features reduce in size. This difference in trend is one of the causes of inhomogeneous deformation.

(2) The size variations of the concave and convex features have been restricted by the deformation coordination of grains and caused a surface roughness (R_a) evolution that first increased and then remained unchanged with increasing stretching strain.

(3) During the stretching process, the large-sized grains in the 270–444 μm range exhibit severe strain localization and the most unfavorable deformation coordination ability, and the small-sized grains in the 37~102 μm range show the most favorable coordination of deformation. The deformation mode of medium-sized grains in the 114–270 μm range tends to grain deflection, and that of others tends to slip.

Future work should elucidate the mechanism of surface roughening under biaxial stress from a deformation perspective by investigating the effect of strain paths on surface roughening behavior.

Author Contributions: Y.C.: methodology, formal analysis, investigation, conducting the SEM and EBSD examinations, and writing—original draft. X.W.: conceptualization, supervision, and project administration. Y.D.: conducting EBSD analysis. All authors have read and agreed to the published version of the manuscript.

Funding: The work was financially supported by the Nature Science Foundation of Heilongjiang Province (LH2020E125), the Scientific Research Foundation for Introducing High-level Talents of Heilongjiang University of Science and Technology.

Data Availability Statement: The data presented in this study are available upon request from the corresponding author.

Acknowledgments: The authors would like to thank the National Key Laboratory of Precision Hot Processing of Metals for Harbin Institute of Technology. This research was partially supported by and partly performed at the Institute of Fluid Pressure Forming Technology, Harbin, China.

Conflicts of Interest: The authors declare no conflict of interest.

References

1. Yuan, S.J.; Liu, W.; Wang, G.F.; He, Z.B.; Fan, X.B. Advance in fluid pressure forming of complex light metal thin-walled components. *Aerosp. Shanghai* **2019**, *36*, 35–41.
2. Yuan, S.J. Fundamentals and Processes of Fluid Pressure Forming Technology for Complex Thin-Walled Components. *Engineering* **2021**, *7*, 358–366. [CrossRef]
3. Liu, C.S.; Xu, W.J.; Liu, M.Y.; Zhang, L. Effect of strain path on roughness evolution of free surface during plastic deformation. *Int. J. Mech. Sci.* **2020**, *173*, 105475. [CrossRef]
4. Romanova, V.A.; Balokhonov, R.R.; Panin, A.V.; Batukhtina, E.E.; Kazachenok, M.S.; Shikhijanov, V.S. Micromechanical Model of Deformation-Induced Surface Roughening in Polycrystalline Materials. *Phys. Mesomech.* **2017**, *20*, 324–333. [CrossRef]
5. Romanova, V.; Balokhonov, R.; Panin, A.; Kazachenok, M.; Kozelskaya, A. Micro-and mesomechanical aspects of deformation-induced surface roughening in polycrystalline titanium. *Mater. Sci. Eng. A* **2017**, *697*, 248–258. [CrossRef]
6. Cheng, C.; Wan, M.; Meng, B.; Fu, M.W. Characterization of the microscale forming limit for metal foils considering free surface roughening and failure mechanism transformation. *J. Mater. Processing Technol.* **2019**, *272*, 111–124. [CrossRef]
7. Kishimoto, T.; Suematsu, S.; Sakaguchi, H.; Tashima, K.; Kajino, S.; Gondo, S.; Suzuki, S. Effect of crystal orientation on inner surface roughness of micro metal tubes in hollow sinking. *Mater. Sci. Eng. A* **2021**, *805*, 140792. [CrossRef]
8. Al-Ghamdi, K.A.; Hussain, G. On the Free-Surface Roughness in Incremental Forming of a Sheet Metal: A Study from the Perspective of ISF Strain, Surface Morphology, Post-Forming Properties, and Process Conditions. *Metals* **2019**, *9*, 553. [CrossRef]
9. Irani, N.; Nicola, L. Modelling surface roughening during plastic deformation of metal crystals under contact shear loading. *Mech. Mater.* **2019**, *132*, 66–76. [CrossRef]
10. Bong, H.J.; Lee, J. Crystal Plasticity Finite Element–Marciniak-Kuczynski Approach with Surface Roughening Effect in Predicting Formability of Ultra-Thin Ferritic Stainless Steel Sheets. *Int. J. Mech. Sci.* **2021**, *191*, 106066. [CrossRef]
11. Cheng, C.; Wan, M.; Meng, B. Size effect on the forming limit of sheet metal in micro-scaled plastic deformation considering free surface roughening. *Procedia Eng.* **2017**, *207*, 1010–1015. [CrossRef]
12. Romanova, V.; Balokhonov, R.; Emelianova, E.; Pisarev, M.; Dymnich, E. Numerical study of the texture effect on deformation-induced surface roughening in titanium polycrystals. *Eng. Fail. Anal.* **2020**, *110*, 104437. [CrossRef]
13. Ma, X.G.; Zhao, J.W.; Du, W.; Zhang, X.; Jiang, Z.Y. Analysis of surface roughness evolution of ferritic stainless steel using crystal plasticity finite element method. *J. Mater. Res. Technol.* **2019**, *8*, 3175–3187.

14. Feng, Z.Y.; Li, H.; Yang, J.C.; Huang, H.; Li, G.J.; Huang, D. Macro-meso scale modeling and simulation of surface roughening: Aluminum alloy tube bending. *Int. J. Mech. Sci.* **2018**, *144*, 696–707. [CrossRef]
15. Panin, A.V.; Kazachenok, M.S.; Romanova, V.A.; Balokhonov, R.R.; Kozelskaya, A.I.; Sinykova, E.A.; Krukovskii, K.V. Strain-Induced Surface Roughening in Polycrystalline VT1-0 Titanium Specimens under Uniaxial Tension. *Phys. Mesomech.* **2018**, *21*, 249–257. [CrossRef]
16. Furushima, T.; Sato, H.; Manabe, K.; Alexandrov, S. Identification of an Empirical Equation for Predicting Free Surface Roughness Evolution in Thin Sheets of Aluminum Alloy and Pure Copper. *J. Manuf. Sci. Eng. Trans. ASME* **2018**, *140*, 034501. [CrossRef]
17. Furushima, T.; Tsunezaki, H.; Hirose, Y. Fracture and surface roughening behaviors in micro metal forming. *Procedia Manuf.* **2018**, *15*, 1481–1486. [CrossRef]
18. Cai, Y.; Wang, X.S.; Yuan, S.J. Analysis of surface roughening behavior of 6063 aluminum alloy by tensile testing of a trapezoidal uniaxial specimen. *Mater. Sci. Eng. A* **2016**, *672*, 184–193. [CrossRef]
19. Emelianova, E.; Romanova, V.; Zinovieva, O.; Balokhonov, R. The effects of surface-layer grain size and texture on deformation-induced surface roughening in polycrystalline titanium hardened by ultrasonic impact treatment. *Mater. Sci. Eng. A* **2020**, *793*, 139896. [CrossRef]
20. Leu, D.K. Free surface roughening under simple tension. *Int. J. Adv. Manuf. Technol.* **2018**, *95*, 2349–2356.
21. Solhjoo, S.; Halbertsma, P.J.; Veldhuis, M.; Toljaga, R.; Pei, Y. Effects of loading conditions on free surface roughening of AISI 420 martensitic stainless steel. *J. Mater. Processing Technol.* **2020**, *275*, 116311. [CrossRef]
22. Romanova, V.; Dymnich, E.; Balokhonov, R.; Zinovieva, O. Mechanical Aspects of Deformation-Induced Surface Roughening in the Presence of Inclusions in a Subsurface Layer. Numerical Modeling. *Front. Mech. Eng.* **2020**, *6*, 66. [CrossRef]
23. Romanova, V.; Balokhonov, R.; Zinovieva, O.; Shakhidjanov, V. Numerical study of the surface hardening effect on the deformation-induced roughening in titanium polycrystals. *Comput. Mater. Sci.* **2016**, *116*, 96–102. [CrossRef]
24. Jung, J.; Yoon, J.I.; Moon, J.H.; Park, H.K.; Kim, H.S. Effect of coarse precipitates on surface roughening of an FCC polycrystalline material using crystal plasticity. *Comput. Mater. Sci.* **2017**, *126*, 121–131.
25. Wouters, O.; Vellinga, W.P.; Tijum, R.V.; de Hosson, J.T.M. On the evolution of surface roughness during deformation of polycrystalline aluminum alloys. *Acta Mater.* **2005**, *53*, 4043–4050.
26. Shi, Y.; Wu, P.D.; Lloyd, D.J.; Embury, D. Numerical study of surface roughening in blow-formed aluminum bottle with crystal plasticity. *Mater. Sci. Eng. A* **2015**, *638*, 97–105.
27. Muhammad, W.; Ali, U.; Brahme, A.P.; Kang, J.; Mishra, R.K.; Inal, K. Experimental analyses and numerical modeling of texture evolution and the development of surface roughness during bending of an extruded aluminum alloy using a multiscale modeling framework. *Int. J. Plast.* **2019**, *117*, 93–121. [CrossRef]
28. Cai, Y.; Wang, X.S.; Yuan, S.J. Quantitative analysis of orange peel during tension of 6063 alloy spun tubes. *Trans. Nonferrous Met. Soc. China* **2018**, *5*, 858–865. [CrossRef]
29. Raabe, D.; Sachtleber, M.; Weiland, H.; Scheele, G.; Zhao, Z. Grain-scale micromechanics of polycrystal surfaces during plastic straining. *Acta Mater.* **2003**, *51*, 1539–1560. [CrossRef]
30. Sachtleber, M.; Raabe, D.; Weiland, H. Surface roughening and color changes of coated aluminum sheets during plastic straining. *J. Mater. Processing Technol.* **2004**, *148*, 68–76.
31. Sabat, R.K.; Muhammad, W.; Mishra, R.K.; Inal, K. Mechanism of microstructure and texture evolution during shear loading of AA6063 alloys. *J. Alloys Compd.* **2021**, *889*, 161607.
32. Wang, S.W.; Zhang, S.H.; Song, H.W.; Chen, Y. Surface roughness improvement of the bent thin-walled copper tube by controlling the microstructure and texture components. *Procedia Manuf.* **2020**, *50*, 613–617.
33. Zhao, Z.; Ramesh, M.; Raabe, D.; Cuitiño, A.M.; Radovitzky, R. Investigation of three-dimensional aspects of grain-scale plastic surface deformation of an aluminum oligocrystal. *Int. J. Plast.* **2008**, *24*, 2278–2297.
34. Abe, T. Surface roughening and formability in sheet metal forming of polycrystalline metal based on r-value of grains. *Int. J. Mech. Sci.* **2014**, *86*, 2–6.
35. Cai, Y.; Wang, X.S.; Yuan, S.J. Surface Roughening Behavior of 6063 Aluminum Alloy during Bulging by Spun Tubes. *Materials* **2017**, *10*, 299.
36. Li, H. Analysis of slip activity and heterogeneous deformation in tension and tension-creep of Ti–5Al–2.5Sn (wt %) using in-situ SEM experiments. *Philos. Mag.* **2012**, *92*, 2923–2946.
37. Le, G.M.; Godfrey, A.; Hansen, N.; Liu, W.; Winther, G.; Huang, X. Influence of grain size in the near-micrometre regime on the deformation microstructure in aluminium. *Acta Mater.* **2013**, *61*, 7072–7086.

materials

MDPI

Article

An Approach for Predicting the Low-Cycle-Fatigue Crack Initiation Life of Ultrafine-Grained Aluminum Alloy Considering Inhomogeneous Deformation and Microscale Multiaxial Strain

Teng Sun [1,2], Lidu Qin [3], Yiji Xie [3], Zhanguang Zheng [1,3,*], Changji Xie [3] and Zeng Huang [3]

1 Key Laboratory of Disaster Prevention and Structural Safety of Ministry of Education, Guangxi Key Laboratory of Disaster Prevention and Engineering Safety, College of Civil Engineering and Architecture, Guangxi University, Nanning 535000, China; suntengqz@163.com

2 Beibu Gulf Key Laboratory of Ocean Engineering Equipment and Technology, College of Mechanical and Marine Engineering, Beibu Gulf University, Qinzhou 535011, China

3 College of Mechanical Engineering, Guangxi University, Nanning 535000, China; qinlidu@126.com (L.Q.); yuyu19971019@126.com (Y.X.); xipiao@163.com (C.X.); 173887560@163.com (Z.H.)

* Correspondence: zhenglight@126.com

Abstract: In this paper, a low-cycle-fatigue (LCF) crack initiation life prediction approach that explicitly distinguishes nucleation and small crack propagation regimes is presented for ultrafine-grained (UFG) aluminum alloy by introducing two fatigue indicator parameters (FIPs) at the grain level. These two characterization parameters, the deformation inhomogeneity measured by the standard deviation of the dot product of normal stress and longitudinal strain and the microscale multiaxial strain considering the non-proportional cyclic additional hardening and mean strain effect, were proposed and respectively regarded as the driving forces for fatigue nucleation and small crack propagation. Then, the nucleation and small crack propagation lives were predicted by correlating these FIPs with statistical variables and cyclic J-integrals, respectively. By constructing a microstructure-based 3D polycrystalline finite element model with a free surface, a crystal plasticity finite element-based numerical simulation was carried out to quantify FIPs and clarify the role of crystallographic anisotropy in fatigue crack initiation. The numerical results reveal the following: (1) Nucleation is prone to occur on the surface of a material as a result of it having a higher inhomogeneous deformation than the interior of the material. (2) Compared with the experimental data, the LCF initiation life of UFG 6061 aluminum alloy could be predicted using the new parameters as FIPs. (3) The predicted results confirm the importance of considering the fatigue behavior of nucleation and small crack propagation with different deformation mechanisms for improving the fatigue crack initiation life prediction accuracy.

Keywords: ultrafine-grained aluminum alloy; fatigue initiation life prediction; fatigue indicator parameter; crystal plasticity

1. Introduction

Aluminum alloy (AA), as a lightweight alloy, has been extensively used in structural and machine components due to its excellent mechanical properties [1]. However, with the development of industry, higher mechanical performance, such as improved hardness and strength, is required of AAs in different situations [2]. Recently, many investigations have focused on improving the mechanical properties of AAs by refining their grain size to nanoscale or UFG through severe plastic deformation techniques, among which equal channel angular pressing has become a popular strategy due to its high strength/weight ratio and promising industrial application prospects. In practice, UFG AA has been used in some key structural components of transport systems, which often endure cyclic loads [3,4]. Therefore, a thorough theoretical investigation of the fatigue properties of UFG AA is crucial in order to promote the application of such materials.

Numerous attempts have been made to identify a rational methodology for fatigue life prediction. The empirical Coffin–Manson law and its modification have been widely used by researchers to characterize the LCF of metallic materials for several decades. The Coffin–Manson law has been experimentally verified but cannot be related to the microstructure features. In order to overcome these limitations, some researchers have tried to interpret the fatigue crack initiation behavior in connection with crystal plasticity theory with the help of FIPs. Chen et al. [5] investigated the accuracy by regarding local slip accumulation and dissipated energy as FIPs to assess fatigue failure; both of these were found to correlate with the crack locations observed in experiments. The investigations conducted by Manonukul and Dunne [6] were the first to point out that the potential crack initiation location experiences the most intense accumulated plastic slip. Later, Cruzado et al. [7] argued that calculating energy dissipation in each particular slip plane is a better approach due to its clear physical background. The same conclusion was also made by Sweeney et al. [8] because the energy dissipation criterion designed in their investigations considered the working hardening induced by dislocation motion. These microstructure-sensitive FIPs identify the "hotspots" of fatigue crack initiation by judging the maximum value of the averaged FIPs, and fatigue failure is deemed to occur when the FIP computed for each cycle reaches a threshold value [7,9]. However, most of these FIPs are applied to coarse grains, which are not fully appropriate for use in UFG metals. In order to overcome these limitations, Zhang et al. [10–12] tried to predict the fatigue crack initiation life using a novel statistics-based criterion, providing new insights into the occurrence of fatigue failure. This research used statistical variables as FIPs to describe the inhomogeneous deformation of the representative volume element (RVE), and the fatigue life could be determined when the statistical variables reached their corresponding critical values. However, inhomogeneous plastic deformation occurring on surfaces has been overlooked in previous investigations. This topic deserves more attention [13,14], as fatigue cracks generally nucleate on the surface of UFG material.

Conventionally, local stress-based criteria for HCF and plastic strain-based criteria for LCF have been applied to estimate fatigue life by dividing the cracks into two stages, i.e., crack initiation and crack propagation [15,16]. Over the last few years, some researchers have further divided total fatigue life into three physical regimes: nucleation (10^{-7}–10^{-5} m), small crack propagation (10^{-5}–10^{-4} m), and long crack propagation (LC > 10^{-3} m) [17]. Additionally, the LCF fatigue crack initiation life of materials can be defined by the number of cycles required to incubate a small crack and propagate it to a macrocrack. These two stages are closely related to the microstructure topology and grain crystal anisotropy, and micromechanics-based modeling is required to accurately predict the fatigue initiation life of materials. In fact, microstructure has a dominant influence in the early stages of fatigue crack formation and growth under HCF conditions [18], while the microstructure morphology of grains/phases influences the mean (50% probability) fatigue response under LCF conditions [19]. However, most of the studies listed above do not explicitly distinguish between the nucleation and small crack propagation phases and predict the fatigue crack initiation life based solely on one FIP, which may introduce errors to the fatigue crack initiation life prediction of UFG AA.

Recently, some efforts have been made to correlate microstructural features with the early stage of fatigue crack behavior based on crystal plasticity theory. For instance, Castelluccio and Mcdowell proposed an FIP based on a modified version of the Fatemi and Socie (F–S) criterion in a high-cycle-fatigue regime [20–22]. In this research, the F–S criterion served as the driving force for small crack propagation. More recently, Yang et al. [23] utilized the F–S model to predict the LCF crack initiation life of GH4169 superalloy. A good prediction accuracy was achieved by assuming the small crack propagation life to account for half of the fatigue crack initiation life. In order to clarify the role of FIP in small crack propagation, a corresponding theoretical analysis and CPFEM simulations were performed by Reddy and Fatemi [18]. Their results showed that FIPs play a similar role to that of ΔK or ΔJ. However, the normal strain on the slip plane, which has been reported to not only accelerate small crack propagation [24,25] but also to lead to non-proportional cyclic additional hardening (NPCAH) [26,27], has been overlooked in previous FIP models. Meanwhile, for a crystal grain under multiaxial cyclic straining loading, the mean strain effect may play a non-negligible role in small crack propagation, as extensive macroscopic multiaxial fatigue experiments have indicated that errors could be made in small crack propagation life if the mean strain effect is not considered [28,29]. Therefore, it might be necessary to search for an FIP that is able to reflect the variation in microscale multiaxial strain, which can help us to characterize the fundamental small crack propagation mechanism.

Recently, some researchers have attempted to assess the fracture toughness of UFG metallic alloys using J-integrals. Additionally, the small crack propagation life can be predicted by associating the cyclic J-integral with the range of crack tip opening displacement. For instance, Ding and Mughrabi [30] proposed an LCF life prediction model for UFG materials based on cyclic J-integrals. In their research, the macroscale indicator parameter, i.e., the plastic strain range, is responsible for the fatigue crack growth. However, the proposed model assumes a very small initial radius of the fatigue crack, without any reference being made to the microstructure information. Similar computational models can also be applied to different kinds of UFG metals [31–33], but they are limited in their rendering of the heterogeneous microstructure and anisotropic mechanical behavior of materials. As outlined in [17], the assessment and prediction of early fatigue behavior (i.e., nucleation and small crack propagation ranging from 10^{-7} to 10^{-5} m) require micromechanics-based models.

In this paper, an LCF fatigue crack initiation life approach that explicitly distinguishes between the nucleation and small crack propagation regime is proposed based on experiments and crystal plasticity modeling. Choosing the UFG 6061AA as a model material, we carried out the following studies: (1) A synthetic polycrystal finite element model with a free surface was constructed and regarded as an analytical model with which to clarify the contribution of the inhomogeneous plastic deformation to the fatigue crack initiation. (2) Though conducting CPFEM simulations, the cycles required for nucleation were assessed based on the inhomogeneity of the material. (3) By introducing the FIP related to the microscale multiaxial strain field into the cyclic J-integral, the small crack propagation rate equation was obtained by associating the cyclic J-integral with the range of crack tip opening displacement. After this, the small crack propagation life was predicted based on a newly deduced Coffin–Manson model for UFG metallic alloys. (4) Comparative studies were conducted between the methods based on only one FIP and the proposed combined approach. (5) The validation of the proposed LCF crack initiation life model was carried out with respect to the errors between the predicted results and the experimental data. Finally, a technical diagram was plotted with an algorithm of all calculation steps (shown in Figure 1).

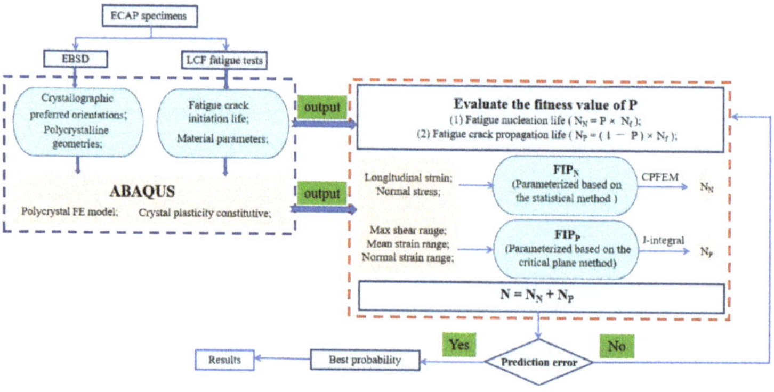

Figure 1. The technical diagram for the proposed method.

2. Material and Strain Fatigue Experiments

The material used in ECAP is AA 6061; its chemical composition is presented in Table 1. The process of ECAP was carried out using a wa-600b electro-hydraulic servo universal testing machine at room temperature. After eight extrusions of ECAP, the samples were machined into cylindrical specimens with a gauge diameter size of 6.25 mm and gauge length of 18 mm, as shown in Figure 2a. Meanwhile, electron backscatter diffraction (EBSD) tests were conducted to obtain the crystallographic orientation and polycrystalline geometry information prior to the LCF tests. The TSL OIM software package was employed to analyze the EBSD data. As shown in Figure 2b, the different colors in the EBSD map represent the orientations of separate grains, and the average grain size is 1.08 μm.

Table 1. Chemical composition of 6061 aluminum alloy (% in weight).

Si	Fe	Cu	Mn	Mg	Cr	Zn	Ti
0.586	0.241	0.264	0.095	0.945	0.07	0.024	0.005

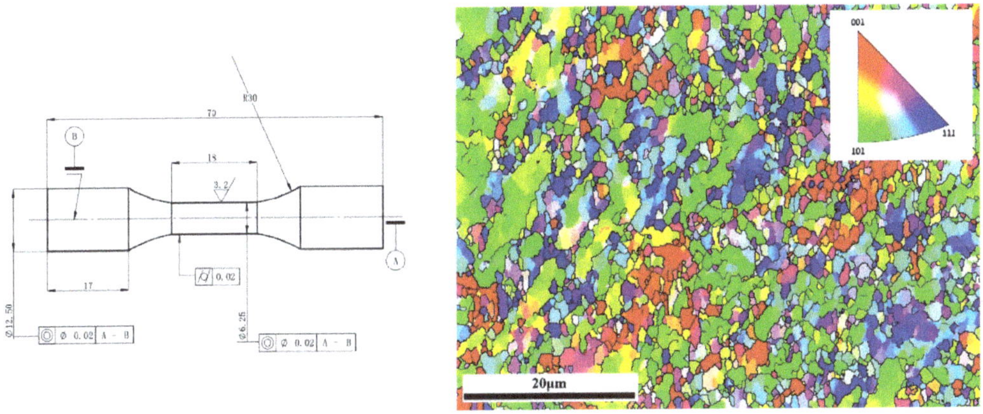

(a) (b)

Figure 2. (**a**) Geometric dimensions of the cylindrical specimen (unit: mm); (**b**) EBSD map of UFG AA6061.

The fatigue test was carried out according to the American Society of Testing Materials (ASTM) E606 (Standard Practice for Strain-Controlled Fatigue Test). A Guanteng PA-20 microcomputer-controlled fatigue testing machine was used, and the gauge distance of the extensometer was 12.5 mm. Meanwhile, all the test specimens were carefully ground and polished using various grit emery papers before fatigue testing, aiming to reduce the influence of surface roughness. The LCF tests were conducted with a sinusoidal loading waveform with a frequency of 0.05 Hz at ambient temperature and carried out on three samples for reproducibility. The ECAP specimens were subjected to cyclic symmetrical tension–compression loading with a constant strain amplitude. The strain amplitudes used in the test were 0.005, 0.006, and 0.007. Finally, the stable hysteresis loops (solid lines) under different strain amplitudes were compared with the simulation results obtained from the CPFEM simulation.

3. Proposed Fatigue Crack Initiation Life Prediction Approach for UFG AA

As discussed in the introduction, the LCF crack initiation life can be estimated by the superposition of fatigue nucleation life (N_N) and small crack propagation life (N_{MSC}), which can be written as [17]:

$$N_f = N_N + N_{MSC} \quad (1)$$

These two physically based regimes are closely related to the microstructure of polycrystalline materials. From a microstructural and physical point of view, the microscopic behavior of material influences the mean (50% probability) fatigue response in low cycle regimes [19]. Following the experimental observation of UFG AA [34–36], we assume that dislocation glide is still the most important deformation mechanism for UFG AA. This assumption allows us to utilize crystal plasticity theory to develop micromechanics-based fatigue criteria that relate FIPs to the phase of nucleation and transgranular small crack propagation.

3.1. Indicator Parameter of Inhomogeneity Considering the Influence of Normal Stress

Previous studies have demonstrated that the meso-inhomogeneous deformation of a material can be described by the standard deviation of the longitudinal strain [11,12]. For materials under LCF loading, it has been reported that a stress-assisted plastic slip-based FIP provides better predictions than one without stress terms [37]. Based on this, we proposed the FIP_N for the nucleation regime by calculating the standard deviation of the dot product of normal stress and mesoscopic longitudinal strain (MLS) for the whole RVE, namely:

$$FIP_N = \hat{f} \tag{2}$$

with

$$\hat{f} = \sqrt{\sum_{k=1}^{n_{RVE}} (f)_k^2 p_k - (\overline{f})^2}, \quad \overline{f} = \sum_{k=1}^{n_{RVE}} (f)_k p_k$$

$$f = \sigma_n \cdot \varepsilon_{ll}$$

$$\sigma_n = \begin{bmatrix} n_x & n_y & n_z \end{bmatrix} \begin{bmatrix} \sigma_x & \tau_{xy} & \tau_{xz} \\ \tau_{xy} & \sigma_y & \tau_{yz} \\ \tau_{xz} & \tau_{yz} & \sigma_z \end{bmatrix} \begin{bmatrix} n_x \\ n_y \\ n_z \end{bmatrix}$$

Here, $\hat{f}$ and $\overline{f}$ represent the standard deviation and mean value of the dot product of normal stress and MLS; n_{RVE} denotes the total number of finite elements; σ_n is the normal stress with respect to the α slip plane, where the maximum plastic shear strain range (MSSR) occurs; and p_k can be calculated by the volume of the kth element divided by the total volume of the RVE.

Furthermore, the FIP_N^{Surf} representing the inhomogeneity on the free surface is proposed as follows:

$$FIP_N^{Surf} = \hat{f}^{Surf} \tag{3}$$

with

$$\hat{f}^{Surf} = \sqrt{\sum_{k=1}^{n_{RVE_{surf}}} (f^{Surf})_k^2 p_k - \left(\overline{f}^{Surf}\right)^2}, \quad \overline{f}^{Surf} = \sum_{k=1}^{n_{RVEsurf}} (f^{Surf})_k p_k$$

$$f^{Surf} = \sigma_n^{Surf} \cdot \varepsilon_{ll}^{Surf}$$

It is noteworthy that all the variables calculated in Equation (3) are defined with respect to the surface layer of the RVE.

3.2. Modeling of Small Crack Propagation Life Based on the Indicator of Microscale Multiaxial Strain

For UFG AA, it has been observed that the micro deformation band is formed from the crack tip and moves along a close-packed plane of atoms [38]. This situation indicates that the physical modeling of small crack propagation in UFG AA should take the microscale slip mechanism into consideration. As shown in Figure 3, there exists a cyclic plastic region (CPR) on the slip plane, within which a fatigue damage region (FDR) is formed ahead of the small crack. Considering the unique deformation mechanism of UFG AA, it is not difficult to extrapolate that dynamic grain coarsening and large-scale shear-dominated deformation are supposed to occur in the FDR due to the complex dislocation interaction. In this work, the microscale control parameter used to correlate small crack propagation with the local field is characterized by FIP utilizing crystal plasticity theory.

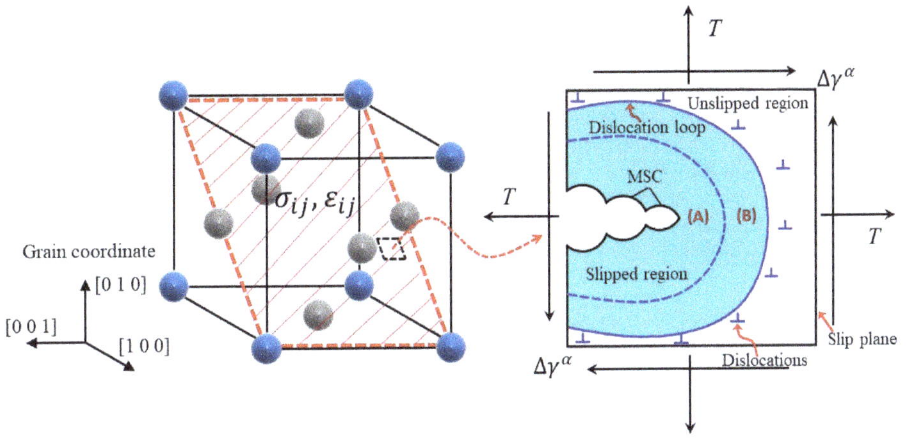

(**A**) (**B**)

Figure 3. Propagation of microstructurally small crack: (**A**) fatigue damage region, (**B**) cyclic plastic region.

As stated above, the local stress–strain field in the FDR evolves in a multiaxial and non-proportional manner, giving rise to NPCAH behavior in the realistic microstructure. Previous macroscale experiments have proven [39] that NPCAH behavior can be described by establishing a multiaxial fatigue damage parameter with respect to the normal strain and maximum shear strain based on the critical plane method. Therefore, based on the assumption that only active slip planes can be potentially considered as critical planes [20], a multiaxial fatigue damage parameter, defined on the specific crystallographic plane, is proposed to reflect the NPCAH effect at the grain level, which can be deduced according to the von Mises yield criterion:

$$\frac{\Delta\varepsilon_{eq,N}}{2} = \left[(\varepsilon_n^*)^2 + \frac{1}{3}\left(\frac{\Delta\gamma_{\max}}{2}\right)^2\right]^{\frac{1}{2}} \tag{4}$$

where $\Delta\gamma_{\max}$ is the MSSR in the α slip system and ε_n^* denotes the normal strain on the crystallographic plane where the MSSR occurs; it can be calculated as $\varepsilon_n^* = \max\limits_{t_A<t<t_E}(\varepsilon_n(t)) - \min\limits_{t_A<t<t_E}(\varepsilon_n(t)) = \varepsilon_n^{\max} - \varepsilon_n^{\min}$. ε_n, representing the normal strain on the slip plane, can be defined as:

$$\varepsilon_n^{(\alpha)} = \begin{bmatrix} n_x & n_y & n_z \end{bmatrix} \begin{bmatrix} \varepsilon_x & \gamma_{xy} & \gamma_{xz} \\ \gamma_{xy} & \varepsilon_y & \gamma_{yz} \\ \gamma_{xz} & \gamma_{yz} & \varepsilon_z \end{bmatrix} \begin{bmatrix} n_x \\ n_y \\ n_z \end{bmatrix} \tag{5}$$

Considering that the mean range on the specified slip plane may play an important role in the small crack growth, the equivalent mean strain range, combined with the multiaxial fatigue damage parameter, is regarded as the local driving force in small crack propagation. Consequently, the new proposed indicator parameter of the microscale multiaxial strain field can be rewritten as:

$$FIP_P = \frac{\Delta\varepsilon_{eq,N}}{2} + \frac{\Delta\varepsilon_{eq,m}}{2} \tag{6}$$

with

$$\frac{\Delta \varepsilon_{eq,m}}{2} = \text{sgn}\left[\text{sgn}\left(\frac{\Delta \varepsilon_{n,m}}{2}\right) \cdot \left(\frac{\Delta \varepsilon_{n,m}}{2}\right)^2 + \frac{1}{3}\text{sgn}\left(\frac{\Delta \gamma_m}{2}\right) \cdot \text{sgn}\left(\frac{\Delta \gamma_m}{2}\right)^2\right] \times \left[\left(\frac{\Delta \varepsilon_{n,m}}{2}\right)^2 + \frac{1}{3}\left(\frac{\Delta \gamma_m}{2}\right)^2\right]^{\frac{1}{2}}$$

where $\Delta\varepsilon_n^m$ and $\Delta\gamma^m$ represent the mean normal strain range and mean shear strain range for the slip plane, respectively.

The relationship between the cyclic J-integral and FIP was clarified in the work of McDowell [40]. For UFG materials, it has been recognized that the cyclic J-integral shows a good correlation with the fatigue crack growth rate under the elastic–plastic field for the case of an LCF condition, and it was suggested as an effective alternative to the application of ΔK in the investigations for both short crack and long crack growth regimes [41,42]. In this paper, the cyclic J-integral is proposed to be responsible for the small crack propagation, and it follows a representative relationship based on the work of Rice [43]:

$$J = -\frac{\partial V_{int}}{\partial r} \tag{7}$$

Here, V denotes the internal energy dissipated in the process of small crack propagation. Previous studies have already demonstrated that the growth rate of fatigue cracks for UFG metals depends on whether the local cyclic peak stress in the fatigue degradation zone ahead of the crack tip approaches the ultimate tensile strength [44–46]. Therefore, the interaction energy dissipated per unit length can be determined using:

$$V_{\text{int}} = -R_{Y,UFG}\widetilde{\varepsilon}_{FDR}\pi(r_{FDR}/2)^2 \tag{8}$$

Here, $R_{Y,UFG}$ denotes the cyclic ultimate tensile strength and $\widetilde{\varepsilon}_{FDR}$ represents the accumulated mean multiaxial damage strain in FDR, which can be calculated by the integration of the strain in CPR (ε_{CPZ}) over the radius of the fatigue damage region (r_{FDR}) [33,47,48]. Then, the mean multiaxial damage strain can be obtained by the integral equations as follows:

$$\widetilde{\varepsilon}_{FDR} = \frac{1}{r_{FDR}}\int_0^{r_{FDR}} \varepsilon_{CPR}(r)dr \tag{9}$$

In order to consider the effect of multiaxial damage strain on the small crack propagation, we extend the model in [30,49] by incorporating the FIP_p into the formulation of ε_{CPR} and r_{FDR}. The newly developed model is described by:

$$\varepsilon_{CPR}(r) = \frac{FIP_P}{2}\left(\frac{r_c}{r}\right)^{1/(n'+1)} \tag{10}$$

$$\text{r}_{FDR} = \frac{\lambda\pi K_{GBS}^2}{16K_{GBC}^2}\left(\frac{(K'_{UFG})^{3+1/n'}}{R_{m,UFG}^{1+1/n'}R_{Y,UFG}^2}\right)\left(\frac{FIP_P}{2}\right)^{3n'+1}\alpha \tag{11}$$

Here, λ is the cyclic plastic zone correction factor, which can be obtained by one experimental fatigue test; K'_{UFG} and n' are material parameters related to the cyclic mechanical response of UFG AA; K_{GBS} and K_{GBC} are, respectively, the GB strengthening (GBS) factor and the GB constraint (GBC) factor. Additionally, both of these can be calculated from the relationship between the cyclic mechanical response and the GB properties of UFG materials [30]

$$\begin{cases} K_{GBS} \equiv R_{Y,0.2}^{UFG}/R_{Y,0.2} \\ K_{GBC} \equiv [(\Delta\sigma/2)/(\Delta\sigma_{eff}/2)]/2 \end{cases} \tag{12}$$

where $R_{P,UFG}$ and R_P represent the effective yield strength of UFG and its CG counterparts and $\Delta\sigma/2, \Delta\sigma_{eff}/2$ are, respectively, the cyclic stress range and equivalent stress range at the steady cycle, which can be obtained by performing CPFEM simulation. These parameters reflect the strengthening effect on the metal matrix and the constraint on the dislocation motion due to the large amount of grain boundaries (GBs) in UFG materials.

After some computation, one can further obtain Equation (13) by substituting Equation (8) into Equation (7), namely:

$$\Delta J = \frac{\lambda \pi^2 K_{GBS}^2}{32K_{GBC}^2}(\frac{n'+1}{n'})\left(\frac{(K'_{UFG})^3}{R_{p,UFG}^2}\right)\left(\frac{FIP_P}{2}\right)^{3n'+1}\alpha \tag{13}$$

Based on the fracture mechanics, the ΔJ parameter and the range of crack tip opening displacement ($\Delta CTOD$) are assumed to satisfy the following relation [50]:

$$\Delta CTOD = \Delta J/\omega R_{P,UFG}, \; with \; \omega = 1.5 \tag{14}$$

Therefore, with the knowledge of Equations (13) and (14), we obtain the fatigue crack growth rate equation for the UFG material as:

$$(\frac{da}{dN})_{MSC} = \frac{\lambda \pi^2 K_{GBS}^2}{96K_{GBC}^2}(\frac{n'+1}{n'})(\frac{K'_{UFG}}{R_{p,UFG}})^3(\frac{FIP_P}{2})^{3n'+1} a \tag{15}$$

Since the cracks will evolve with the loading cycles and finally reach a critical value, one can obtain the fatigue life formula by integrating Equation (15) from an assumed small crack r_i to a critical crack size r_f. The expression of FCP life for the considered material is then:

$$N_f = \left[\frac{96K_{GBC}^2}{\lambda \pi^2 K_{GBS}^2}\left(\frac{n'}{n'+1}\right)\left(\frac{R_{p,UFG}}{K'_{UFG}}\right)^3 \ln\left(\frac{r_f}{r_i}\right)\right]\left(\frac{FIP_P}{2}\right)^{-(3n'+1)} \tag{16}$$

After rearranging the equation above, one can find that Equation (16) can be easily translated into the standard Coffin–Manson model, i.e., Equation (17), with the consideration of the deformation mechanism at the microscale and its corresponding mechanical response at the macroscale:

$$\frac{FIP_P}{2} = \left[\frac{192K_{GBC}^2}{\lambda \pi^2 K_{GBS}^2}\left(\frac{n'}{n'+1}\right)\left(\frac{R_{p,UFG}}{K'_{UFG}}\right)^3 \ln\left(\frac{r_f}{r_i}\right)\right]^{\frac{1}{3n'+1}} \left(2N_f\right)^{-1/(3n'+1)} \tag{17}$$

4. Simulation Methodology

4.1. Dislocation-Based Constitutive Model

The following formulations of the dislocation-based constitutive model for UFG material were based on the theory established in [51]. In order to explore the cyclic plasticity behavior of UFG AA, the flow rule was modified by introducing a nonlinear kinematic hardening term to the slip system, i.e.,

$$\dot{\gamma}^{(\alpha)} = v_0 \frac{b}{d} \exp\left\{-\frac{\Delta G_0}{KT}\left[1 - \left\langle \frac{\left|\tau^{(\alpha)} - \chi^{(\alpha)}\right| - \tau_{ath}^{(\alpha)}}{\hat{\tau}_{UFG}} \right\rangle^p\right]^q\right\} \mathrm{sgn}(\tau^{(\alpha)} - \chi^{(\alpha)}) \tag{18}$$

Here, the brackets $\langle \ldots \rangle$ denote the Macaulay bracket, meaning that $\langle x \rangle \equiv x$ for $x \geq 0$; otherwise $\langle x \rangle \equiv 0$. P and q are two material parameters. v_0 denotes the attempt frequency and b and d represent the Burgers vector and the mean grain size of UFG metal, respectively. The physical meaning of "b/d" corresponds to the plastic shear strain generated in the progress of dislocation sliding from GBs to GBs. $\Delta G^{(\alpha)}$ denotes the reduced activation energy; α refers to the total amount of slip in the system, which is equal to 12 for FCC polycrystalline metal; $\dot{\gamma}^{(a)}$ is the inelastic strain rate for the active a-slip systems, i.e., a {110}

<110> system for an FCC crystal structure; and $\tau^{(\alpha)}$, representing the resolved shear stress, can be computed as:

$$\tau^{(\alpha)} = \mathbf{M}^{e} : \mathbf{P}^{(a)},\ with\ \mathbf{M}^{e} = \mathbf{F}^{eT}\mathbf{F}^{e} \cdot \mathbf{T}^{e} \quad (19)$$

Here, $\mathbf{M}^{e}$ denotes the Mandel stress tensor; $\mathbf{T}^{e}$ represents the second-order Piola–Kirchhoff stress tensor; $\mathbf{E}_{e}$ is the elastic Green–Lagrange strain tensor; and $\mathbf{C}$ is the fourth-order elastic tensor. Based on the Hooke's law, their relationship can be written as:

$$\mathbf{T}^{e} = \mathbf{C} : \mathbf{E}^{e} \text{ with } \mathbf{E}^{e} = \frac{1}{2}\left(\mathbf{F}^{eT}\mathbf{F}^{e} - \mathbf{I}\right) \quad (20)$$

The lattice friction stress, $\hat{\tau}_{UFG}$, is the critical shear stress used to generate crystal plastic slip at 0 K, which can be obtained based on a previous investigation into UFG metals [51]:

$$\hat{\tau} = 2\Delta G_0/Lb\omega_0 \quad (21)$$

where ω_0 represents the average travel distance of the dislocation and L represents the distance of two pinning obstacles, which can be determined from the product of volume fraction (c) and grain size as $L = md$ $(0 < m < 1)$.

To account for the evolution of back-stress at the grain level, a kinematic hardening rule including a strain hardening term and a dynamic recovery term is introduced based on the work of Busso and McClintock [52], namely:

$$\chi^{(\alpha)} = h_b\dot{\gamma}^{(\alpha)} - \frac{r_D}{\tau_{ath}^{(\alpha)}}\chi^{(\alpha)}\left|\dot{\gamma}^{(\alpha)}\right| \quad (22)$$

Here, h_b and r_D are two material parameters and $\tau_{ath}^{(\alpha)}$ represents the slip resistance law in the α slip system [53]:

$$\tau_{ath}^{(\alpha)} = \mu Gb\sqrt{\sum_{\beta=1}^{12} A^{\alpha\beta}\rho^{\beta}}, \quad A^{\alpha\beta} = h_0\left[\omega_1 + (1-\omega_2)\sigma^{\alpha\beta}\right] \quad (23)$$

where μ is a material parameter that is equal to 0.7, G is the shear modulus, and h_0 is a scalar function describing the strength of dislocation pair interaction. Generally, $A_{\alpha\beta}$ is a 12×12 dislocation interaction matrix that describes the degree of obstruction between different slip systems. In this investigation, an interaction matrix consisting of two elements corresponding to weak interactions and strong interactions is adopted for simplicity.

Considering the competition between the multiplication and annihilation of dislocations for the α slip system, the overall dislocation density evolution equation is governed by the balance law, which is suggested as:

$$\dot{\rho}^{(\alpha)} = \frac{\left|\dot{\gamma}^{(\alpha)}\right|}{b}\left(\frac{1}{d} + \frac{\sqrt{\sum_{a\neq\beta}\rho^{\beta}}}{K} - y_k\exp\left(-\frac{\Delta GB}{k_BT}\right)\rho^{(\alpha)}\right) \quad (24)$$

where K is a material constant that controls the size of the mean free path [54]. For UFG materials, the process of the absorption of dislocations by GBs is proportional to the diffusivity along the GBs over the course of cyclic loading. Therefore, the absorption rate of dislocations at GBs is proposed to be related to the temperature and thermal activation energy of GBs.

4.2. Construction of the Finite Element Model

4.2.1. Applied Boundary Conditions

The loading direction applied to the polycrystalline finite element model is shown in Figure 4a. Additionally, the specific boundary conditions can be implemented using the following steps: (1) Normal displacement was applied with U3 = U for the positive surface of the third axis, while U3 = 0 was applied for the negative surface of the third axis. (2) On the negative surface of axes 1 and 2, the displacement in the normal direction of all nodes was equal to zero. (3) For the positive surface of the first axis, node A at the lower right corner of the surface was taken as the reference point, while the other nodes on the surface were recorded as A′ set. In order to meet the condition of macro-uniaxial loading and continuous material deformation, the constraint Equation (25) was embedded into ABAQUS to satisfy the boundary conditions above.

$$-U_1^{A}+U_1^{A'}=0 \tag{25}$$

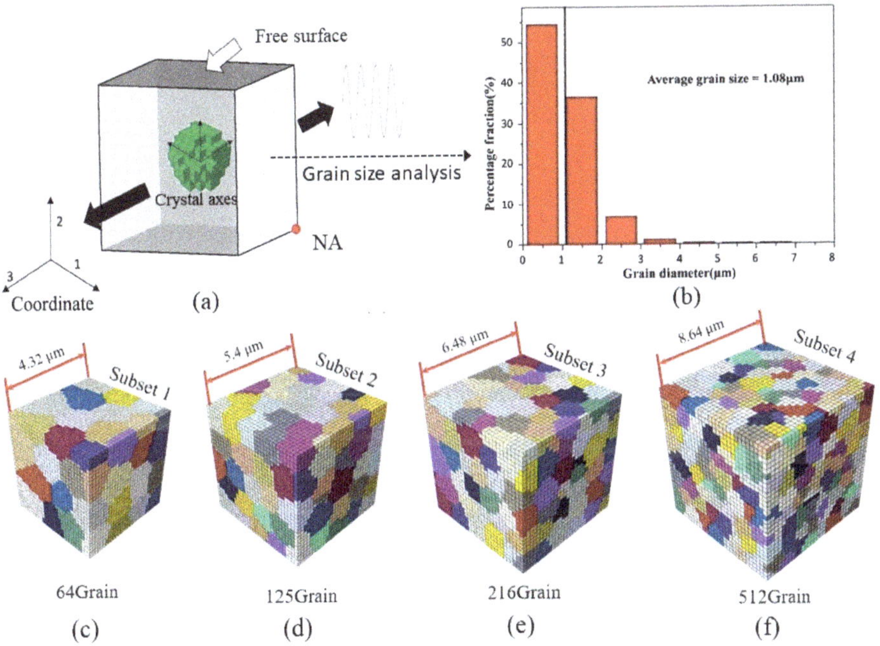

Figure 4. (**a**) Boundary condition and loading direction applied to RVE; (**b**) grain size analysis; (**c**) subset 1–64 grains; (**d**) subset 2–125 grains; (**e**) subset 3–216 grains; (**f**) subset 4–512 grains.

4.2.2. Validation of the RVE

Based on the microscopic experiment of UFG 6061AA, it is acknowledged that Goss texture is formed for UFG 6061AA in the process of ECAP. In other words, the polycrystalline aggregate will show obvious anisotropy. Therefore, the number of grains in RVE should be sufficient to reflect the plasticity heterogeneities of the real microstructure. However, numerical simulations with a large number of grains under a crystal plasticity framework are very costly from a computational viewpoint. Thus, a reasonable number of grains with the reduced crystallographic texture should be used for the accurate prediction of mechanical response at the polycrystal level. In this investigation, four RVEs with the same number of elements (27,000) but containing 27, 64, 216, and 512 grains (Figure 4c–f, different colors represent different grains) were constructed by grain size analysis (as shown in Figure 4b) and assigned a reduced crystallographic texture extracted from the pole map (Figure 5). The computationally obtained results and their local enlarged drawing at the peak stress are plotted in Figure 6. The difference in macro hysteresis loop shows that the peak stress identified for 216 grains is slightly higher than that for 512 grains, but less than 2 MPa. Therefore, a 3D polycrystalline RVE established using 27,000 elements and 216 grains was adopted in the following CPFE simulations.

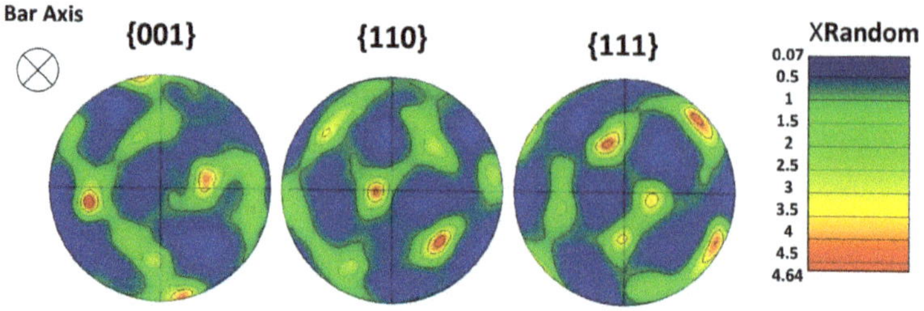

Figure 5. The pole figure of UFG 6061AA.

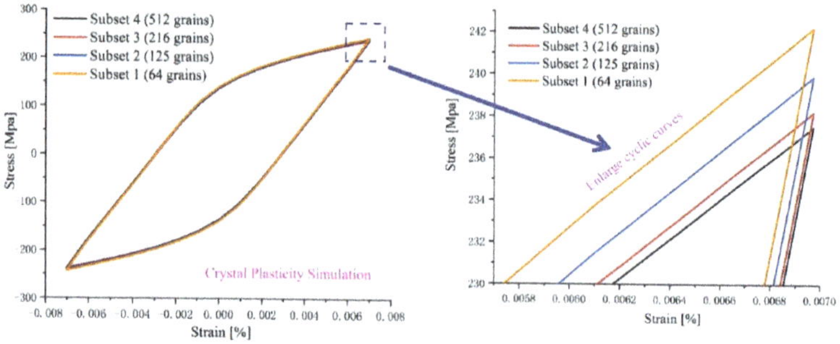

Figure 6. Comparison between the steady hysteresis loops under repeated load and the mechanical response from the polycrystal FE model simulated with different numbers of grains.

4.3. Parameter Identification

4.3.1. Identification of Material Parameters and Simulating Cyclic Deformation Behavior

The material parameters used in the crystal plasticity constitutive equation were classified into three categories. The first category contained the microstructural features of UFG metals, including grain size and grain orientation, which were consistent with those observed in the EBSD experiment. The second category was mainly physical material parameters, such as the GB thermal activation energy, Boltzmann constant, and Burger vector, all of which were related to the unique properties of UFG metallic material. These parameters controlling the development of isotropic and kinematic hardening were initialized based on previously published studies on UFG AA at room temperature [47,55–57]. Lastly, the remaining parameters, such as the three independent single-crystal elastic constants and dynamic recovery rate, were adjusted from those in the symmetric strain-controlled test using a "trial-and-error" method. All the calibrated parameters discussed above are listed in Table 2.

Finally, the rationality of the material parameters was verified in terms of the stable stress–strain hysteresis loops and their corresponding numerical simulation. Figure 7 shows a comparative analysis between the simulated stable stress–strain hysteresis loops and the experimental ones at the strain amplitudes of 0.5%, 0.6%, and 0.7%. The maximum stress difference between the test data and the micro-level simulations is less than 3 MPa, which indicates that the obtained simulated stable hysteresis curves are in good agreement with the experimental results. Meanwhile, all the simulated results predict a slight cyclic softening, which is consistent with the experimental observation. Moreover, in order to reflect the reduction in dislocation density observed in the cyclic loading process, two internal variables, i.e., initial dislocation density and critical shear stress, are reduced with the microstructure evolution. In summary, all the above results show that the calibrated material parameters for the crystal plasticity model are reasonable and acceptable for use in further investigations.

Table 2. Material parameters for CPFEM simulation.

Elastic Constants and Flow Parameters			Hardening Parameters		
Parameter	Dimension	Value	Parameter	Dimension	Value
C_{11}	MPa	77,159	d_{mean}	μm	1.08
C_{12}	MPa	55,706	G	MPa	26,209
C_{44}	MPa	35,910	ω_1	-	1.0
v_0	s^{-1}	1013	ω_2	-	1.1
ρ_0	1014 mm^{-2}	1.5	h_0	-	0.014
c	-	0.014	y_c	-	0.9×10^{-25}
b	nm	0.286 [47]	r_D	MPa	118
ΔG_0	ev	1.4 [51]	h_b	MPa	8831
m	-	0.28	ΔG_B	J/mol	87,000 [51]
k_B	J/K	1.38×10^{-23}	u	-	3.55
p	-	0.99			
q	-	1.01			
T	K	298			

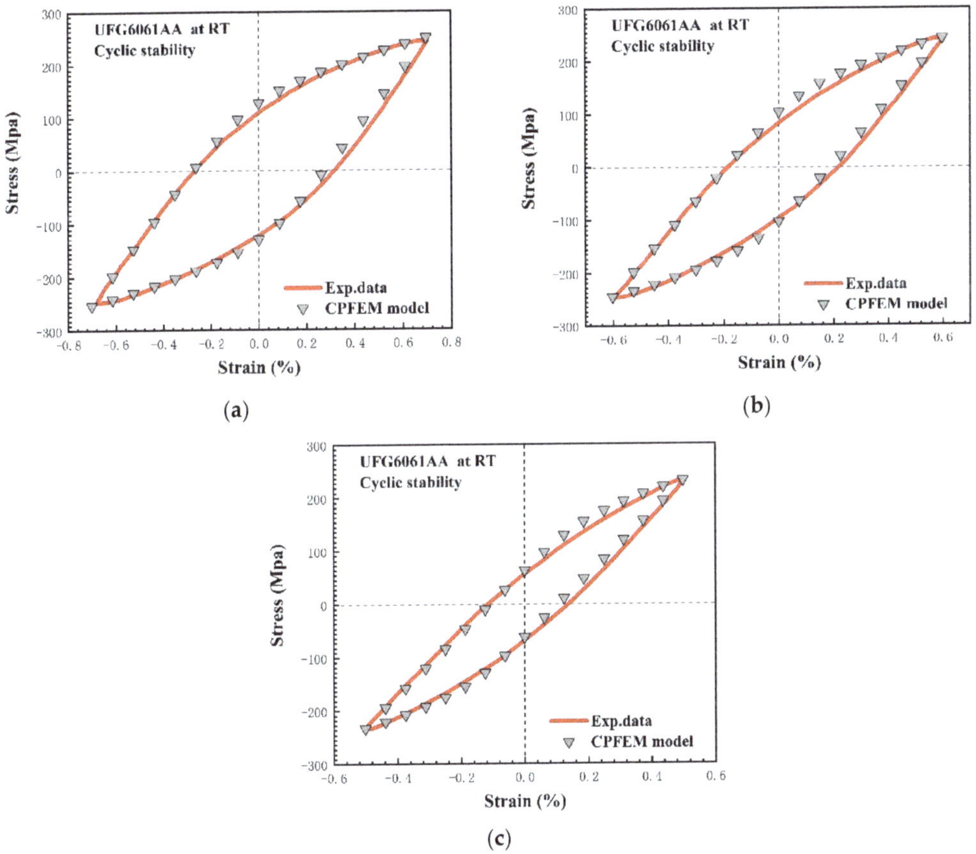

Figure 7. Comparison between the experimental and computational cyclic stress–strain curves in terms of hysteresis loops: (**a**) 0.005; (**b**) 0.006; (**c**) 0.007.

4.3.2. Identification of Material Parameters for Small Crack Propagation Equation

The small crack propagation life of UFG 6061AA was predicted according to Equation (16). Obviously, the main factors that affect the small crack propagation life of UFG 6061AA are the defined initial fatigue crack length, the material parameters, and the investigated material. Thus, a series of tension–compression symmetric strain-controlled experiments were conducted carefully to identify these key parameters prior to the application of UFG 6061AA. However, very limited information about the fatigue crack initiation life of UFG AA is available in the literature. Additionally, it is hard to measure the corresponding real-time crack length of a smooth cylindrical specimen using the currently available observation equipment. Therefore, in this work, an empirical approach outlined in the ASTM E606 standard was adopted; i.e., the total fatigue crack initiation life was estimated when the maximum stress of the cyclic stress–strain curve had dropped below 5% of the stabilized hysteresis loop peaks. Then, the critical crack size was determined based on the fracture appearance using SEM observation. In this work, the critical crack size for UFG 6061AA was taken as 25 μm, which is approximately 20 times larger than the initial mean grain size. In order to obtain the material parameters related to the cyclic plasticity behavior, a set of cyclic stress–strain curves for UFG and CG 6061AA were plotted based on a series of symmetrical cyclic strain-controlled experiments (shown in Figure 8). It should be noted that the cyclic yield strength was calculated by the formulation fitted by the approximation $R_{P,0.2} = K'(0.002)^{n}$ [30]. In addition, the GBS factor can be obtained by the experimental method reported in [58] or the CPFEM simulation. Finally, all the parameters used in our calculation are listed in Table 3.

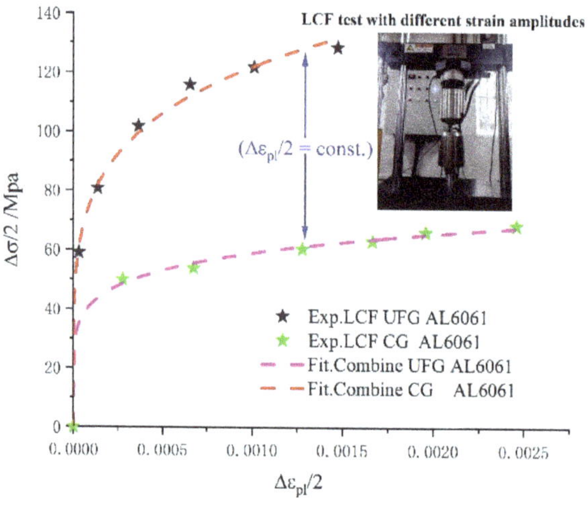

Figure 8. Stress amplitude versus plastic strain amplitude curves of UFG 6061AA.

Table 3. Characteristic parameters of UFG 6061AA and CG 6061AA.

Materials	n'	K'_{UFG}	**Cyclic Yield Strength** $R^{UFG}_{P,0.2}$ **(MPa)**	**Strengthening Factor** K_{GBS}
UFG 6061AA/RT State	0.21	484	139	2.1
CG AA6061/RT State	0.15	168	66	

5. Results and Discussion

5.1. Determining the Statistical Variable Describing the Inhomogeneity

According to the previous experimental observations of fatigue experiments, microcracks tend to nucleate on the surface of metal rather than in the bulk [59]. The reason for this is that the surface of samples generally suffers from manufacturing defects and a poor finish, leading to the accumulation of damage that accelerates the fatigue failure of components under cyclic loading. Based on the use of an RVE with preferred orientation according to the EBSD, Figure 9a,b compare the distribution of mesoscale longitudinal strain at the maximum tension point of the 3rd and 300th cycles, respectively. It can be seen that the free surface becomes obviously uneven as the number of cycles increases. Aiming to quantitatively depict the degree of deformation of the RVE and its free surface, the inhomogeneity of the polycrystalline aggregation and its free surface are described by Equations (2) and (3), respectively.

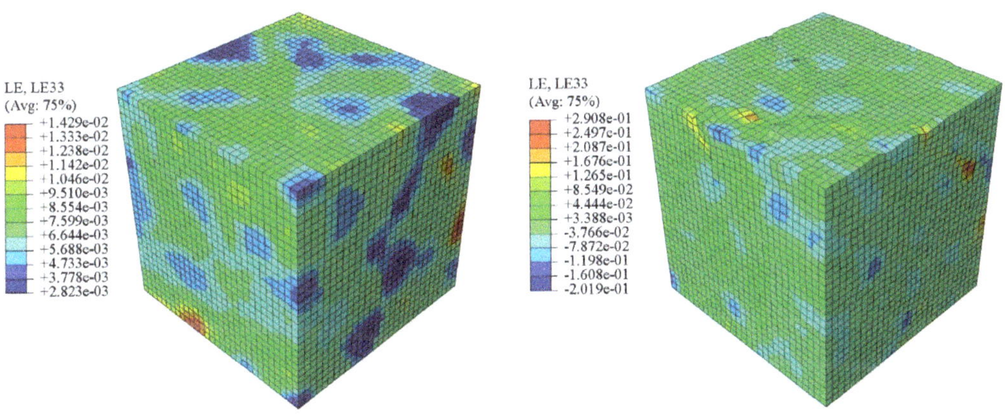

(**a**) (**b**)

Figure 9. Longitudinal strain contours of the RVE at the maximum tension moment with respect to the (**a**) 3rd cycle and (**b**) 300th cycle under a strain amplitude of 0.7%.

Figure 10 displays the histograms of the frequency distribution for f and f^{Surf} under various strain amplitudes. It is evident that the statistical distributions for f and f^{Surf} are Gaussian-like, similar to those reported in [11,12]. Therefore, it is reasonable to adopt these two parameters to signify the level of dispersity of the inhomogeneous local strain field. As shown in Figure 10a–c, the SD of f presents a trend of growth, which implies that the inhomogeneity of the material increases with the increase in the applied strain amplitude. Figure 10c,d compare the SD of f for the whole RVE at an amplitude of 0.7%, and it is found that the SD of f at the 300th cycle is 20 times larger than that at the 3rd cycle. Furthermore, Figure 10e,f investigate the inhomogeneity on the free surface of the RVE by calculating the f^{Surf} at the 300th and 3rd cycle with respect to the strain amplitude of 0.7%. A very similar evolution was found but with a 28.25 times difference. When comparing Figure 10c,d with e,f, the SD of f^{Surf} is obviously higher than that of f in the 300th and 3rd cycle. Based on the analysis above, we can conclude the following: (1) The cycle numbers and the applied loading conditions are the two key external factors that affect the inhomogeneity of materials. (2) Even if there is no damage on the surface, the metal surface suffers from more inhomogeneous plastic deformation than the interior of the material, which further clarifies the contribution of the plasticity heterogeneities and micromechanical interactions between neighbor grains to the fatigue damage mechanism. (3) The SD of f^{Surf} calculated with respect to the free surface layer of the RVE is more suitable for use as a fatigue criterion to judge the fatigue nucleation occurrence.

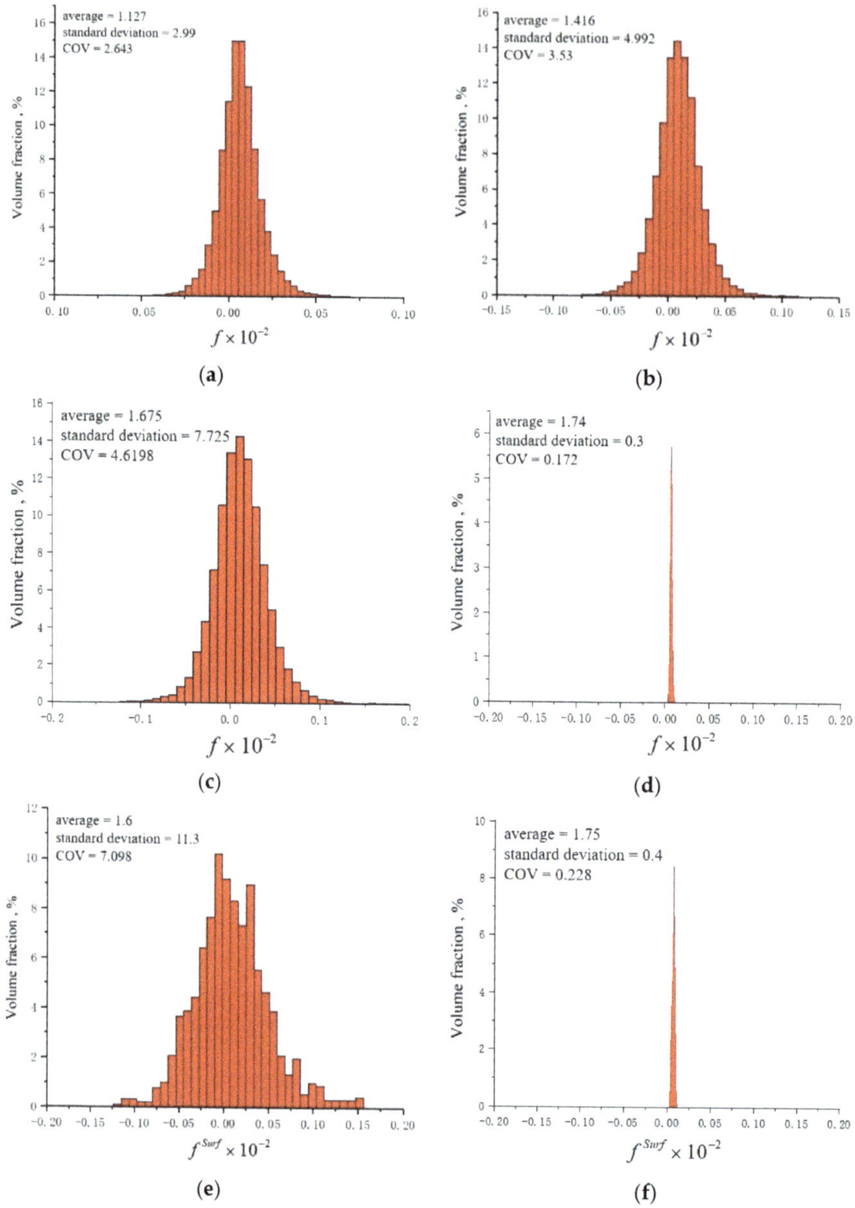

Figure 10. The histograms of the frequency distribution for f at the maximum moment of tension under the strain amplitude of (**a**) 0.5%, (**b**) 0.6%, and (**c**) 0.7% in the 300th cycle. (**d**) The f in the 3rd cycle under the strain amplitude of 0.7%; (**d**) the distribution of f^{Surf} on the free surface in terms of the (**e**) 300th cycle and (**f**) 3rd cycle.

5.2. Analysis of Local Stress and Strain Field on the Free Surface

From the microscopic viewpoint, the inhomogeneity of polycrystalline aggregation can be described by the Schmid factor of individual grains. The contours of the Mises stress and maximum principal strain are plotted in Figure 11; they were calculated by ABAQUS after computational stabilization at the 10th cycle with respect to the tension peak. The heterogeneous state of plastic deformation at the grain level was mainly caused by the preferred orientation processed by ECAP. Since the condition of stress equilibrium and strain compatibility needs to be satisfied during deformation, the misorientation of the neighboring grains makes the GBs "hot ports" in the polycrystalline aggregation. As displayed in Figure 11a,b, more details of the local stress and strain response on the free surface of the RVE were investigated along two horizontal lines. Moreover, two types of grain combinations, that is to say, the hard–soft combination and the soft–soft combination, were selected to analyze the effect of crystal orientation on the inhomogeneous local stress and strain field. It can be observed from Figure 12a,b that the maximum Mises stress appears at hard grains, while the minimum value appears at soft grains. On the other hand, it can be seen that the transition process of stress and strain between two soft grains tends to be relatively smooth. For the two adjacent soft grains, the elastic and plastic behavior at the micro-level is anisotropic. Therefore, it can be inferred that this inhomogeneous deformation process firstly induces a grain shape change to satisfy the strain compatibility and then stress equilibrium along the GBs. Finally, it should be noted that the combination of hard grain and hard grain was been discussed in this investigation, but a similar evolution could be inferred from the evolution law of the soft–soft combination.

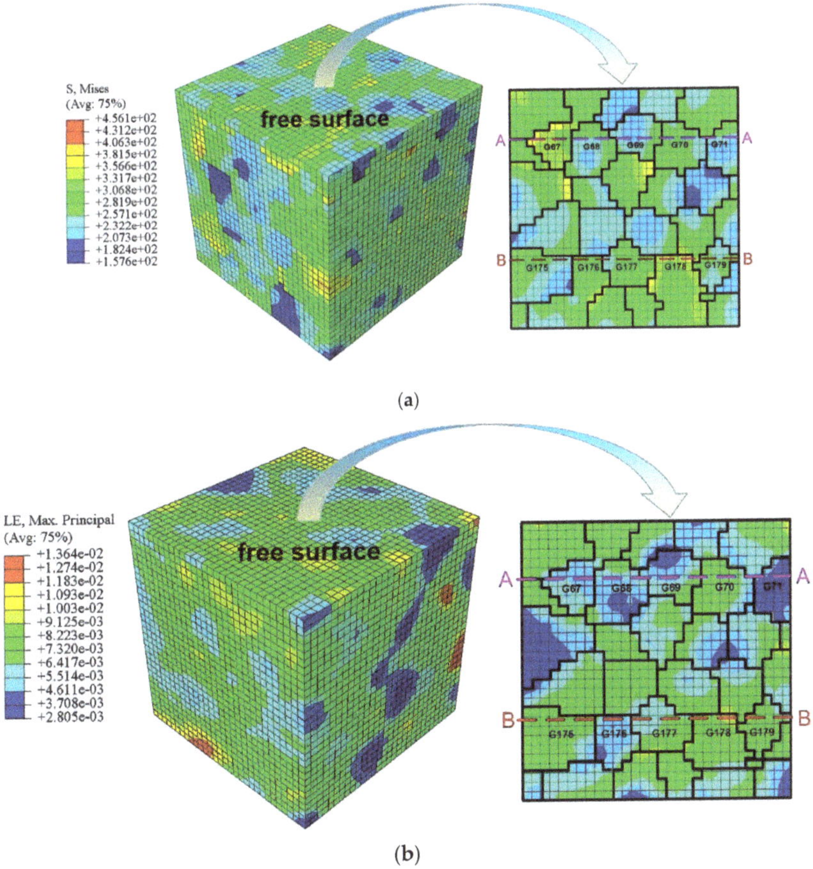

Figure 11. Contours of (**a**) Mises and (**b**) LE maximum principal strain for the whole RVE and its free surface.

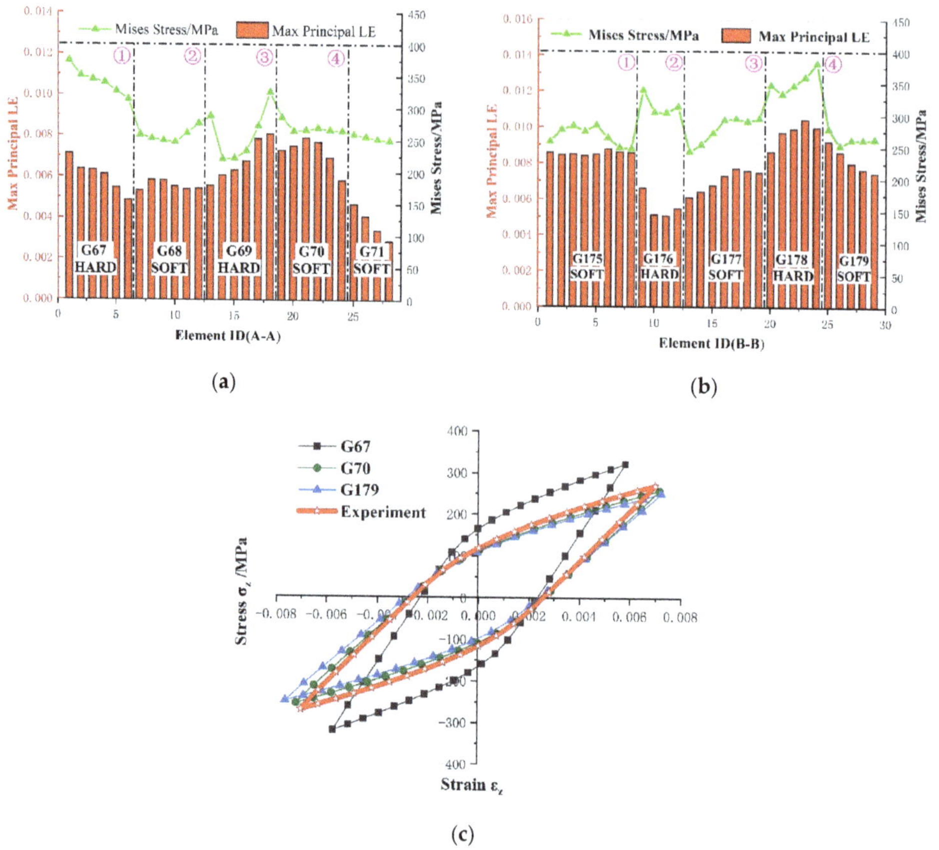

Figure 12. Local stress and strain analysis arising from the preferred crystallographic orientation and neighbor grain constraint effect: along the (**a**) horizontal line A; (**b**) horizontal line B; (**c**) cyclic response for specified grains.

In order to further validate the aforementioned deduction, two soft grains and one hard grain were selected to explore the microscale cyclic plastic response at the grain level. Figure 12c shows the mechanical responses of individual grains obtained from soft grains 67 and 70 and hard grain 169 with respect to the 10th cycle. It can be seen clearly that the two soft grains exhibit a higher strain response and a lower stress peak as compared to the hysteresis loops obtained from the experiment, while the hard grain exhibits a higher stress response and a lower strain response. These results further demonstrate that the material properties of hard grains and soft grains are anisotropic in terms of elastic and plastic deformation, which are, respectively, linked to elastic lattice distortion and the plastic slip accumulated over the course of cyclic loading. Moreover, one can see that the yield strength, Young's modulus, and cyclic response for the soft grains are smaller than those obtained from the experiments, which is in contrast to the situation seen in hard grains. These well-oriented grains, which are prone to slip, will contribute to a larger accumulation of micro-level plastic deformation and higher hysteresis energy dissipation.

5.3. Prediction of Low-Cycle-Fatigue Crack Initiation Life

5.3.1. Nucleation Life Based on the Evolutions of the Inhomogeneity on the Free Surface

Further investigations of the relationship between the inhomogeneous meso-plastic deformation and fatigue nucleation were conducted by tracking the evolution of FIP_N^{Surf}

through CPFEM simulation. For UFG AA, the inclusion of large amounts of precipitates distributed dispersedly within the matrix was observed; this contributes to the localized crystallographic slip concentration, accelerating the nano-void nucleation and growth [35,37] and resulting in the scatter characteristic of nucleation life in UFG materials. At present, there are technical difficulties in tracking the real-time evolution of nucleation and small cracks in smooth cylindrical fatigue specimens using the currently available observation equipment. Therefore, we made the assumption that the fatigue nucleation regime may account for 30%, 50%, 70%, and 100% of the fatigue initiation life for simplicity.

Figure 13 shows the evolution curves of the simulated FIP_N^{Surf}, characterizing the inhomogeneous micro-plastic deformation that has occurred under three different strain amplitudes in four probabilities as the number of cycles increases. It can clearly be seen that the FIP_N^{Surf} increases as the number of cycles increases, and the evolution law is similar within the four assumed probability ranges. The simulation results also indicate that the inhomogeneous deformation on the free surface became increasingly serious during the cyclic loading. The relationship between the experimental data and their averaged values is marked by red solid dots and blue crosses distributed along the FIP evolution curves. The variation ranges of FIP_N^{Surf}, i.e., 2.6∼15.8, 6.7∼28, 9∼40.2, and 14∼57.8, were divided into six equal parts by seven critical values. Subsequently, a group of horizontal lines, including four black dotted lines and three solid lines plotted in red, pink, and green, was drawn though these critical values. The values of FIP_N^{Surf} corresponding to the intersections of these critical lines and FIP evolution curves are listed in Table 4.

Selecting the limiting values in Table 4, the comparisons between the experiment life and the predicted nucleation life are shown in Figure 14. The abscissa is the nucleation life calculated from the product of the total fatigue initiation life and the assumed probability. The ordinate is the predicted nucleation life obtained by utilizing the proposed FIP_N^{Surf}. The bold diagonal solid orange line represents perfect fatigue crack initiation life prediction. Compared with Figure 14a,b, it can be seen that all the predicted data points are within the ±3.0 error band. However, compared with the experimental results, this approach overestimates the nucleation life at larger values of strain amplitude, while showing conservative predicted results at lower values of strain amplitude. The reason for the lower predictability of this model can be attributed to the fact that it exclusively focuses on the inhomogeneous plastic deformation of the material, but ignores the effect of the inclusion of precipitates dispersed in large amounts of UFG aluminum alloy matrix, which may trigger early micro-crack nucleation due to deformation incompatibility. This causes the nucleation life to be lower than expected in the case of higher strain amplitudes. On the other hand, the predicted results tend to be conservative under lower strain amplitudes. This is mainly due to the smaller grain size in the multiaxis stress state being conducive to more homogeneous slipping plastic deformation in the UFG regime. As a result, the stress concentration is reduced at the micro level, which delays the propagation of small cracks to macroscopic ones [60,61]. This is manifested as an improvement in the fatigue strength of UFG metals at the macro level. In addition, it is necessary to point out that the statistical predicted method based on inhomogeneity can better capture the scatter characteristic of nucleation behavior to an extent. Although the approach based on inhomogeneous plastic deformation is capable of assessing the fatigue initiation life of UFG metallic alloys, more thorough investigations that take into particular consideration the unique microstructure characteristics and the macroscopic mechanical response of UFG material are required to establish a more physically based equation for this kind of material.

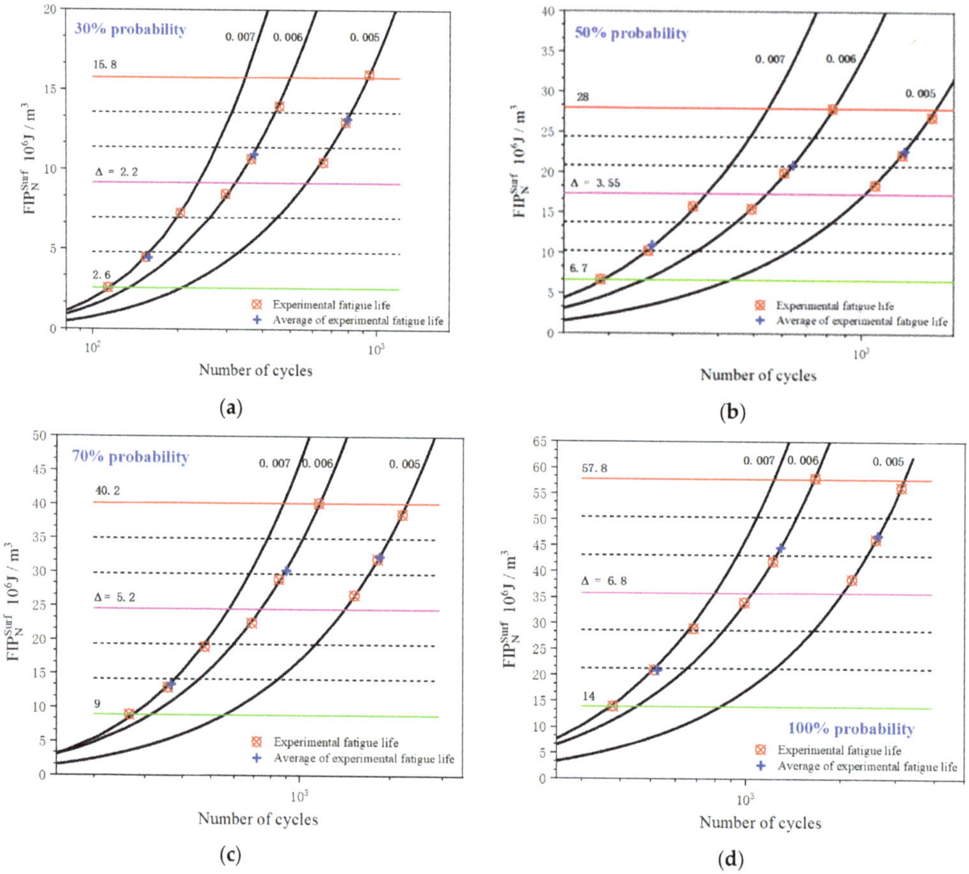

Figure 13. The evolution of FIP_N^{Surf} under different probabilities.

Table 4. Nucleation life prediction for various FIP_N^{Surf} with respect to their different probabilities.

	30% Probability							50% Probability						
FIP_N^{Surf} / $\Delta E_t/2$	2.6	4.8	7	9.2	11.4	13.6	15.8	6.7	10.25	13.8	17.35	20.9	24.45	28
0.5%	258	313	450	568	680	804	908	438	621	820	1013	1201	1406	1589
0.6%	131	198	260	302	386	426	502	245	336	443	527	642	718	827
0.7%	113	167	194	237	271	302	360	189	256	313	380	430	495	543
	70% Probability							**100% Probability**						
FIP_N^{Surf} / $\Delta E_t/2$	9	14.2	19.4	24.6	29.8	35	40.2	14	21.3	28.6	35.9	43.2	50.5	57.8
0.5%	551	846	1118	1406	1693	1981	2265	830	1223	1615	2010	2400	2818	3185
0.6%	310	450	596	728	889	1026	1157	445	652	851	1054	1245	1436	1653
0.7%	265	380	497	428	667	765	880	378	512	670	802	933	1081	1228

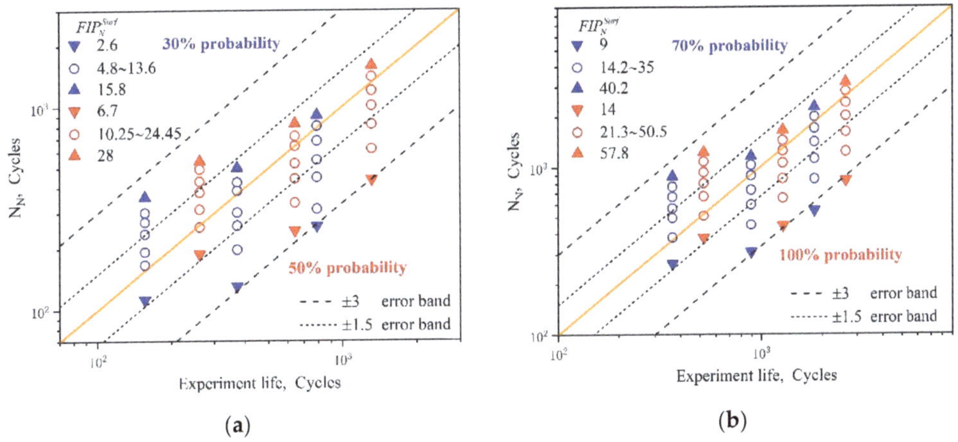

Figure 14. Comparisons between the experimental and predicted loading cycles for nucleation by using FIP_N^{Surf} on the free surface: (**a**) 30% probability and 50% probability; (**b**) 70% probability and 100% probability.

5.3.2. Small Crack Propagation Life Prediction Based on Microscale Multiaxial Strain

Figure 15a shows the distributions of the FIP_P for 207 grain sets in the RVE by contours at a strain amplitude of 0.005 after numerical cyclic stability. The results show the FIP_P presents a strong inhomogeneous nature with values varying in the range of -6.614×10^{-7} to 1.535×10^{-2}. In order to characterize the FIP_P with respect to the strain amplitude of 0.5%, 0.6%, and 0.7%, the maximum value of FIP_P was extracted from the corresponding grain sets after cyclic deformation. The data points obtained are displayed in the scatter diagram shown in Figure 15b–d. It can be seen the maximum FIP_P for the crystal plasticity model obtained under different strain amplitudes varies in the range of 2.38×10^{-3} to 1.54×10^{-2}, 3.58×10^{-3} to 1.84×10^{-2}, and 4.11×10^{-3} to 2.14×10^{-2}, respectively. The homogenized values of the largest FIP_{MSC} are, respectively, 7.46×10^{-3}, 8.94×10^{-3}, and 1.05×10^{-2}. Subsequently, these homogenized FIP_{PS} are introduced into Equation (16) to calculate the small crack propagation life. Furthermore, through the analysis of the distribution of the maximum FIP_P value of crystal sets, it can be determined that the stress–strain states at the critical sites evolve in a multiaxial and non-proportional manner. From the data shown in Figure 16, the errors between the predicted results and the experimental data can be seen to be acceptable when the error band is set to the range of ±3.0. It is noteworthy that the range of four probabilities corresponds to different cyclic plastic zone correction factors for different deformation mechanisms. In addition, it is not difficult to find that the fatigue initiation life approach based on the microscale multiaxial strain, corresponding to 100% probability, is obviously conservative as a result of negligence in the nucleation regime.

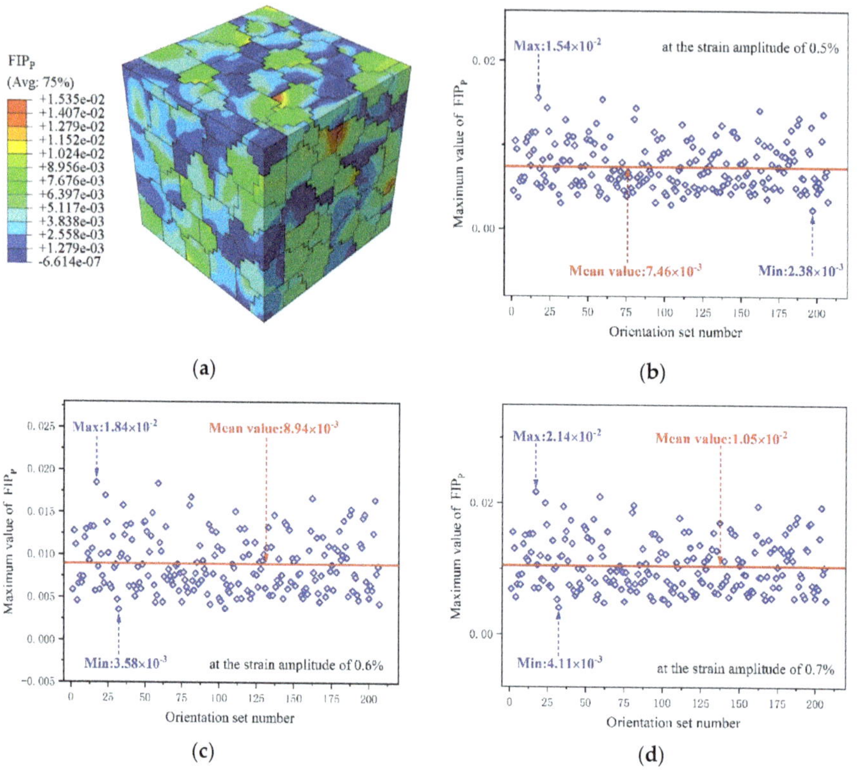

Figure 15. (**a**) Distributions of FIP_p by contours with respect to the strain amplitude of 0.5%. Effect of the grain orientation on the maximum FIP_p at a strain amplitude of (**b**) 0.5%, (**c**) 0.6%, and (**d**) 0.7%.

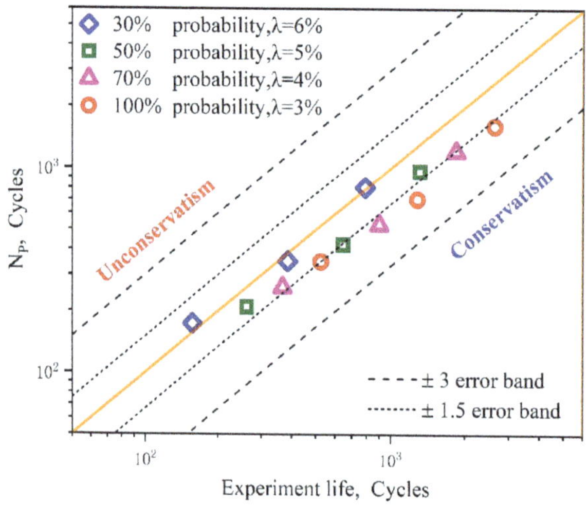

Figure 16. Comparisons between the experimental and predicted small crack propagation life based on the microscale multiaxial FIP_P with respect to four probabilities.

5.3.3. Fatigue Crack Initiation Life Prediction Based on the Combination of Inhomogeneity and Microscale Multiaxial Strain

As shown in Figure 17, the fatigue crack initiation life predicted by the proposed methodology is more reasonable than that based on only one FIP (corresponding to 100% probability) and the previously published method [61] under various macroscopic strain amplitudes. When the percentage of nucleation regime varies from 50% to 70% probability, all the predicted data points are within ±1.5 error bands. The best prediction accuracy was achieved when the phase of nucleation accounted for 70% of the LCF crack initiation life. The prediction data and their errors with the averaged fatigue lives for different prediction models with respect to the corresponding half strain range, $\Delta E_t/2 = 0.5\%$, 0.6%, 0.7%, are also listed in Tables 5 and 6. To describe the relationship between the probability of FN, applied strain amplitude, and fatigue initiation life, a three-dimensional fatigue initiation life histogram was drawn, as shown in Figure 18. It can be clearly seen that the fatigue crack initiation life increases along with the probability of FN, while decreasing as the applied strain amplitude increases. Meanwhile, it can also be seen that the fatigue initiation life is dominated by the applied strain amplitude. The findings shown above further indicate that the influences of the inhomogeneous plastic strain for fatigue initiation life are significant at lower applied strain ranges, while the influences of the multiaxial strain field for small crack propagation are remarkable only at larger strain ranges. The proposed method, which explicitly considers the phase of nucleation and small cracks, can be used to provide an accurate description of the early behavior of UFG metallic alloys through CPFEM simulation. Since our approach considers the heterogeneity characteristics arising from crystallographic orientation and the microscale LCF multiaxial strain field, it can capture the grain-scale deformation localization well and therefore improve the precision of fatigue crack initiation life prediction.

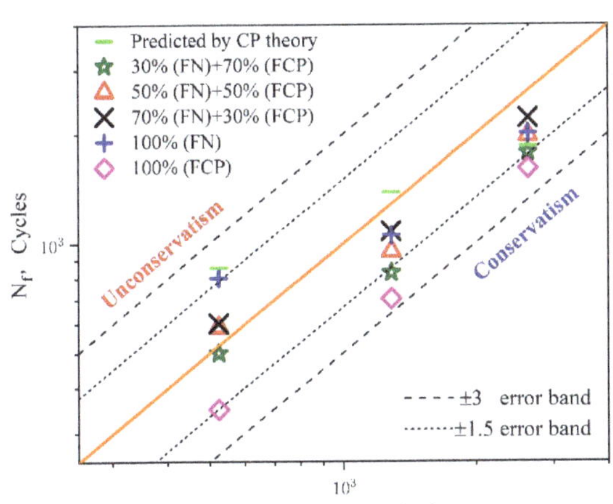

Figure 17. Comparisons between the experiments, predicted results using the proposed method under different probability combinations, and the method based on the crystal plasticity (CP) theory.

Table 5. Tested LCF lives of the UFG AA6061 vs. the predicted results considering only one FIP.

$\Delta E_t/2$ (%)	N_f	$\overline{N}_f$	N_N	$\frac{N_N - \overline{N}_f}{\overline{N}_f}$	N_P	$\frac{N_P - \overline{N}_f}{\overline{N}_f}$
0.5	3125/2173/2589	2629	2010	−0.24	1608	−0.39
0.6	1653/986/1213	1284	1054	−0.18	704	−0.45
0.7	512/378/681	524	802	0.53	349	−0.33

Table 6. Computation results and prediction errors made by N1, N2, N3, and N4 schemes.

N_1: 30%(FN) + 70%(FCP)		N_2: 50%(FN) + 50%(FCP)		N_3: 70%(FN) + 30%(FCP)		N_4: Previous Published Method	
N_1	$\frac{N_1 - \overline{N}_f}{\overline{N}_f}$	N_2	$\frac{N_2 - \overline{N}_f}{\overline{N}_f}$	N_3	$\frac{N_3 - \overline{N}_f}{\overline{N}_f}$	N_4	$\frac{N_4 - \overline{N}_f}{\overline{N}_f}$
1774	−0.33	1978	−0.25	2210	−0.16	1853	−0.30
830	−0.35	949	−0.26	1079	−0.16	1385	0.08
499	−0.05	589	0.12	602	0.15	856	0.63

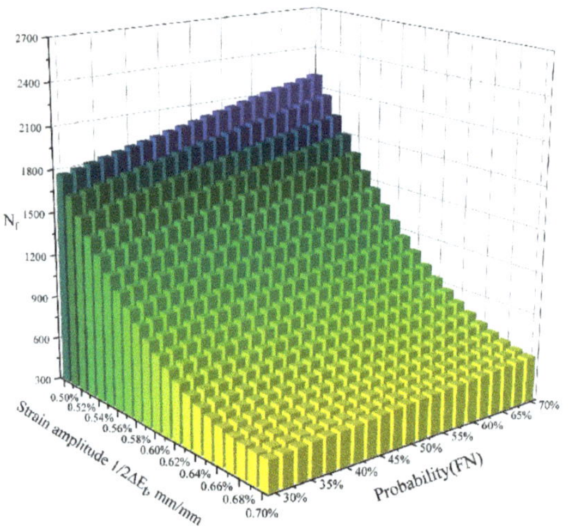

Figure 18. Comparisons between the experiments, predicted results using the proposed method under different probability combinations, and a previously published method based on the crystal plasticity (CP) theory.

6. Conclusions

In this work, an LCF crack initiation life prediction approach was proposed that explicitly distinguishes the nucleation and small crack propagation regime; it contains two newly proposed FIPs relating to inhomogeneity and the microscale multiaxial strain field. Micro-level numerical simulations and a series of strain-controlled fatigue experiments were conducted for UFG 6061AA in order to examine the prediction capacities of the proposed model. The main findings of this study can be summarized as follows:

(1) By performing a statistical analysis of the RVE and its free surface, it was proven that, even if there are no machining defects or damage on the free surface of the specimen, this is still the most dangerous place for fatigue nucleation to occur due to the evolution of inhomogeneous deformation. Additionally, stress-assisted FIP based on the statistical method is capable of expressing the degree of inhomogeneity of UFG material.

(2) Regarding the two newly proposed FIPs as the driving force for the nucleation and small crack propagation, we predicted the nucleation life and small crack propagation life with respect to different probability factors. The predicted accuracy of the fatigue crack initiation life based on only one FIP was acceptable when the error bands were set in the range of ±3.

(3) When the phase of nucleation accounted for 50% to 70% of the LCF crack initiation life for three different strain amplitudes, the predicted accuracy of the developed numerical process was improved, with all the predicted data points lying within the ±1.5 error band. The proposed methodology accompanied by the two FIPs provides new insights into the early stage of the LCF fatigue behavior of UFG AA.

However, the criterion used to judge fatigue failure may warrant further research to take into account the effect of the surface roughness and material internal defects in engineering materials and structural components. Moreover, in situ experimental observation will be conducted in future work to better understand the LCF properties of UFG aluminum alloy and provide a more accurate verification of the proposed methods.

Author Contributions: Conceptualization, T.S. and Z.Z.; methodology, T.S.; software, L.Q. and Y.X.; validation, T.S. and Z.Z.; formal analysis, T.S.; investigation, T.S.; resources, Z.Z.; data curation, Z.H.; writing—original draft preparation, T.S. and Z.Z.; writing—review and editing, T.S. and Z.Z.; visualization, C.X.; supervision, T.S.; project administration, Z.Z.; funding acquisition, Z.Z. All authors have read and agreed to the published version of the manuscript.

Funding: This research was funded by the Natural Science Foundation of Guangxi Province of China (2021GXNSFAA220119), the Guangxi Key R&D Program (guike AB20159012), Interdisciplinary Scientific Research Foundation of GuangXi University (Grant No. 2022JCC023), the Guangxi Science and Technology Base and Special Talents (guike AD21220101), the National Natural Science Foundation of China (51675110, 11962001), and the Project of Improving the Basic Scientific Research Ability of Young- and Middle-aged Teachers in Guangxi Universities in 2021 (2021KY0426).

Conflicts of Interest: The authors declare no conflict of interest. The funders had no role in the design of the study; in the collection, analyses, or interpretation of data; in the writing of the manuscript; or in the decision to publish the results.

Nomenclature

UFG AA	Ultrafine-grained aluminum alloy
CPFE	Crystal plasticity finite element
FIP	Fatigue indicator parameter
RVE	Representative volume element
EBSD	Electron backscatter diffraction
LCF	Low cycle fatigue
HCF	High cycle fatigue
F–S	Fatemi and Socie
NPCAH	Non-proportional cyclic additional hardening
N_N	Fatigue nucleation life
N_P	Small crack propagation life
FIP_N	FIP for the whole RVE
FIP_N^{Surf}	FIP on the free surface
FIP_p	FIP for small crack propagation
$R_{P,UFG}, R_P$	Effective yield strength of UFG and its CG counterparts
$\hat{f}, f^{Surf}$	The dot product of normal stress and mesoscopic longitudinal strain for the RVE and its free surface
n', K'_{UFG}	Cyclic stress–strain hardening exponent and coefficient
$\varepsilon_n^{(\alpha)}$	Normal strain on the α slip system
$\Delta\varepsilon_{eq,N}/2$	Equivalent multiaxial strain reflecting non-proportional cyclic additional hardening
$\Delta\varepsilon_{eq,m}/2$	Mean equivalent normal strain amplitude
$\Delta\gamma_m/2$	Mean shear strain amplitude
$\Delta\varepsilon_{n,m}/2$	Mean normal strain amplitude
V_{int}	Interaction energy
$\tilde{\varepsilon}_{FDR}$	The accumulated mean multiaxial damage strain in FDR
ε_{CPZ}	The strain in cyclic plastic region
r_{FDR}	The radius of the fatigue damage region
C_{11}, C_{12}, C_{44}	Elastic constants
α, r_i, r_f	Crack length, initial crack length and critical crack length

$\chi^{(\alpha)}$	Back stress on the αth slip system
$\tau^{(\alpha)}$	Resolved shear stress on the αth slip system
$\Delta\sigma/2$, $\Delta\sigma_{eff}/2$	Cyclic stress amplitude, effective stress amplitude
ρ_0	Initial dislocation density
b	Burgers vector
ΔG_0	Activation energy
m	Activation length parameter
k_B	Boltzmann constant
p, q	Exponential constants
T	Tested temperature
d_{mean}	Mean grain size
G	Shear modulus
ω_1, ω_2	Latent hardening ratio
v_0	Attempt frequency
$A^{\alpha\beta}$	Hardening matrix
y_c	Material constant
r_D	Dynamic recovery parameter
h_b	Hardening constant
ΔGB	Activation energy of grain boundary diffusion
ΔG_0	Activation energy
u	Material parameter
$\varepsilon_{ll}, \varepsilon_{ll}^{Surf}$	Mesoscopic longitudinal strain for the RVE and its free surface
$\sigma_n, \sigma_n^{Surf}$	Normal stress for the RVE and its free surface
$\Delta\gamma_{max}/2$	Maximum plastic shear strain amplitude
N_f	Total fatigue initiation life
N_N	Fatigue nucleation life
$\hat{\tau}_{UFG}$	Lattice friction stress at the current temperature
r_{FDR}	Fatigue damage region
K_{GBS}	Grain boundary strengthening factor
K_{GBC}	Grain boundary constraint factor
$\Delta CTOD$	Crack tip opening displacement
$\tau_{ath}^{(\alpha)}$	Slip resistance on the αth slip system
N_P	Small crack propagation life
FCP	Fatigue crack propagation
FN	Fatigue nucleation

References

1. Stojanovic, B.; Milan, B.; Epler, I. Application of Aluminum and Aluminum Alloys in Engineering. *Appl. Eng. Lett. J. Eng. Appl. Sci.* **2018**, *3*, 52–62. [CrossRef]
2. Li, H.; Yan, Z.; Cao, L. Bake hardening behavior and precipitation kinetic of a novel Al-Mg-Si-Cu aluminum alloy for lightweight automotive body. *Mater. Sci. Eng. A* **2018**, *728*, 88–94. [CrossRef]
3. Lipinska, M.; Chrominski, W.; Olejnik, L.; Golinski, J.; Rosochowski, A.; Lewandowska, M. Ultrafine-Grained Plates of Al-Mg-Si Alloy Obtained by Incremental Equal Channel Angular Pressing: Microstructure and Mechanical Properties. *Metall. Mater. Trans. A* **2017**, *48*, 4871–4882. [CrossRef]
4. Malekjani, S.; Hodgson, P.D.; Cizek, P.; Sabirov, I.; Hilditch, T.B. Cyclic deformation response of UFG 2024 Al alloy. *Int. J. Fatigue* **2011**, *33*, 700–709. [CrossRef]
5. Chen, B.; Jiang, J.; Dunne, F.P.E. Is stored energy density the primary meso-scale mechanistic driver for fatigue crack nucleation? *Int. J. Plast.* **2018**, *101*, 213–229. [CrossRef]
6. Manonukul, A.; Dunne, F. High- and low-cycle fatigue crack initiation using polycrystal plasticity. *Proc. R. Soc. A Math. Phys. Eng. Sci.* **2004**, *460*, 1881–1903. [CrossRef]
7. Cruzado, A.; Lucarini, S.; Llorca, J.; Segurado, J. Microstructure-based fatigue life model of metallic alloys with bilinear Coffin-Manson behavior. *Int. J. Fatigue* **2018**, *107*, 40–48. [CrossRef]
8. Sweeney, C.A.; O'Brien, B.; Dunne, F.P.E.; McHugh, P.E.; Leen, S.B. Strain-gradient modelling of grain size effects on fatigue of CoCr alloy. *Acta Mater.* **2014**, *78*, 341–353. [CrossRef]
9. Mughrabi, H. Microstructural fatigue mechanisms: Cyclic slip irreversibility, crack initiation, non-linear elastic damage analysis. *Int. J. Fatigue* **2013**, *57*, 2–8. [CrossRef]
10. Zhang, K.S.; Wu, M.S.; Feng, R. Simulation of microplasticity-induced deformation in uniaxially strained ceramics by 3-D Voronoi polycrystal modeling. *Int. J. Plast.* **2005**, *21*, 801–834. [CrossRef]

11. Zhang, K.-S.; Ju, J.W.; Li, Z.; Bai, Y.-L.; Brocks, W. Micromechanics based fatigue life prediction of a polycrystalline metal applying crystal plasticity. *Mech. Mater.* **2015**, *85*, 16–37. [CrossRef]
12. Zhang, K.-S.; Shi, Y.-K.; Ju, J.W. Grain-level statistical plasticity analysis on strain cycle fatigue of a FCC metal. *Mech. Mater.* **2013**, *64*, 76–90. [CrossRef]
13. Mughrabi, H.; Höppel, H.W. Cyclic deformation and fatigue properties of very fine-grained metals and alloys. *Int. J. Fatigue* **2010**, *32*, 1413–1427. [CrossRef]
14. Macek, W.; Branco, R.; Costa, J.D.; Trembacz, J. Fracture Surface Behavior of 34CrNiMo6 High-Strength Steel Bars with Blind Holes under Bending-Torsion Fatigue. *Materials* **2021**, *15*, 80. [CrossRef]
15. Kujawski, N.D. Sequence effects of small amplitude cycles on fatigue crack initiation and propagation in 2024-T351 aluminum. *Int. J. Fatigue* **2001**, *23*, 807–815.
16. Zhuge, H. Experimental and Numerical Study of Combined High and Low Cycle Fatigue Performance of Low Alloy Steel and Engineering Application. *Materials* **2021**, *14*, 3395.
17. Yuan, H.; Zhang, W.; Castelluccio, G.M.; Kim, J.; Liu, Y. Microstructure-sensitive estimation of small fatigue crack growth in bridge steel welds. *Int. J. Fatigue* **2018**, *112*, 183–197. [CrossRef]
18. Moshtaghi, M.; Safyari, M. Effect of Work-Hardening Mechanisms in Asymmetrically Cyclic-Loaded Austenitic Stainless Steels on Low-Cycle and High-Cycle Fatigue Behavior. *Steel Res. Int.* **2020**, *92*, 2000242. [CrossRef]
19. Pineau, A.; McDowell, D.L.; Busso, E.P.; Antolovich, S.D. Failure of metals II: Fatigue. *Acta Mater.* **2016**, *107*, 484–507. [CrossRef]
20. Castelluccio, G.M.; McDowell, D.L. Mesoscale modeling of microstructurally small fatigue cracks in metallic polycrystals. *Mater. Sci. Eng. A* **2014**, *598*, 34–55. [CrossRef]
21. Castelluccio, G.M.; McDowell, D.L. Microstructure-sensitive small fatigue crack growth assessment: Effect of strain ratio, multiaxial strain state, and geometric discontinuities. *Int. J. Fatigue* **2016**, *82*, 521–529. [CrossRef]
22. Przybyla, C.P.; Musinski, W.D.; Castelluccio, G.M.; McDowell, D.L. Microstructure-sensitive HCF and VHCF simulations. *Int. J. Fatigue* **2013**, *57*, 9–27. [CrossRef]
23. Yang, M.; Lu, S. A microstructure-based homogenization model for predicting the low-cycle fatigue initiation life of GH4169 superalloy. *Fatigue Fract. Eng. Mater. Struct.* **2021**, *44*, 1546–1561. [CrossRef]
24. Sonsino, C.M. Multiaxial fatigue life response depending on proportionality grade between normal and shear strains/stresses and material ductility. *Int. J. Fatigue* **2020**, *135*, 105468. [CrossRef]
25. Albinmousa, J. Investigation on multiaxial fatigue crack path using polar stress–strain representation. *Int. J. Fatigue* **2016**, *92*, 406–414. [CrossRef]
26. Tong, J.; Lin, B.; Lu, Y.W.; Madi, K.; Tai, Y.H.; Yates, J.R.; Doquet, V. Near-tip strain evolution under cyclic loading: In situ experimental observation and numerical modelling. *Int. J. Fatigue* **2015**, *71*, 45–52. [CrossRef]
27. Zhao, B.; Xie, L.; Song, J.; Zhao, Z.; Xu, G. Prediction of multiaxial fatigue life for complex three-imensional stress state considering effect of additional hardening. *Fatigue Fract. Eng. Mater. Struct.* **2019**, *42*, 2558–2578. [CrossRef]
28. Itoh, T.; Nakamura, H.; Takanashi, M.; Wu, M. Multiaxial low cycle fatigue life of Ti-6Al-4V under non-proportional loading with mean strain. *Theor. Appl. Fract. Mech.* **2017**, *90*, 165–173. [CrossRef]
29. Jiang, W.; Yang, P. Experimental studies on crack propagation and accumulative mean strain of cracked stiffened plates under low-cycle fatigue loads. *Ocean. Eng.* **2020**, *214*, 107744. [CrossRef]
30. Ding, H.Z.; Mughrabi, H.; Höppel, H.W. Alow-cycle fatigue life prediction model of ultrafine-grained metals. *Fatigue Fract. Eng. Mater. Struct.* **2002**, *25*, 975–984. [CrossRef]
31. Sga, B.; Sm, B. Fracture toughness characteristics of ultrafine grained Nb–Ti stabilized microalloyed and interstitial free steels processed by advanced multiphase control rolling. *Mater. Charact.* **2020**, *159*, 110003.
32. Hohenwarter, A.; Pippan, R. Fracture toughness evaluation of ultrafine-grained nickel. *Scr. Mater.* **2011**, *64*, 982–985. [CrossRef]
33. Kumar, N.; Owolabi, G.M.; Jayaganthan, R. Al 6082 alloy strengthening through low strain multi-axial forging. *Mater. Charact.* **2019**, *155*, 109761. [CrossRef]
34. Kumar, N.; Goel, S.; Jayaganthan, R.; Owolabi, G.M. The influence of metallurgical factors on low cycle fatigue behavior of ultra-fine grained 6082 Al alloy. *Int. J. Fatigue* **2018**, *110*, 130–143. [CrossRef]
35. Hockauf, K.; Wagner, M.F.X.; Halle, T.; Niendorf, T.; Hockauf, M.; Lampke, T. Influence of precipitates on low-cycle fatigue and crack growth behavior in an ultrafine-grained aluminum alloy. *Acta Mater.* **2014**, *80*, 250–263. [CrossRef]
36. Kato, M. Thermally activated dislocation depinning at a grain boundary in nanocrystalline and ultrafine-grained materials. *Mater. Sci. Eng. A* **2009**, *516*, 276–282. [CrossRef]
37. Li, D.-F.; Barrett, R.A.; O'Donoghue, P.E.; O'Dowd, N.P.; Leen, S.B. A multi-scale crystal plasticity model for cyclic plasticity and low-cycle fatigue in a precipitate-strengthened steel at elevated temperature. *J. Mech. Phys. Solids* **2017**, *101*, 44–62. [CrossRef]
38. Meng, L.; Goyal, A.; Doquet, V.; Ranc, N.; Couzinié, J.-P. Ultrafine versus coarse grained Al 5083 alloys: From low-cycle to very-high-cycle fatigue. *Int. J. Fatigue* **2019**, *121*, 84–97. [CrossRef]
39. Wang, C.H.; Brown, M.W. A path-independent parameter for fatigue under proportional and non-proportional loading. *Fatigue Fract. Eng. Mater. Struct.* **1993**, *16*, 1285–1297. [CrossRef]
40. McDowell, D.L.; Berard, J.Y. A ΔJ-based approach to biaxial fatigue. *Fatigue Fract. Eng. Mater. Struct.* **1992**, *15*, 719–741. [CrossRef]
41. Kinoshita, K.; Shimokawa, T.; Kinari, T. Grain Boundary Structure Dependence of Extrinsic Grain Boundary Dislocation Emission Phenomena: A Molecular Dynamics Study. *Mater. Trans.* **2012**, *53*, 147–155. [CrossRef]

42. Geng, J.; Li, Y.; Xiao, H.; Wang, Z.; Wang, M.; Chen, D.; Wang, H. Deformation-induced ultrafine grains near fatigue crack tip and correlative fatigue damage in Al matrix composite. *Scr. Mater.* **2021**, *193*, 49–54. [CrossRef]
43. Rice, J.R. A Path Independent Integral and the Approximate Analysis of Strain Concentration by Notches and Cracks. *J. Appl. Mech.* **1968**, *35*, 379–386. [CrossRef]
44. Esmaeili, A.; Shaeri, M.H.; Noghani, M.T.; Razaghian, A. Fatigue behavior of AA7075 aluminium alloy severely deformed by equal channel angular pressing. *J. Alloys Compd.* **2018**, *757*, 324–332. [CrossRef]
45. Fintová, S.; Arzaghi, M.; Kuběna, I.; Kunz, L.; Sarrazin-Baudoux, C. Fatigue crack propagation in UFG Ti grade 4 processed by severe plastic deformation. *Int. J. Fatigue* **2017**, *98*, 187–194. [CrossRef]
46. Goto, M.; Yamamoto, T.; Han, S.Z.; Lim, S.H.; Kim, S.; Ahn, J.H.; Lee, S.J.; Yakushiji, T.; Lee, J. Crack initiation mechanism in ultrafine-grained copper fabricated by severe plastic deformation in the high-cycle fatigue regime. *Mater. Sci. Eng. A* **2020**, *788*, 139569. [CrossRef]
47. Rao, P.N.; Singh, D.; Brokmeier, H.G.; Jayaganthan, R. Effect of ageing on tensile behavior of ultrafine grained Al 6061 alloy. *Mater. Sci. Eng. A* **2015**, *641*, 391–401. [CrossRef]
48. Ma, K.; Wen, H.; Tao, H.; Topping, T.D.; Schoenung, J.M. Mechanical behavior and strengthening mechanisms in ultrafine grain precipitation-strengthened aluminum alloy. *Acta Mater.* **2014**, *62*, 141–155. [CrossRef]
49. Rice, J.R. Mechanics of Crack Tip Deformation and Extension by Fatigue. In *Fatigue Crack Propagation*; American Society for Testing and Materials: West Conshohocken, PA, USA, 1967.
50. Broek, D. *Elementary Engineering Fracture Mechanics*; Springer Science & Business Media: Berlin/Heidelberg, Germany, 1982.
51. Khan, A.S.; Liu, J. A deformation mechanism based crystal plasticity model of ultrafine-grained/nanocrystalline FCC polycrystals. *Int. J. Plast.* **2016**, *86*, 56–69. [CrossRef]
52. Busso, E.P.; McClintock, F.A. A dislocation mechanics-based crystallographic model of a B2-type intermetallic alloy. *Int. J. Plast.* **1996**, *12*, 1–28. [CrossRef]
53. Dong, Y.; Kang, G.; Yu, C. A dislocation-based cyclic polycrystalline visco-plastic constitutive model for ratchetting of metals with face-centered cubic crystal structure. *Comput. Mater. Sci.* **2014**, *91*, 75–82. [CrossRef]
54. Le Pécheur, A.; Curtit, F.; Clavel, M.; Stephan, J.M.; Rey, C.; Bompard, P. Polycrystal modelling of fatigue: Pre-hardening and surface roughness effects on damage initiation for 304L stainless steel. *Int. J. Fatigue* **2012**, *45*, 48–60. [CrossRef]
55. Fujita, T.; Horita, Z.; Langdon, T.G. Using grain boundary engineering to evaluate the diffusion characteristics in ultrafine-grained Al–Mg and Al–Zn alloys. *Mater. Sci. Eng. A* **2004**, *371*, 241–250. [CrossRef]
56. Koju, R.K.; Mishin, Y. Atomistic study of grain-boundary segregation and grain-boundary diffusion in Al-Mg alloys. *Acta Mater.* **2020**, *201*, 596–603. [CrossRef]
57. Padmanabhan, K.A.; Basariya, M.R. Mesoscopic grain boundary sliding as the rate controlling process for high strain rate superplastic deformation. *Mater. Sci. Eng. A* **2009**, *527*, 225–234. [CrossRef]
58. Zha, M.; Li, Y.; Mathiesen, R.H.; Bjørge, R.; Roven, H.J. Microstructure evolution and mechanical behavior of a binary Al–7Mg alloy processed by equal-channel angular pressing. *Acta Mater.* **2015**, *84*, 42–54. [CrossRef]
59. Davidson, D.L.; Mcclung, R.C. Local constraint near fatigue cracks in alloys and particulate composites. *Int. J. Fract.* **1997**, *84*, 81–98. [CrossRef]
60. Hultgren, G.; Mansour, R.; Barsoum, Z.; Olsson, M. Fatigue probability model for AWJ-cut steel including surface roughness and residual stress. *J. Constr. Steel Res.* **2021**, *179*, 106537. [CrossRef]
61. Das, P.; Jayaganthan, R.; Chowdhury, T.; Singh, I.V. Fatigue behaviour and crack growth rate of cryorolled Al 7075 alloy. *Mater. Sci. Eng. A* **2011**, *528*, 7124–7132. [CrossRef]

materials

MDPI

Article

Analytical Solution of Thermo–Mechanical Properties of Functionally Graded Materials by Asymptotic Homogenization Method

Dan Chen [1], Lisheng Liu [2,*], Liangliang Chu [2,*] and Qiwen Liu [2]

[1] Department of Mechanics and Engineering Structure, Wuhan University of Technology, Wuhan 430070, China; 292562@whut.edu.cn

[2] Hubei Key Laboratory of Theory and Application of Advanced Materials Mechanics, Wuhan University of Technology, Wuhan 430070, China; qiwen_liu@whut.edu.cn

* Correspondence: liulish@whut.edu.cn (L.L.); chull_620@whut.edu.cn (L.C.)

Abstract: In this work, a general mathematical model for functionally graded heterogeneous equilibrium boundary value problems is considered. A methodology to find the local problems and the effective properties of functionally graded materials (FGMs) with generalized periodicity is presented, using the asymptotic homogenization method (AHM). The present models consist of the matrix metal Mo and the reinforced phase ceramic ZrC, the constituent ratios and the property gradation profiles of which can be described by the designed volume fraction. Firstly, a new threshold segmentation method is proposed to construct the gradient structure of the FGMs, which lays the groundwork for the subsequent research on the properties of materials. Further, a study of FGMs varied along a certain direction and the influence of the varied constituents and graded structures in the behavior of heterogeneous structures are investigated by the AHM. Consequently, the closed–form formulas for the effective thermo–mechanical coupling tensors are obtained, based on the solutions of local problems of FGMs with the periodic boundary conditions. These formulas provide information for the understanding of the traditional homogenized structure, and the results also be verified the correctness by the Mori–Tanaka method and AHM numerical solution. The results show that the designed structure profiles have great influence on the effective properties of the present inhomogeneous heterogeneous models. This research will be of great reference significance for the future material optimization design.

Keywords: functionally graded materials; asymptotic homogenization method; effective properties; thermo–mechanical coupling

1. Introduction

The growing demand and optimization of industrial applications reveal the limitations of traditional materials. The development of functionally graded materials (FGMs) solves these limitations to a large extent by integrating mutually exclusive properties [1–3]. FGMs are a kind of advanced composite materials and show a local characteristic dependence on the spatial distribution of their constituent phases [4,5], which were first designed for the Japanese space shuttle project in 1983 to reduce the thermal stress caused by the high temperature of the metal and ceramic interface. This advanced heterogeneous composite improves the thermal shock resistance of the material along the gradient direction [6,7]. The superior material properties of FGMs had been widely applied in different areas, such as in electronic packaging, automobile and aerospace industries [8]. Many studies had been done on the fabrication, application, and mechanical properties [9–12] of FGMs. How to predict and determine accurately the properties of such inhomogeneous FGMs is vital research problem for obtaining the thermo–mechanical ability under the conditions of some extreme environment [13,14]. However, the properties of FGMs depend not only on the constituent materials, but also on their microstructure. The complexity of the microstructure of FGMs, i.e., the content of components, distribution of inclusions and so on, makes the prediction of their equivalent thermo–mechanical properties very untoward [15].

In recent years, a great deal of research has focused its attention on the prediction of the effective properties of composite materials. Theoretical boundary methods include Hashin–Shtrikman bounds [16–19] and Budiansky bounds [20], but these methods do not consider the complexity of microstructure. For composites, it is very vital to understand their accurate properties after the microstructure is determined. Based on Eshelby's equivalent inclusion theory [21], combined with the average stress, self–consistent method [22], generalized self–consistent method [23] and Mori–Tanaka method [24–26] are proposed to predict the effective properties of composites. For example, Tran et al. [25] derived the solution of Eshelby's spherical heterogeneous problem by using the Mori–Tanaka method to predict the effective properties of gradient composites with spherical inclusions. However, because Eshelby's equivalent inclusion theory is based on ellipsoid, these methods are initially only applicable to composites with ellipsoidal inclusions. For composites with complex shapes, the finite element analysis method (FEM) with periodic boundary conditions can be used [27,28]. Kundalwal et al. [27] studied the effective properties of fiber–reinforced composites by using the FEM, and calculated the effective elastic constants of the materials. Zhang et al. [28] studied the multi field properties of electromagnetic thermoelastic composites by using the FEM based on micromechanics, and obtained the effective thermal expansion coefficient and other parameters of intelligent composites under periodic boundary conditions. However, these methods lack strict mathematical framework to clarify the relationship between the effective properties and the material layout. Recently, based on strict mathematical derivation, a novel homogenization method is proposed, asymptotic homogenization method (AHM), and shows good accuracy compared with experiments [29–31]. The method of AHM can be incorporated into the coupling between the micro and macro behaviors of heterogeneous materials [32–36].

It can be seen from the existing literature that AHM has been widely used in the prediction and analysis of the effective properties of composites, with good calculation results. In terms of prediction of the effective mechanical properties [37–40], for example, Santana et al. [38] predicted the effective constitutive coefficient tensor of the composite by using two–scale AHM considering the effect of debonding between fiber and matrix. The comparison of homogenization and non–homogenization numerical models shows that almost all effective coefficients have good convergence. Nasirov et al. [39] further used the three–scale formulas of AHM to predict Young's modulus of fiber from micro scale to mesoscale and from mesoscale to macro scale, which is in good agreement with the tensile test. The study of multi physical field coupling has further developed the application of AHM, especially the prediction of effective thermo–mechanical properties [41–45]. Muhammad et al. [42] established AHM for three–scale composite analysis considering the thermomechanical effect, and compared the Young's modulus and Poisson's ratio obtained from the analysis with the experimental results. The results show that the two methods have good consistency in Young's modulus. Zhao et al. [43] studied the thermo–mechanical properties of particle–filled polymers using AHM. The results show that the numerical results of progressive homogenization method are in good agreement with the experimental results. For periodic composites, the progressive homogenization method is more reasonable and accurate than the representative volume element method. Dirichlet formulation overestimates the effective tensor and Neumann formulation underestimates the effective tensor. AHM can also be used to study multi physical field coupling such as mechano–chemical coupling [46], thermo–electric coupling [47], and magneto–electric coupling [48]. However, the research on the effective properties of FGMs by AHM has not been reported.

Because AHM is based on strict mathematical framework, it finally leads to a complex implementation, which is mainly related to the complex microstructure of heterogeneous materials [49]. In general, the relationship between macro and micro structure is realized through the concept of the representative volume unit (RVE) [50–52]. RVE needs to characterize accurately the microstructure of composites. Tian et al. [53,54] proposed the RVE generation method of spatially randomly distributed fibers in combination with the random sequential absorption (RSA) technique, and then Tian et al. [55] used RSA technique to generate the RVE of composites containing cylindrical inclusions. At the same time, Dutra et al. [15] believe that the steps of AHM using numerical simulation are not simple, such as applying the first boundary condition and periodic boundary condition. When Ramos et al. [56] studied two–phase parallel fiber reinforced periodic viscoelastic composites, a simple closed–form formula for the effective properties of square and hexagonal element composites was obtained by using AHM. Based on the elastic viscoelastic correspondence principle and assuming that there are perfect contact conditions at each split interface, the local problems and global viscoelastic properties are obtained in explicit form. Compared with the traditional Maxwell and Kelvin models, its effectiveness is verified. It can be seen that the closed–form formula of effective performance not only has good accuracy, but also does not need to build a complex microstructure model for numerical solution, which can greatly reduce the demand for computers and can be better used in practice. Therefore, when studying the effective thermo–mechanical properties of FGMs, it is very important to deduce the closed–form formula of effective thermo–mechanical properties by AHM on the premise of ensuring the accuracy of calculation results.

In this study, a methodology to solve the local problems and the effective properties of FGMs with generalized periodicity is presented base on the method of AHM. For FGMs with special distribution, i.e., power–law distribution with two parameters, the regulating effects of different gradient models on FGMs thermo–mechanical performance are analyzed. Gaussian random field method and our new threshold segmentation algorithm are used to establish a two–phase random gradient structure that obeys the power–law distribution of two parameters, so as to show the distribution characteristics of different gradient model components. The theoretical closed–form formulas for calculating the performance of FGMs are derived from the gradual homogenization method and its generalized periodicity. For the thermo–mechanical coupling field, these properties include elastic tensor's components, thermal expansion tensor's components and thermal conduction tensor's components. The Mori–Tanaka method and numerical solution of AHM in literature have been used to verify the correctness of the present method. The quantitative relationship between the gradient structure and thermo–mechanical properties of FGMs and the resulting analysis is an important part of our research.

2. The Establishment of the Gradient Microstructure

In this section, the Gaussian random field method combined with a new threshold segmentation algorithm is mainly considered to construct the FGM model. For various random media, various construction methods based on finite morphological and statistical information extracted from 2D/3D data are proposed, including random–set method [57], Gaussian random field method [58,59] etc. The random–set method can be used for composites with spherical or polygonal inclusions, but for inclusions with arbitrary shape, the calculation is complex and difficult to realize. In the Gaussian random field method, the field–field correlation function is used to construct the Gaussian field, and the horizontal cutting is used to obtain the two–phase micro model. According to [59], the 2D original gray random distribution microstructure model of heterogeneous materials can be obtained. Generally, in order to obtain the two–phase microstructure model, it is necessary to use an image segmentation algorithm to separate the target phase and background matrix. The determination of the cut–off value in the segmentation algorithm depends on whether the volume fraction of the target phase of the obtained two–phase model is consistent with the target model.

When a cut–off value is selected in the whole gray model, a uniform two–phase model with a certain volume fraction will be obtained. Obviously, this cannot establish the FGM model with gradient distribution of the volume fraction of the component phase. In order to construct the gradient model, we propose a new threshold segmentation algorithm, that is, the whole model is divided into several regions, the number of regions depends on the number of layers of the gradient model itself, and different cut–off values are selected in each block according to the required volume fraction.

It is assumed that the thickness of FGM models studied is h, which is composed of metal phase and ceramic phase. The properties of the studied model are assumed to be isotropic and homogeneous in the cross section while following a power law distribution with two parameters in the thickness direction i.e., $V_p(z)$ is the volume fraction of the component at the distance z from the bottom according to Ref. [60], which can be expressed as,

$$V_p(z) = \begin{cases} \frac{c_0}{2}\left(1+\frac{2z}{h}\right)^m, -\frac{h}{2} \le z \le 0 \\ c_0\left[1-\frac{1}{2}\left(1-\frac{2z}{h}\right)^m\right], 0 \le z \le \frac{h}{2} \end{cases} \quad (1)$$

and the volume fraction of the component of the top surface satisfies,

$$V_b(z) = 1 - V_p(z) \quad (2)$$

where c_0 denotes the component ratio coefficient to adjust the constituent ratio of FGMs, $0 \le c_0 \le 1.0$; m is the gradient index of material property to describe the property gradation profiles of FGMs, i.e., m = 0 corresponds to the homogeneous traditional composition. The volume fraction V_p(z) varies with the structure of FGMs (c_0, m), as shown Figure 1. Once the composition of the material composition ratio is determined, the current structure can only be adjusted by the structural design method through the gradient index (Figure 1b). Given the existence of two parameters, component ratio coefficient and gradient coefficient, this gradient structure is more scientific, which is beneficial to explore the influence of different volume component phase distribution on the thermo–mechanical properties of FGM structure.

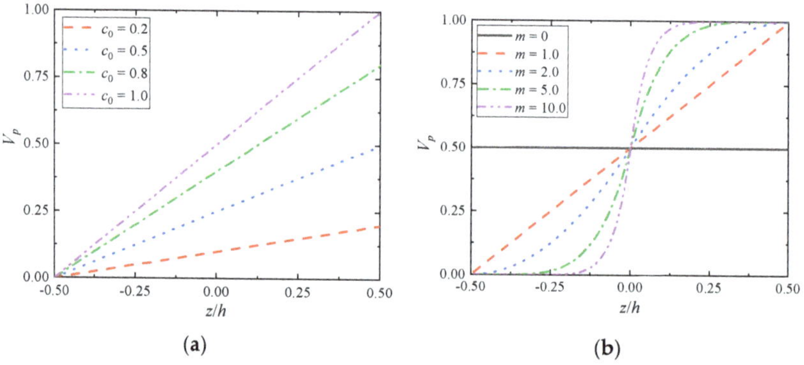

Figure 1. Variation of volume fraction V_p vs. FGMs structure: (**a**) the variational material component ratio coefficients c_0 (m = 1.0); (**b**) the variational gradient indexes m (c_0 = 1.0).

According to the new threshold segmentation algorithm, the FGMs model with the volume fraction of component phase following Equations (1) and (2) in the thickness direction is constructed. The gradient model shows different gradient distributions according to the component ratio coefficient c_0 and gradient index m, as shown in Figures 2 and 3. It is observed that the dark phase and the light phase of the model show the opposite change trend in the longitudinal direction, i.e., they represent two different component phases, respectively. The volume proportion of the light phase from bottom to top is increasing, showing a gradient change.

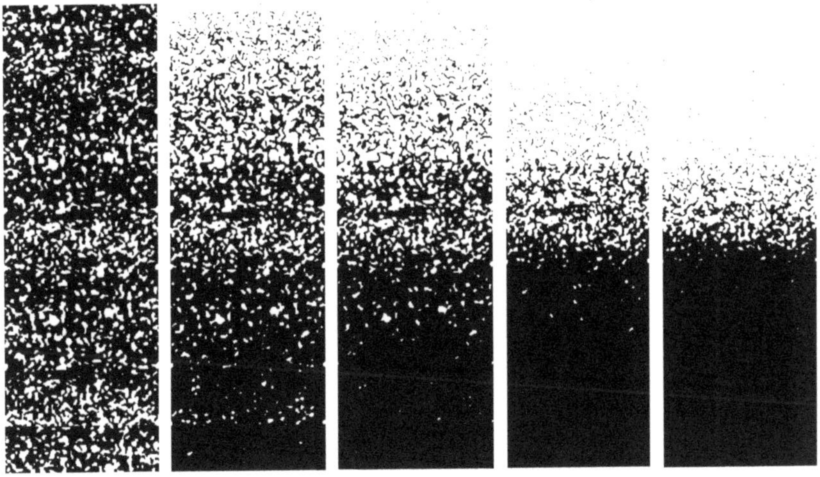

(a) (b) (c) (d) (e)

Figure 2. Establishment of the gradient model with (**a**) $m = 0$; (**b**) $m = 1.0$; (**c**) $m = 2.0$; (**d**) $m = 5.0$; (**e**) $m = 10.0$ ($c_0 = 1.0$).

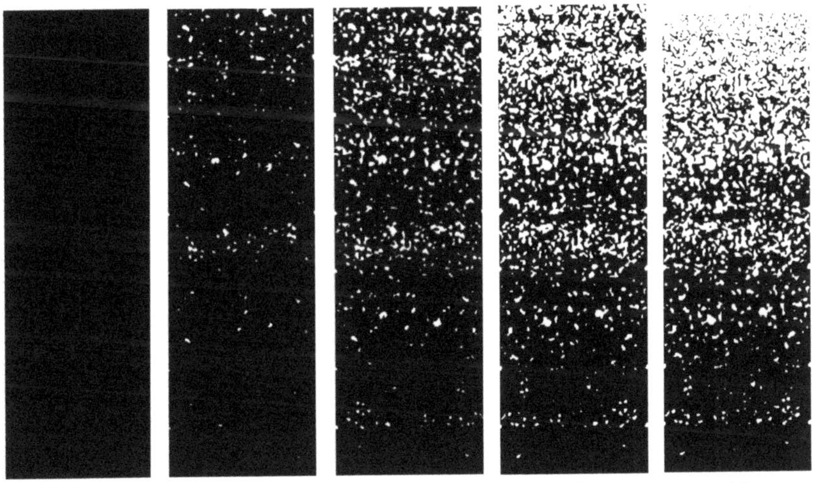

(a) (b) (c) (d) (e)

Figure 3. Establishment of the gradient model with (**a**) $c_0 = 0$; (**b**) $c_0 = 0.2$; (**c**) $c_0 = 0.5$; (**d**) $c_0 = 0.8$; (**e**) $c_0 = 1.0$ ($m = 1.0$).

In Figure 2, the component ratio coefficient $c_0 = 1.0$ of the gradient models have different gradient indices m. For different m, the models have little difference near $z/h = 0$, but the farther away from this position, the greater the difference. With the increase of m, the closer the position where the light phase begins to appear to the top surface, the narrower the position area where the two–phase components form the interactive structure. When the gradient index $m = 1.0$, the gradient model with different component ratio coefficient c_0 is constructed as shown in Figure 3. There is only the dark phase when $c_0 = 0$. The volume fraction of the light phase increased gradually from bottom to top in the thickness direction, ($c_0 \neq 0$), but there was no drastic change trend. At the same time, the two–phase components almost form an interactive structure in the whole thickness direction. With the increase of c_0, the proportion of the overall volume of the light phase increases continuously. When $c_0 = 1.0$, only the light phase is on the top surface.

It should be noted that what are constructed here are not the actual microstructure, but the microstructure model similar to the actual FGMs. Strictly speaking, the gradient change of the gradient model is not continuous, but actually presents a stepped gradient change according to the region. In fact, the real FGMs is not a material with continuous gradient change in the ideal sense.

3. Asymptotic Homogenization Method

In this section, the local problems and effective properties of FGMs with generalized periodicity are derived and solved by AHM, and the theoretical closed calculation formulas for calculating the thermo–mechanical properties of FGMs are obtained, including effective elastic tensor's components, effective thermal expansion tensor's components and effective thermal conduction tensor's components.

Let $\Omega \subset R^3$ be a three–dimensional open connected bounded domain with an infinitely smooth boundary $\partial\Omega$. The equilibrium problem on a thermal mechanical coupling structure Ω is given by the kinetic equation and heat conduction equation,

$$\frac{\partial \sigma_{ij}}{\partial x_j} = -f_i \text{ on } \Omega, \text{ with } \mathbf{u} = 0 \text{ on } \partial\Omega \tag{3}$$

$$\frac{\partial q_i}{\partial x_i} = 0 \text{ on } \Omega, \text{ with } \mathbf{u} = 0 \text{ on } \partial\Omega \tag{4}$$

where, $\boldsymbol{\sigma}$ denotes the stress acting on the domain Ω; $\mathbf{f}$ is the body force; $\mathbf{x}$ is macroscopic (global) coordinate system; $\mathbf{q}$ is the thermal circulation vector.

Stress–strain relationship in linear thermomechanical problem can be expressed as follows,

$$\delta_{ij} = C_{ijkl}(\varepsilon_{kl} - \varepsilon^t_{kl}) \tag{5}$$

where, $\mathbf{C}$ is the four–order elasticity tensor of the material; ε is the total strain, and ε^t denotes thermal strain.

The mechanical strain–displacement relationship and thermal strain be stated as, respectively,

$$\varepsilon_{ij} = \frac{1}{2}\left(\frac{\partial u_i}{\partial x_j} + \frac{\partial u_j}{\partial x_i}\right) \tag{6}$$

$$\varepsilon^t_{ij} = \alpha_{ij}\Delta T \tag{7}$$

where, $\mathbf{u}$ denotes the actual displacement; $\boldsymbol{\alpha}$ is the second–order thermal expansion tensor; T is the temperature field, and $\Delta T = T - T_0$, T_0 is the initial temperature.

By substituting Equation (7) into Equation (5), and considering the symmetry of elastic tensor, the stress can be expressed as follows,

$$\sigma_{ij} = C_{ijkl}\varepsilon_{kl} - \beta_{ij}\Delta T \tag{8}$$

where $\beta_{ij} = C_{ijkl}\cdot\alpha_{kl}$.

For the thermal circulation vector $\mathbf{q}$,

$$q_i = -\kappa_{ij}\frac{\partial T}{\partial x_j} \tag{9}$$

where $\boldsymbol{\kappa}$ is the second–order thermal conductivity tensor.

These coefficients hold the following symmetry relationships, $C_{ijkl} = C_{jikl} = C_{klij}$, $\beta_{ij} = \beta_{ji}$. To complete our problem formulation, we compliment the boundary conditions with the following interphase conditions, $||\sigma_{ij}\, n_j|| = 0$, $||q_i n_i|| = 0$, where $||A||$ denotes a discontinuity in the value of A.

3.1. Asymptotic Analysis and Model Development

It is apparent that the problem at hand is characterized by two scales: the macroscopic scale of order **x** which accounts for the variation of the dependent variables from one unit–cell to the next, and the microscopic scale of order **y** which accounts for periodicity. Appropriately, the first step in the asymptotic homogenization technique is the definition of a new microscopic variable y_i as follows,

$$y_i = \frac{x_i}{\varepsilon} \tag{10}$$

with a constant $0 < \varepsilon << 1$. Equation (10) represents the ratio of the length unit vector in the microscopic coordinates to the length unit vector in the macroscopic coordinates.

And the partial derivative becomes,

$$\nabla_i \mathbf{y} = \frac{\partial}{\partial x_i} + \frac{1}{\varepsilon}\frac{\partial}{\partial y_i} \tag{11}$$

Then introduction of **y** necessitates the transformation, through the application of the chain rule, Equation (3) becomes the following Equation (12) and Equation (4) becomes the following Equation (13),

$$\frac{\partial \sigma_{ij}(\mathbf{x},\mathbf{y})}{\partial x_j} + \frac{1}{\varepsilon}\frac{\partial \sigma_{ij}(\mathbf{x},\mathbf{y})}{\partial y_j} = -f_i \text{ on } \Omega, \text{ with } \mathbf{u}(\mathbf{x},\mathbf{y}) = 0 \text{ on } \partial\Omega \tag{12}$$

$$\frac{\partial q_i(\mathbf{x},\mathbf{y})}{\partial x_i} + \frac{1}{\varepsilon}\frac{\partial q_i(\mathbf{x},\mathbf{y})}{\partial y_i} = 0 \text{ on } \Omega, \text{ with } \mathbf{u}(\mathbf{x},\mathbf{y}) = 0 \text{ on } \partial\Omega \tag{13}$$

And Equation (8) becomes the following Equation (14), Equation (9) becomes the following Equation (15),

$$\sigma_{ij}(\mathbf{x},\mathbf{y}) = \frac{1}{2}C_{ijkl}(\mathbf{y})\left\{\frac{\partial u_k(\mathbf{x},\mathbf{y})}{\partial x_l} + \frac{1}{\varepsilon}\frac{\partial u_k(\mathbf{x},\mathbf{y})}{\partial y_l}\right\} - \beta_{ij}(\mathbf{y})\Delta T(\mathbf{x},\mathbf{y}) \tag{14}$$

$$q_i(\mathbf{x},\mathbf{y}) = -\kappa_{ij}(\mathbf{y})\frac{\partial T(\mathbf{x},\mathbf{y})}{\partial x_j} \tag{15}$$

The next step in the model development is to asymptotically expand the stress and displacement fields as well as the thermal circulation into infinite series in terms of integral powers of the small parameter ε. However, According to [61], the first and second order expansion terms of the AHM are necessary to obtain more micro information regarding composite structures for the thermo–mechanical problem. This means that the second order expansion term must be considered to ensure the accuracy of AHM for the thermo–mechanical problem. Moreover, according to [61], the first two order expansions are sufficiently accurate for most composite problems in general. Thus, those parameters are expanded depend on ε,

(1) Basic Expansions

$$u_i(\mathbf{x},\mathbf{y}) = u_i^{(0)}(\mathbf{x},\mathbf{y}) + \varepsilon u_i^{(1)}(\mathbf{x},\mathbf{y}) + \varepsilon^2 u_i^{(2)}(\mathbf{x},\mathbf{y}) + O(\varepsilon^3) \tag{16}$$

$$T(\mathbf{x},\mathbf{y}) = T^{(0)}(\mathbf{x},\mathbf{y}) + \varepsilon T^{(1)}(\mathbf{x},\mathbf{y}) + \varepsilon^2 T^{(2)}(\mathbf{x},\mathbf{y}) + O(\varepsilon^3) \tag{17}$$

(2) Derived Expansions

$$\sigma_{ij}(\mathbf{x},\mathbf{y}) = \sigma_{ij}^{(0)}(\mathbf{x},\mathbf{y}) + \varepsilon \sigma_{ij}^{(1)}(\mathbf{x},\mathbf{y}) + \varepsilon^2 \sigma_{ij}^{(2)}(\mathbf{x},\mathbf{y}) + O(\varepsilon^3) \tag{18}$$

$$q_i(\mathbf{x},\mathbf{y}) = q_i^{(0)}(\mathbf{x},\mathbf{y}) + \varepsilon q_i^{(1)}(\mathbf{x},\mathbf{y}) + \varepsilon^2 q_i^{(2)}(\mathbf{x},\mathbf{y}) + O(\varepsilon^3) \tag{19}$$

The first items $u_i^{(0)}(\mathbf{x},\mathbf{y})$, $T^{(0)}(\mathbf{x},\mathbf{y})$, $\sigma_{ij}^{(0)}(\mathbf{x},\mathbf{y})$, $q_i^{(0)}(\mathbf{x},\mathbf{y})$ in Equations (16)–(19) represent the physical quantities in macroscale; The second items $u_i^{(1)}(\mathbf{x},\mathbf{y})$, $T^{(1)}(\mathbf{x},\mathbf{y})$, $\sigma_{ij}^{(1)}(\mathbf{x},\mathbf{y})$, $q_i^{(1)}(\mathbf{x},\mathbf{y})$ are the microscale physical quantities; The items $u_i^{(n)}(\mathbf{x},\mathbf{y})$, $T^{(n)}(\mathbf{x},\mathbf{y})$, $\sigma_{ij}^{(n)}(\mathbf{x},\mathbf{y})$, $\sigma_{ij}^{(n)}(\mathbf{x},\mathbf{y})\quad(n=2,3,\cdots)$ are the physical quantities in eventual smaller scales.

To obtain the asymptotic expansion for the mechanical stress and heat flux, substituting the asymptotic expansion Equations (16)–(19) into the constitutive equations Equations (14) and (15) and get,

$$\sigma_{ij}^{(n)}=C_{ijkl}\left(\frac{\partial u_k^{(n)}}{\partial x_l}+\frac{\partial u_k^{(n+1)}}{\partial y_l}\right)-\beta_{ij}\Delta T^{(n)},\quad(n=0,1,2\cdots)\tag{20}$$

$$q_i^{(n)}=-\kappa_{ij}\left(\frac{\partial T^{(n)}}{\partial x_j}+\frac{\partial T^{(n+1)}}{\partial y_j}\right),\quad(n=0,1,2\cdots)\tag{21}$$

where, $\Delta T^{(n)}\quad(n=0,1,2\cdots)$ represents the change of temperature compared with the initial temperature at the different scales, similar to $T^{(n)}\quad(n=0,1,2\cdots)$.

Taking the above relations Equations (18) and (19) into the governing Equations (12) and (13), one can get,

$$\frac{\partial\sigma_{ij}^{(0)}}{\partial x_j}+\varepsilon\frac{\partial\sigma_{ij}^{(1)}}{\partial y_j}+\frac{1}{\varepsilon}\frac{\partial\sigma_{ij}^{(0)}}{\partial x_j}+\frac{\partial\sigma_{ij}^{(1)}}{\partial y_j}+\varepsilon^2\frac{\partial\sigma_{ij}^{(2)}}{\partial x_j}+\varepsilon\frac{\partial\sigma_{ij}^{(2)}}{\partial y_j}=-f_i\tag{22}$$

$$\frac{\partial q_i^{(0)}}{\partial x_j}+\varepsilon\frac{\partial q_i^{(1)}}{\partial y_j}+\frac{1}{\varepsilon}\frac{\partial q_i^{(0)}}{\partial x_j}+\frac{\partial q_i^{(1)}}{\partial y_j}+\varepsilon^2\frac{\partial q_i^{(2)}}{\partial x_j}+\varepsilon\frac{\partial q_i^{(2)}}{\partial y_j}=0\tag{23}$$

Subsequently, equating like powers of ε, we obtain a series of differential equations. For the terms of the stress field expansion,

$$O(\varepsilon^{-1}):\frac{\partial\sigma_{ij}^{(0)}}{\partial y_j}=0\tag{24}$$

$$O(\varepsilon^{0}):\frac{\partial\sigma_{ij}^{(0)}}{\partial x_j}+\frac{\partial\sigma_{ij}^{(1)}}{\partial y_j}=-f_i\tag{25}$$

$$O(\varepsilon^{n}):\frac{\partial\sigma_{ij}^{(n)}}{\partial x_j}+\frac{\partial\sigma_{ij}^{(n+1)}}{\partial y_j}=0,\quad(n=1,2,3\cdots)\tag{26}$$

For the terms of the thermal balance field expansion,

$$O(\varepsilon^{-1}):\frac{\partial q_k^{(0)}}{\partial y_k}=0\tag{27}$$

$$O(\varepsilon^{n}):\frac{\partial q_k^{(n)}}{\partial x_k}+\frac{\partial q_k^{(n+1)}}{\partial y_k}=0,\quad(n=0,1,2\cdots)\tag{28}$$

3.2. *Asymptotic Homogenization and Governing Equations*

(1) For the displacement field problem

The determination of displacement field problem is Equations (24) and (25). We obtain $\sigma_{ij}^{(0)}, \sigma_{ij}^{(1)}$ from Equation (20),

$$\sigma_{ij}^{(0)} = C_{ijkl}\left(\frac{\partial u_k^{(0)}}{\partial x_l} + \frac{\partial u_k^{(1)}}{\partial y_l}\right) - \beta_{ij}\Delta T^{(0)} \tag{29}$$

$$\sigma_{ij}^{(1)} = C_{ijkl}\left(\frac{\partial u_k^{(1)}}{\partial x_l} + \frac{\partial u_k^{(2)}}{\partial y_l}\right) - \beta_{ij}\Delta T^{(1)} \tag{30}$$

Then substituting Equation (29) into Equation (24), we can get,

$$\frac{\partial}{\partial y_j}\left\{C_{ijkl}\frac{\partial u_k^{(1)}}{\partial y_l}\right\} = \frac{\partial \beta_{ij}}{\partial y_j}\Delta T^{(0)} - \frac{\partial C_{ijkl}}{\partial y_j}\frac{\partial u_k^{(0)}}{\partial x_l} \tag{31}$$

The separation of variables on the right–hand side of the above equation prompts us to write down the solution of $u_n^{(1)}$ as,

$$u_n^{(1)}(\mathbf{x},\mathbf{y}) = V_n(\mathbf{x}) + N_n^{kl}(\mathbf{y})\frac{\partial u_k^{(0)}(\mathbf{x})}{\partial x_l} + L_n(\mathbf{y})\Delta T^{(0)}(\mathbf{x}) \tag{32}$$

where, $N^{kl}(\mathbf{y})$ and $L(\mathbf{y})$ are the local functions of equilibrium problem to be solved, $V(\mathbf{x})$ is the homogeneous part of the solution.

Substituting Equation (32) into (31), the functions satisfy,

$$\frac{\partial}{\partial y_j}\left\{C_{ijkl}\frac{\partial V_n(\mathbf{x})}{\partial y_l}\right\} = 0 \tag{33}$$

$$\frac{\partial}{\partial y_j}\left\{C_{ijkl}(\mathbf{y})\frac{\partial N_n^{kl}(\mathbf{y})}{\partial y_l}\right\} = -\frac{\partial C_{ijkl}(\mathbf{y})}{\partial y_j} \tag{34}$$

$$\frac{\partial}{\partial y_j}\left\{C_{ijkl}(\mathbf{y})\frac{\partial L_n(\mathbf{y})}{\partial y_l}\right\} = \frac{\partial \beta_{ij}(\mathbf{y})}{\partial y_j} \tag{35}$$

Substituting Equation (32) into (29), then,

$$\sigma_{ij}^{(0)} = \left(C_{ijkl}(\mathbf{y}) + C_{ijmn}(\mathbf{y})\frac{\partial N_m^{kl}(\mathbf{y})}{\partial y_n}\right)\frac{\partial u_k^{(0)}(\mathbf{x})}{\partial x_l} - \left(\beta_{ij}(\mathbf{y}) - C_{ijmn}(\mathbf{y})\frac{\partial L_m(\mathbf{y})}{\partial y_n}\right)\Delta T^{(0)}(\mathbf{x}) \tag{36}$$

Substituting Equation (36) into (25), and subsequently average over the volume of the unit–cell. Thus,

$$\begin{aligned} &\tfrac{1}{|Y|}\int_Y \tfrac{\partial \sigma_{ij}^{(1)}(\mathbf{x},\mathbf{y})}{\partial y_j}\mathrm{d}v + \tfrac{1}{|Y|}\int_Y\left(C_{ijkl}(\mathbf{y}) + C_{ijmn}(\mathbf{y})\tfrac{\partial N_m^{kl}(\mathbf{y})}{\partial y_n}\right)\mathrm{d}v\tfrac{\partial^2 u_k^{(0)}(\mathbf{x})}{\partial x_j \partial x_l} \\ &-\tfrac{1}{|Y|}\int_Y\left(\beta_{ij}(\mathbf{y}) - C_{ijmn}(\mathbf{y})\tfrac{\partial L_m(\mathbf{y})}{\partial y_n}\right)\mathrm{d}v\tfrac{\partial(\Delta T^{(0)}(\mathbf{x}))}{\partial x_j} = -f_i \end{aligned} \tag{37}$$

where, Y is the dimension vector of the unit–cell, and $|Y|$ is the volume of the unit–cell.

Keeping the periodicity of the associated functions in mind and remembering to treat x_i as a parameter as far as integration with respect to $\mathbf{y}$ is concerned results in an expression of the form,

$$\overline{C}_{ijkl}\frac{\partial^2 u_k^{(0)}(\mathbf{x})}{\partial x_j \partial x_l} - \overline{\beta}_{ij}\frac{\partial(\Delta T^{(0)}(\mathbf{x}))}{\partial x_j} = -f_i \tag{38}$$

With

$$\overline{C}_{ijkl} = \frac{1}{|Y|}\int_Y \left(C_{ijkl}(\mathbf{y}) + C_{ijmn}(\mathbf{y})\frac{\partial N_m^{kl}(\mathbf{y})}{\partial y_n} \right) \mathrm{d}v \tag{39}$$

$$\overline{\beta}_{ij} = \frac{1}{|Y|}\int_Y \left(\beta_{ij}(\mathbf{y}) - C_{ijmn}(\mathbf{y})\frac{\partial L_m(\mathbf{y})}{\partial y_n} \right) \mathrm{d}v \tag{40}$$

The first integral vanishes as a consequence of the periodicity of $\sigma_{ij}^{(1)}$ and following the application of the divergence theorem.

(2) For the heat conduction problem

According to Equation (28), we can obtain,

$$\frac{\partial q_k^{(0)}}{\partial x_k} + \frac{\partial q_k^{(1)}}{\partial y_k} = 0 \tag{41}$$

Thus, the heat conduction problem can be determined by Equations (27) and (41).

According to Equation (21), substituting and equating like powers of ε, we obtain, for $q_k^{(0)}, q_k^{(1)}$,

$$q_k^{(0)}(\mathbf{x},\mathbf{y}) = -\kappa_{ik}(\mathbf{y})\left(\frac{\partial T^{(0)}(\mathbf{x})}{\partial x_i} + \frac{\partial T^{(1)}(\mathbf{x},\mathbf{y})}{\partial y_i} \right) \tag{42}$$

$$q_k^{(1)}(\mathbf{x},\mathbf{y}) = -\kappa_{ik}(\mathbf{y})\left(\frac{\partial T^{(1)}(\mathbf{x},\mathbf{y})}{\partial x_i} + \frac{\partial T^{(2)}(\mathbf{x},\mathbf{y})}{\partial y_i} \right) \tag{43}$$

Substituting Equation (42) into (27), and then,

$$\frac{\partial}{\partial y_k}\left\{ \kappa_{ik}(\mathbf{y})\frac{\partial T^{(1)}(\mathbf{x},\mathbf{y})}{\partial y_i} \right\} = -\frac{\partial \kappa_{ik}(\mathbf{y})}{\partial y_k}\frac{\partial T^{(0)}(\mathbf{x})}{\partial x_i} \tag{44}$$

The separation of variables on the right–hand side of the above equation prompts us to write down the solution of $T^{(1)}$ as,

$$T^{(1)}(\mathbf{x},\mathbf{y}) = L'(\mathbf{x}) + M^k(\mathbf{y})\frac{\partial T^{(0)}(\mathbf{x})}{\partial x_k} \tag{45}$$

where, $M(\mathbf{y})$ is the local function of equilibrium problem to be solved likes $N^{kl}(\mathbf{y})$ and $L(\mathbf{y})$, $L'(\mathbf{x})$ is the homogeneous part of the solution.

Substituting Equation (45) into (44), the functions satisfy,

$$\frac{\partial}{\partial y_k}\left\{ \kappa_{lk}(\mathbf{y})\frac{\partial L'(\mathbf{x})}{\partial y_l} \right\} = 0 \tag{46}$$

$$\frac{\partial}{\partial y_k}\left\{ \kappa_{lk}(\mathbf{y})\frac{\partial M^i(\mathbf{y})}{\partial y_l} \right\} = -\frac{\partial \kappa_{ik}(\mathbf{y})}{\partial y_k} \tag{47}$$

Substituting Equation (45) into (42), then,

$$q_k^{(0)}(\mathbf{x},\mathbf{y}) = -\left(\kappa_{ik}(\mathbf{y}) + \kappa_{lk}(\mathbf{y})\frac{\partial M^i(\mathbf{y})}{\partial y_l} \right)\frac{\partial T^{(0)}(\mathbf{x})}{\partial x_i} \tag{48}$$

Substitute Equation (48) into (41), and subsequently average over the volume of the unit–cell. Thus,

$$\frac{1}{|Y|}\int_Y \frac{\partial q_k^{(1)}(\mathbf{x},\mathbf{y})}{\partial y_k}\mathrm{d}v - \frac{1}{|Y|}\int_Y \left(\kappa_{ik}(\mathbf{y}) + \kappa_{lk}(\mathbf{y})\frac{\partial M^i(\mathbf{y})}{\partial y_l} \right)\mathrm{d}v\frac{\partial T^{(0)}(\mathbf{x})}{\partial x_i \partial x_k} = 0 \tag{49}$$

Keeping the periodicity of the associated functions in mind and remembering to treat x_i as a parameter as far as integration with respect to **y** is concerned results in an expression of the form,

$$-\overline{\kappa}_{ik}\frac{\partial T^{(0)}(\mathbf{x})}{\partial x_i \partial x_k}=0 \tag{50}$$

With

$$\overline{\kappa}_{ik}=\frac{1}{|Y|}\int_Y\left(\kappa_{ik}(\mathbf{y})+\kappa_{lk}(\mathbf{y})\frac{\partial M^i(\mathbf{y})}{\partial y_l}\right)dv \tag{51}$$

The first integral vanishes as a consequence of the periodicity of $q_k^{(1)}$ and following the application of the divergence theorem.

Prior to calculating the effective coefficients by Equations (39), (40) and (51), one must first determine the local functions N_m^{kl}; L_m; M^i by Equations (34), (35) and (47) from the appropriate unit–cell problems.

3.3. Determination of Effective Coefficients

In this section, the equivalent performance of a graded composite considering the thermo–mechanical coupling effect is studied. Assuming that the graded composite, where the gradient layers are orthogonal to the axis Ox_3 and the constituents of the structure have an arbitrary isotropic. The graded composite material satisfies the generalized periodicity. The local function of the composite material with generalized periodicity satisfies a set of partial differential equations, which is called the local problem on the periodic element.

The coefficients of the equilibrium problem on a thermal mechanical coupling structure are rapidly oscillating functions. Then, Equations (34), (35) and (47) become,

$$\frac{\partial}{\partial y_3}\left\{C_{i3k3}(\mathbf{y})\frac{\partial N_n^{kl}(\mathbf{y})}{\partial y_3}\right\}=-\frac{\partial C_{i3kl}(\mathbf{y})}{\partial y_3} \tag{52}$$

$$\frac{\partial}{\partial y_3}\left\{C_{i3k3}(\mathbf{y})\frac{\partial L_n(\mathbf{y})}{\partial y_3}\right\}=\frac{\partial \beta_{i3}(\mathbf{y})}{\partial y_3} \tag{53}$$

$$\frac{\partial}{\partial y_3}\left\{\kappa_{l3}(\mathbf{y})\frac{\partial M^i(\mathbf{y})}{\partial y_3}\right\}=-\frac{\partial \kappa_{i3}(\mathbf{y})}{\partial y_3} \tag{54}$$

Integrating once with respect to y_3 yields,

$$\frac{\partial N_m^{kl}}{\partial y_3}=-C_{m3i3}^{-1}C_{i3kl}+C_{m3i3}^{-1}A_{ikl} \tag{55}$$

$$\frac{\partial L_k}{\partial y_3}=C_{k3i3}^{-1}\beta_{i3}+C_{k3i3}^{-1}B_i \tag{56}$$

$$\frac{\partial M^i}{\partial y_3}=-\kappa_{33}^{-1}\kappa_{3i}+\kappa_{33}^{-1}C_i \tag{57}$$

where A_{ikl}, B_i, C_i are constants of integration.

Based on the above relations, and integrating one more time with respect to y_3, remembering at the same time that the functions $N_m^{kl}(y_3)$, $L_k(y_3)$ and $M^i(y_3)$ are 1–periodic in y_3, results in the following expression for the constants of integration,

$$A_{ikl}=\left\langle C_{m3i3}^{-1}\right\rangle^{-1}\left\langle C_{m3i3}^{-1}C_{i3kl}\right\rangle \tag{58}$$

$$B_i=-\left\langle C_{k3i3}^{-1}\right\rangle^{-1}\left\langle C_{k3i3}^{-1}\beta_{i3}\right\rangle \tag{59}$$

$$C_i=\left\langle \kappa_{33}^{-1}\right\rangle^{-1}\left\langle \kappa_{33}^{-1}\kappa_{3i}\right\rangle \tag{60}$$

Substituting Equations (58)–(60) into Equations (55)–(57), then,

$$\frac{\partial N_m^{kl}}{\partial y_3} = -C_{m3i3}^{-1}C_{i3kl} + C_{m3i3}^{-1}\left\langle C_{m3i3}^{-1}\right\rangle^{-1}\left\langle C_{m3i3}^{-1}C_{i3kl}\right\rangle \tag{61}$$

$$\frac{\partial L_k}{\partial y_3} = C_{k3i3}^{-1}\beta_{i3} - C_{k3i3}^{-1}\left\langle C_{k3i3}^{-1}\right\rangle^{-1}\left\langle C_{k3i3}^{-1}\beta_{i3}\right\rangle \tag{62}$$

$$\frac{\partial M^i}{\partial y_3} = -\kappa_{33}^{-1}\kappa_{3i} + \kappa_{33}^{-1}\left\langle \kappa_{33}^{-1}\right\rangle^{-1}\left\langle \kappa_{33}^{-1}\kappa_{3i}\right\rangle \tag{63}$$

Substituting Equations (61)–(63) into Equations (34), (35) and (47), we can get the effective elasticity/stiffness coefficients,

$$C_{ijkl}^H = \left\langle C_{ijkl}\right\rangle - \left\langle C_{ijm3}C_{m3p3}^{-1}C_{p3kl}\right\rangle + \left\langle C_{ijm3}C_{m3p3}^{-1}\right\rangle\left\langle C_{p3s3}^{-1}\right\rangle^{-1}\left\langle C_{s3q3}^{-1}C_{q3kl}\right\rangle \tag{64}$$

$$\beta_{ij}^H = \left\langle \beta_{ij}\right\rangle - \left\langle C_{ijm3}C_{m3p3}^{-1}\beta_{p3}\right\rangle + \left\langle C_{ijm3}C_{m3p3}^{-1}\right\rangle\left\langle C_{p3s3}^{-1}\right\rangle^{-1}\left\langle C_{s3q3}^{-1}\beta_{q3}\right\rangle \tag{65}$$

$$\kappa_{ik}^H = \left\langle \kappa_{ik}\right\rangle - \left\langle \kappa_{3k}\kappa_{33}^{-1}\kappa_{3i}\right\rangle + \left\langle \kappa_{3k}\kappa_{33}^{-1}\right\rangle\left\langle \kappa_{33}^{-1}\right\rangle^{-1}\left\langle \kappa_{33}^{-1}\kappa_{3i}\right\rangle \tag{66}$$

where, C^H, β^H, κ^H are the effective elasticity tensor, the tensor related to effective thermal expansion tensor $\boldsymbol{\alpha}^H$ and the effective thermal conductivity tensor, respectively. It is noted that $\boldsymbol{\alpha}^H$ can be calculated by $\beta_{ij} = C_{ijkl}\cdot\alpha_{kl}$.

For the isotropic materials, its elastic constant matrix $\mathbf{C}_{6\times6}$, thermal expansion coefficient matrix $\boldsymbol{\alpha}_{3\times3}$, thermal conductivity coefficient matrix $\boldsymbol{\kappa}_{3\times3}$ are given by,

$$\begin{gathered}\mathbf{C}_{6\times6} = \begin{bmatrix} C_{11} & C_{12} & C_{12} & 0 & 0 & 0 \\ C_{12} & C_{11} & C_{12} & 0 & 0 & 0 \\ C_{12} & C_{12} & C_{11} & 0 & 0 & 0 \\ 0 & 0 & 0 & C_{66} & 0 & 0 \\ 0 & 0 & 0 & 0 & C_{66} & 0 \\ 0 & 0 & 0 & 0 & 0 & C_{66} \end{bmatrix}, \\ \boldsymbol{\alpha}_{3\times3} = \begin{pmatrix} \alpha_{11} & 0 & 0 \\ 0 & \alpha_{11} & 0 \\ 0 & 0 & \alpha_{11} \end{pmatrix}, \boldsymbol{\kappa}_{3\times3} = \begin{pmatrix} \kappa_{11} & 0 & 0 \\ 0 & \kappa_{11} & 0 \\ 0 & 0 & \kappa_{11} \end{pmatrix}\end{gathered} \tag{67}$$

where, $C_{11} = E(1-v)/[(1+v)(1-2v)]$, $C_{12} = Ev/[(1+v)(1-2v)]$, $C_{66} = (C_{11} - C_{12})/2$.

Thus, the analytical formulas of the effective properties for FGMs are assumed to be made by a mixture of two isotropic elastic constituents that are derived and we can obtain the following closed–form formulas,

(1) For the effective elastic tensor's components,

$$\begin{aligned} &C_{11}^H = \langle C_{11}\rangle - \left\langle C_{12}^2 C_{11}^{-1}\right\rangle + \left\langle C_{12}C_{11}^{-1}\right\rangle^2\left\langle C_{11}^{-1}\right\rangle^{-1}, \\ &C_{12}^H = \langle C_{12}\rangle - \left\langle C_{12}^2 C_{11}^{-1}\right\rangle + \left\langle C_{12}C_{11}^{-1}\right\rangle^2\left\langle C_{11}^{-1}\right\rangle^{-1}, \\ &C_{13}^H = C_{23}^H = \left\langle C_{12}C_{11}^{-1}\right\rangle\left\langle C_{11}^{-1}\right\rangle^{-1}, \\ &C_{33}^H = \left\langle C_{11}^{-1}\right\rangle^{-1}, \\ &C_{44}^H = C_{55}^H = \tfrac{1}{2}\left\langle (C_{11} - C_{12})^{-1}\right\rangle^{-1}, \\ &C_{66}^H = \tfrac{1}{2}\left\langle C_{11}^H - C_{12}^H\right\rangle \end{aligned} \tag{68}$$

where, <·> is the average. It can be discovered that the FGMs are transversely isotropic and have 5 independent constants (C_{66}^H is not an independent constant), the matrix is symmetric.

(2) For the effective thermal expansion tensor's components, due to

$$\begin{aligned} \beta_{11}^H = \beta_{22}^H &= \langle\beta_{11}\rangle - \left\langle C_{12}C_{11}^{-1}\beta_{11}\right\rangle + \left\langle C_{12}C_{11}^{-1}\right\rangle\left\langle C_{11}^{-1}\right\rangle^{-1}\left\langle C_{11}^{-1}\beta_{11}\right\rangle, \\ \beta_{33}^H &= \left\langle C_{11}^{-1}\right\rangle^{-1}\left\langle C_{11}^{-1}\beta_{11}\right\rangle \end{aligned} \quad (69)$$

thus, we can calculate the effective thermal expansion tensor's components α_{kl} in terms of the $\beta_{ij} = C_{ijkl}\cdot\alpha_{kl}$.

(3) For the effective thermal conductivity tensor's components,

$$\kappa_{11}^H = \kappa_{22}^H = \kappa_{33}^H = \langle\kappa_{11}\rangle \quad (70)$$

4. Numerical Results and Discussion

4.1. Validation of the Present Method

This section mainly verifies the validity of the analytical method based on the asymptotic homogenization method proposed in this paper. We compare the calculation results of this method with the theoretical and numerical results respectively, and the correctness is verified.

Firstly, the composites we studied are FGMs composed of Mo and ZrC. Due to the lack of relevant experimental data, the results of the present method are compared with the classical method of Mori–Tanaka for predicting the effective coefficients of composites. The relevant material parameters at room temperature are given in Table 1. The material properties calculated by the methods of present and Mori–Tanaka are plotted as a function of the volume fraction of ZrC in Figure 4. It is observed that for this special material combination, the homogenized material properties are almost the same as those obtained by Mori–Tanaka method. Compared with Mori–Tanaka method, C_{11}^H and C_{66}^H calculated by the analytical method are larger, while C_{12}^H is smaller. On the whole, however, the analytical results of the AHM are in good agreement with those of the Mori–Tanaka method.

Table 1. The elastic modulus E, Poisson's ratio v, thermal expansion coefficient α and thermal conductivity coefficient κ of materials at room temperature (25 °C).

Materials	E [Gpa]	v	α [10^{-6}/K]	κ [W/(m·K)]
ZrC	400 [10]	0.20 [62]	6.7 [62]	20.52 [63]
Mo	327 [64]	0.293 [65]	4.8 [64]	128.0 [64]

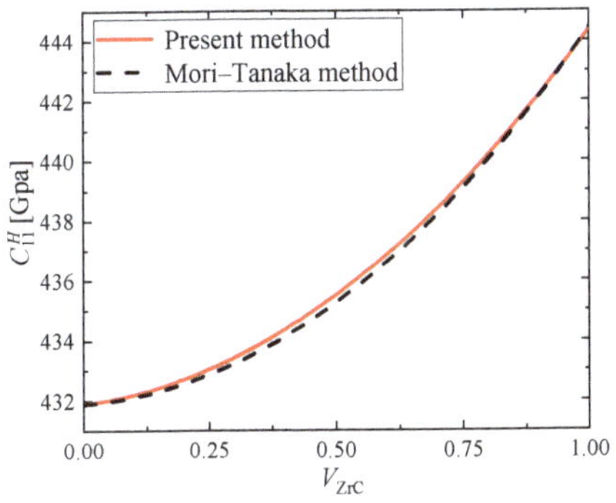

(a)

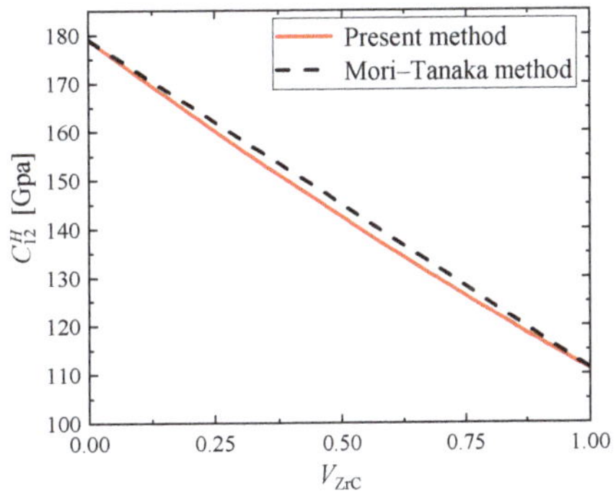

(b)

Figure 4. *Cont.*

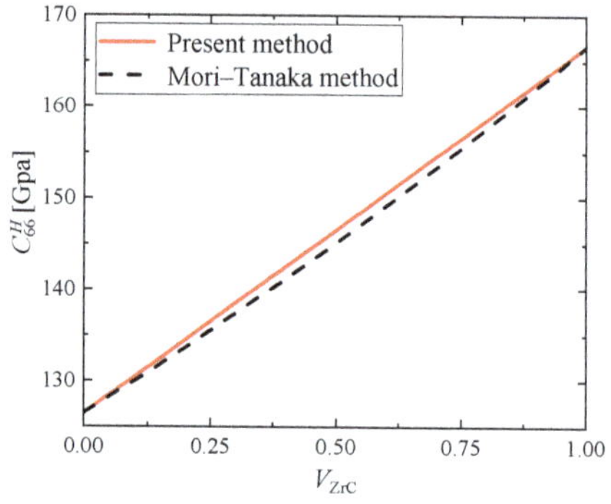

(c)

Figure 4. Comparison of two methods for effective elastic tensor's components (**a**) C_{11}^{H}; (**b**) C_{12}^{H}; (**c**) C_{66}^{H}.

On the other hand, the equivalent coefficient of Al/SiC fiber reinforced metal matrix composites was calculated. The elastic modulus of Al and SiC are 70.0 and 450.0 Gpa, and Poisson's ratio are 0.30, 0.17, respectively. The homogenized results obtained using the present method are compared to the results obtained using the AHM on Abaqus [15], as shown in Figure 5. Three different fiber volume fractions were investigated: 20%, 35% and 55%. The results obtained by [10] are also shown in Figure 5. It is possible to observe that the homogenized coefficients obtained using the present method are practically the same as those obtained in [15]. For all the results the differences are less than 1.5%.

In summary, the effectiveness of our present method has been proved. Therefore, the thermo–mechanical properties of FGMs with different gradient changes will be discussed according to this method.

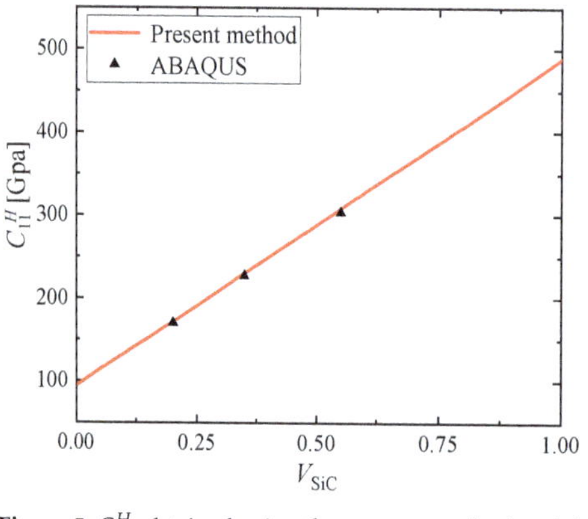

Figure 5. C_{11}^{H} obtained using the present method and the AHM on Abaqus [15].

4.2. *Effective Elastic Tensor's Components*

In Figure 6, the effective elastic tensor's components perturbation along the gradient direction are plotted by selecting distinct component ratio coefficient c_0 while fixing the gradient index as $m = 1.0$. Assume that the component ratio coefficient varies from $c_0 = 0$ to $c_0 = 1.0$. For FGMs ($c_0 \neq 0$), the variation of effective elastic tensor's components depends on the gradient distribution of the component volume fraction, and presents a nonlinear change trend. With the increase of component ratio coefficient c_0, the performance of the bottom material remains unchanged because there is only the material Mo when $z = 0$; However, the material properties in other places except the bottom are changing more and more sharply, and the difference of material properties between the top and bottom is also increasing. This indicates that the larger the component ratio coefficient c_0, the more obvious the difference of local material properties of FGMs; the volume fraction of the material phase ZrC increases with the increase of c_0, which affects the properties of the material to a great extent.

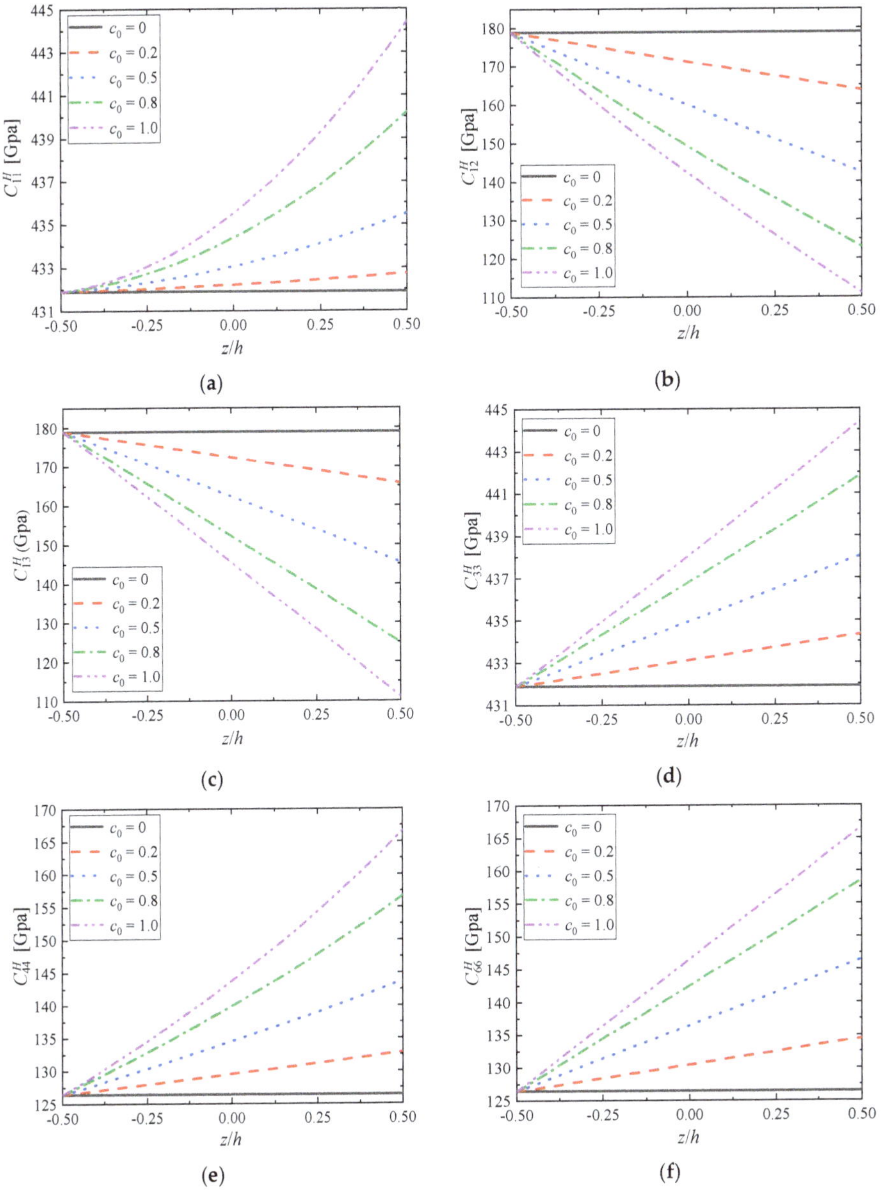

Figure 6. Distributions of (**a**) C_{11}^H; (**b**) C_{12}^H; (**c**) C_{13}^H; (**d**) C_{33}^H; (**e**) C_{44}^H; (**f**) C_{66}^H along the gradient direction when $m = 1.0$.

In Figure 7, when the component ratio coefficient $c_0 = 1.0$ [V_{ZrC}:V_{Mo} = 0.5:0.5], the influences of gradient index m on the perturbation of the effective elastic tensor's component along the gradient direction is shown. It is assumed that the gradient index of volume fraction varies from $m = 0$ to $m = 10.0$. For FGMs ($m \neq 0$), the variation of the effective elastic tensor's components depends on the gradient distribution of volume fraction while exhibiting similar or anti–similar trends with that shown in Figure 1b. With the increase of gradient index m, the difference of material properties on the top and bottom surfaces remains constant, because the materials of these location do not vary with the change of gradient index m, and they are only related to the volume ratio c_0; However, with the increase of gradient index m, the variation of material properties near the top and bottom surfaces is slower and slower, but the variation of material properties is more severe when approaching the middle surface, which indicates that the inhomogeneity of material phase gradient changes leads to the inhomogeneity of material property changes, mainly determined by the different volume fraction at the current position z/h caused by gradient index m. The distribution of the material properties of FGMs along the gradient direction is asymmetric and with a turning point when $z/h = 0$ which means V_{ZrC}:V_{Mo} = 0.5:0.5 in this section.

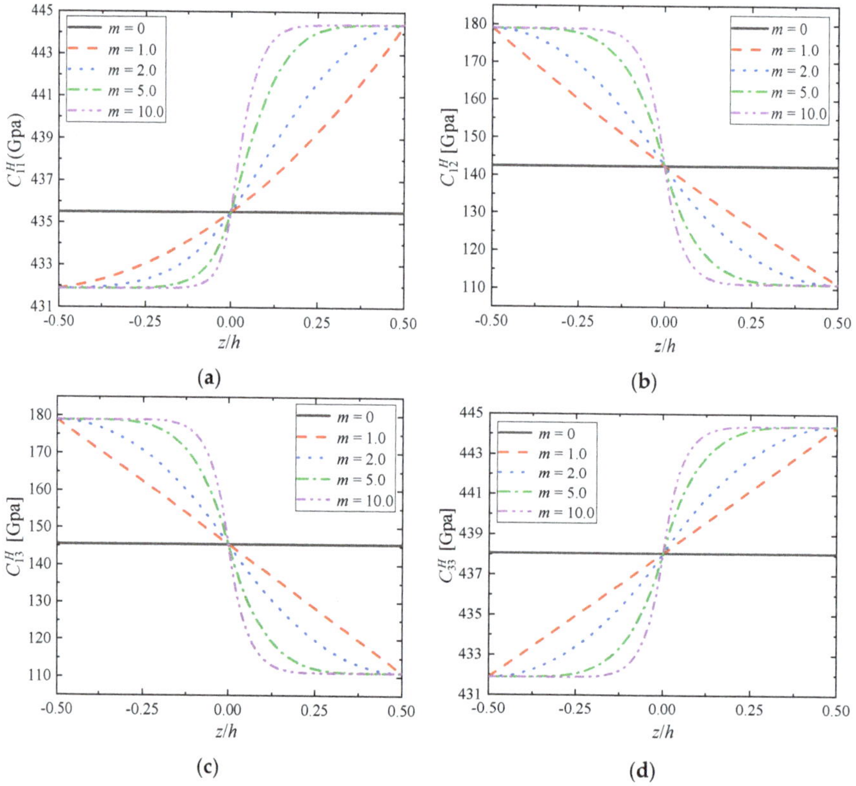

Figure 7. *Cont.*

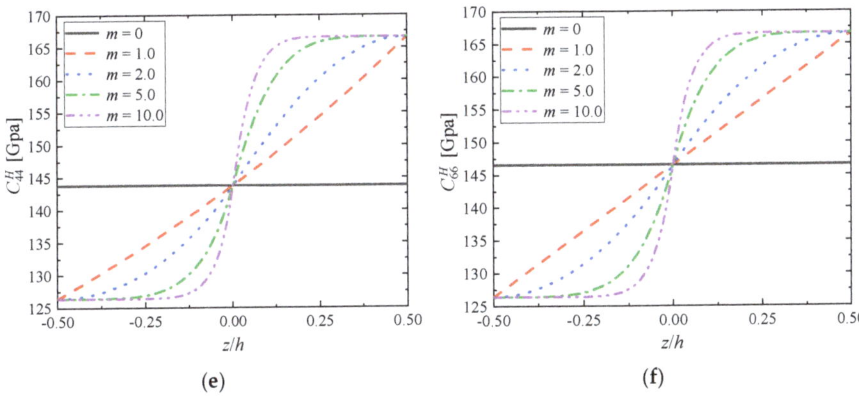

Figure 7. Distributions of (**a**) C^H_{11}; (**b**) C^H_{12}; (**c**) C^H_{13}; (**d**) C^H_{33}; (**e**) C^H_{44}; (**f**) C^H_{66} along the gradient direction when $c_0 = 1.0$.

4.3. Effective Thermal Expansion Tensor's Components

The graphs present changes in the effective thermal expansion coefficient tensor's components depending on the component ratio coefficient c_0 when fixing the gradient index as $m = 1.0$ (Figure 8) and the gradient index m when fixing the component ratio coefficient as $c_0 = 1.0$ (Figure 9).

As can be seen from Figure 8, for FGMs ($c_0 \neq 0$), the change of the effective thermal expansion tensor's components depends on the gradient distribution of the volume fraction of the component, and presents a nonlinear change trend. The effective thermal expansion tensor's components increase with the increase of ZrC volume fraction, which seems to indicate that FGMs are easier to deform after thermal shock than Mo, and the thermal shock resistance of FGMs is not optimized compared with Mo. At the same time, with the increase of the component ratio coefficient c_0, the effective performance of the bottom surface remains unchanged, but the top surface is larger and larger, i.e., the difference between the effective thermal expansion tensor's components of the top surface and the bottom surface is larger and larger, and the overall difference of FGMs is obvious. When the structural FGMs change linearly, the nonlinear changes of the effective tensor's components of elasticity and thermal expansion are observed, which indicates that the gradient change not only affects the volume ratio of the component phases, but also leads to the influence of other aspects on the overall material performance after the two materials are compounded. Thus, considering the microstructures of FGMs is vital when studying their effective properties.

In Figure 9, the change of the effective thermal expansion tensor's components with the gradient position exhibits a trend similar to FGMs ($m \neq 0$) shown in Figure 1b. With the increase of the gradient index m, the effective thermal expansion tensor's components are the same as the effective elastic tensor's components, and the performance difference between the top surface and the bottom surface remains unchanged, because the materials at these positions do not change with the change of the gradient index m, but only relate to the volume ratio c_0; However, with the increase of gradient index m, the material properties near the top and bottom surface change more slowly, but the material properties change more and more sharply near the middle surface. The effective thermal expansion tensor's components of FGMs is asymmetrically distributed along the gradient direction, and there is a turning point when $z/h = 0$, i.e., $V_{ZrC}:V_{Mo} = 0.5:0.5$.

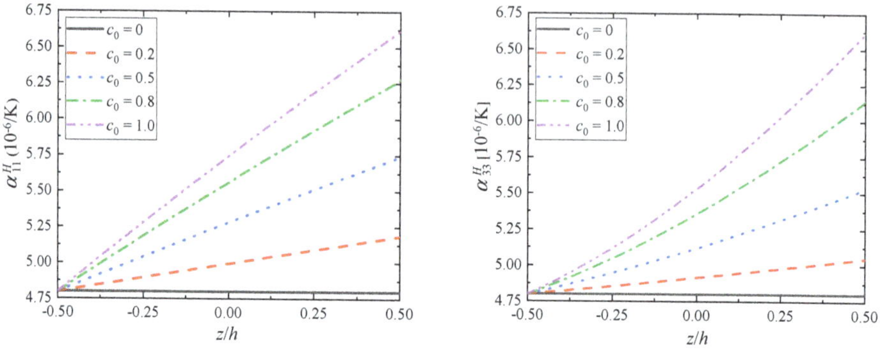

(**a**) (**b**)

Figure 8. Distributions of (**a**) α_{11}^{H} and (**b**) α_{33}^{H} along the gradient direction (m = 1.0).

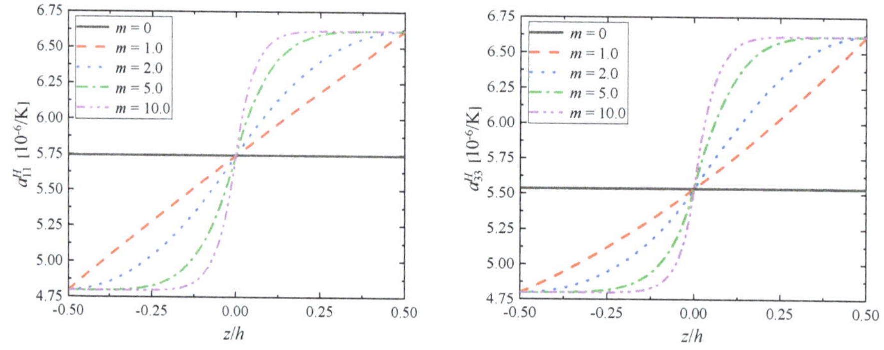

(**a**) (**b**)

Figure 9. Distributions of (**a**) α_{11}^{H} and (**b**) α_{33}^{H} along the gradient direction (c_0 = 1.0).

4.4. Effective Thermal Conductivity Tensor's Components

The graphs present the changes in the thermal conductivity tensor's components depending on the component ratio coefficient c_0 when the gradient index m = 1.0 (Figure 10a) and the gradient index c_0 when the component ratio coefficient c_0 = 1.0 (Figure 10b).

From Figure 10, the variation of thermal conductivity tensor's components depends on the gradient distribution of the component volume fraction, and presents a linear change trend. The thermal conductivity tensor's components of the gradient material decrease with the increase of ZrC volume fraction, which indicates that the internal heat transfer efficiency of the material is decreasing. When the component ratio coefficient c_0 increases, this decrease is more obvious in Figure 10a; with the increase of the gradient index m, the range of the overall thermal conductivity does not change, but there are local differences, and the smaller m, the smaller the local differences in Figure 10b.

Considering the effective tensor's components of thermal expansion and thermal conductivity, the change of the effective performance tensor's components is closely related to the volume fraction of the component phase. With the increase of ZrC volume fraction, the thermal expansion tensor's components of FGMs increases, i.e., FGMs are more likely to deform when subjected to thermal shock. However, the thermal conductivity tensor's components are decreasing, that is to say, it is more difficult to conduct heat in FGMs. However, in general, the deformation effect of FGMs caused by the increase of thermal expansion tensor's components is less than that caused by the decrease of thermal conductivity tensor's components, which means that under the same heating conditions,

FGMs have smaller thermal expansion degree and better thermal shock resistance than single metal material Mo.

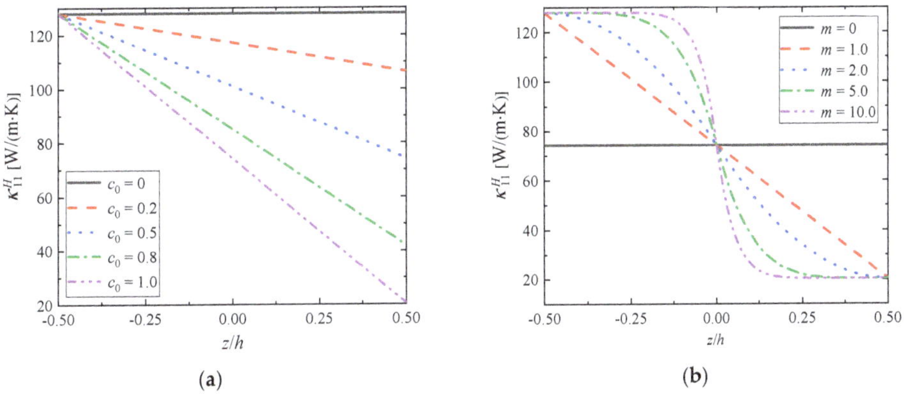

Figure 10. Distributions of κ_{11}^H along the gradient direction when (**a**) $m = 1.0$; (**b**) $c_0 = 1.0$.

5. Conclusions

In summary, a methodology to find the local problems and the effective thermo–mechanical coefficients equations for structures with generalized periodicity is presented based on the method of AHM. Following the proposal, the homogenized problem for FGMs is derived, the relationship between the structure and properties of Mo/ZrC FGMs is studied. The constituent ratio and the property gradation profiles of Mo/ZrC FGMs are described by design volume fraction. From the calculation results, it can be observed that the component ratio coefficient and gradient index have important influences on the thermo–mechanical properties of FGMs. The component ratio coefficient affects the overall performance difference of FGMs, while the gradient index affects the local performance difference. This work opens up the possibility to predict the changing trend of FGMs performance with different gradient structures, and provides guidance for further innovation and unique design of new FGMs structures.

Author Contributions: Conceptualization, D.C. and L.L.; methodology and validation investigation, D.C. and L.C.; writing–original draft preparation, D.C. and L.C.; writing–review and editing, Q.L. All authors have read and agreed to the published version of the manuscript.

Funding: Lisheng Liu acknowledges the support from the National Natural Science Foundation of China (No. 11972267), Liangliang Chu acknowledges the support from the Fundamental Research Funds for the Central Universities (No. 2021IVA060).

Data Availability Statement: All data generated or analysed during this study are included in this published article.

Conflicts of Interest: The authors have no conflict of interest to declare that are relevant to the content of this article.

References

1. Sam, M. Microstructural, mechanical and tribological analysis of functionally graded copper composite. *Int. J. Cast Met. Res.* **2020**, *33*, 123–133.
2. Kazemzadeh–Parsi, M.-J.; Francisco, C.; Amine, A. Proper generalized decomposition for parametric study and material distribution design of multi–directional functionally graded plates based on 3D elasticity solution. *Materials* **2021**, *14*, 6660. [CrossRef] [PubMed]
3. Huang, W.; Xue, K.; Li, Q. Functionally graded rectangular plate with/without cutouts subject to general boundary conditions. *Materials* **2021**, *14*, 7088. [CrossRef] [PubMed]
4. Nejad, M.Z.; Alamzadeh, N.; Hadi, A. Thermoelastoplastic analysis of FGM rotating thick cylindrical pressure vessels in linear elastic–fully plastic condition. *Compos. Part B Eng.* **2018**, *154*, 410–422. [CrossRef]

5. Saleh, B.; Jiang, J.; Ma, A.; Song, D.; Xu, Q. Review on the influence of different reinforcements on the microstructure and wear behavior of functionally graded aluminum matrix composites by centrifugal casting. *Met. Mater. Int.* **2020**, *26*, 933–960. [CrossRef]
6. Birsan, M.; Altenbach, H.; Sadowski, T.; Eremeyev, V.A.; Pietras, D. Deformation analysis of functionally graded beams by the direct approach. *Compos. Part B Eng.* **2012**, *43*, 1315–1328. [CrossRef]
7. Moita, J.S.; Araújo, A.L.; Correia, V.F. Material distribution and sizing optimization of functionally graded plate–shell structures. *Compos. Part B Eng.* **2018**, *142*, 263–272. [CrossRef]
8. Zhou, W.; Zhang, R.; Ai, S.; He, R.; Pei, Y.; Fang, D. Load distribution in threads of porous metal–ceramic functionally graded composite joints subjected to thermomechanical loading. *Compos. Struct.* **2015**, *134*, 680–688. [CrossRef]
9. Leon, L.M. Functionally gradient metal matrix composites: Numerical analysis of the microstructure–strength relationships. *Compos. Sci. Technol.* **2006**, *66*, 1873–1887.
10. Zhou, W.; Ai, S.; Chen, M.; Zhang, R.; He, R.; Pei, Y.; Fang, D. Preparation and thermodynamic analysis of the porous $ZrO_2/(ZrO_2$ + Ni) functionally graded bolted joint. *Compos. Part B Eng.* **2015**, *82*, 13–22. [CrossRef]
11. Fan, F.; Lei, B.; Sahmani, S.; Safaei, B. On the surface elastic–based shear buckling characteristics of functionally graded composite skew nanoplates. *Thin Wall. Struct.* **2020**, *154*, 106841. [CrossRef]
12. Fan, F.; Kiani, K. A rigorously analytical exploration of vibrations of arbitrarily shaped multi–layered nanomembranes from different materials. *Int. J. Mech. Sci.* **2021**, *206*, 106603. [CrossRef]
13. Wu, D.; Liu, A.; Huang, Y.; Huang, Y.; Pi, Y.; Gao, W. Mathematical programming approach for uncertain linear elastic analysis of functionally graded porous structures with interval parameters. *Compos. Part B Eng.* **2018**, *152*, 282–291. [CrossRef]
14. Malikan, M.; Eremeyev, V.A. A new hyperbolic–polynomial higher–order elasticity theory for mechanics of thick FGM beams with imperfection in the material composition. *Compos. Struct.* **2020**, *249*, 112486. [CrossRef]
15. Dutra, T.A.; Ferreira, R.T.L.; Resende, H.B.; Guimaraes, A.; Guedes, J.M. A complete implementation methodology for asymptotic homogenization using a finite element commercial software: Preprocessing and postprocessing. *Compos. Struct.* **2020**, *245*, 112305. [CrossRef]
16. Fernández, M.; Böhlke, T. Hashin–Shtrikman bounds with eigenfields in terms of texture coefficients for polycrystalline materials. *Acta Mater.* **2019**, *165*, 686–697. [CrossRef]
17. Salah, S.B.E.; Nait-Ali, A.; Gueguen, M.; Nadot-Martin, C. Non–local modeling with asymptotic expansion homogenization of random materials. *Mech. Mater.* **2020**, *147*, 103459. [CrossRef]
18. Brown, J.M. Determination of Hashin–Shtrikman bounds on the isotropic effective elastic moduli of polycrystals of any symmetry. *Comput. Geosci.* **2015**, *80*, 95–99. [CrossRef]
19. Hashin, Z.; Shtrikman, S. A variational approach to the theory of the elastic behaviour of multiphase materials. *J. Mech. Phys. Solids* **1963**, *11*, 127–140. [CrossRef]
20. Budiansky, B. On the elastic moduli of some heterogeneous materials. *J. Mech. Phys. Solids* **1965**, *13*, 223–227. [CrossRef]
21. Eshelby, J.D. The determination of the elastic field of an ellipsoidal inclusion, and related problems. *Proc. Math. Phy. Eng. Sci.* **1957**, *241*, 376–396.
22. Hill, R. A self–consistent mechanics of composite materials. *J. Mech. Phys. Solids* **1965**, *13*, 213–222. [CrossRef]
23. Christensen, R.M.; Lo, K.H. Solutions for effective shear properties in three phase sphere and cylinder models. *J. Mech. Phys. Solids* **1979**, *27*, 315–330. [CrossRef]
24. Yun, G.J.; Zhu, F.Y.; Lim, H.J.; Choi, H. A damage plasticity constitutive model for wavy CNT nanocomposites by incremental Mori–Tanaka approach. *Compos. Struct.* **2021**, *258*, 113178. [CrossRef]
25. Tran, V.P.; Brisard, S.; Guilleminot, J.; Sab, K. Mori–Tanaka estimates of the effective elastic properties of stress–gradient composites. *Int. J. Solids Struct.* **2018**, *146*, 55–68. [CrossRef]
26. Peng, X.; Hu, N.; Long, X.; Zheng, H. Extension of combined self–consistent and Mori–Tanaka approach to evaluation of elastoplastic property of particulate composites. *Acta Mech. Solida Sin.* **2013**, *26*, 71–82. [CrossRef]
27. Kundalwal, S.I.; Ray, M.C. Effective properties of a novel continuous fuzzy–fiber reinforced composite using the method of cells and the finite element method. *Eur. J. Mech.-A/Solids* **2012**, *36*, 191–203. [CrossRef]
28. Zhang, Z.; Wang, X. Effective multi–field properties of electro–magneto–thermoelastic composites estimated by finite element method approach. *Acta Mech. Solida Sin.* **2015**, *28*, 145–155. [CrossRef]
29. Hassani, B.; Hinton, E. A review of homogenization and topology opimization II–analytical and numerical solution of homogenization equations. *Comput. Struct.* **1998**, *69*, 719–738. [CrossRef]
30. Macedo, R.Q.; Ferreira, R.T.L.; Guedes, J.M.; Donadon, M.V. Intraply failure criterion for unidirectional fiber reinforced composites by means of asymptotic homogenization. *Compos. Struct.* **2017**, *159*, 335–349. [CrossRef]
31. Macedo, R.Q.D.; Ferreira, R.T.L.; Donadon, M.V.; Guedes, J.M. Elastic properties of unidirectional fiber–reinforced composites using asymptotic homogenization techniques. *J. Braz. Soc. Mech. Sci.* **2018**, *40*, 255–265. [CrossRef]
32. Brito-Santana, H.; Medeiros, R.D.; Rodriguez-Ramos, R.; Tita, V. Different interface models for calculating the effective properties in piezoelectric composite materials with imperfect fiber–matrix adhesion. *Compos. Struct.* **2016**, *151*, 70–80. [CrossRef]
33. Medeiros, R.D.; Moreno, M.E.; Marques, F.D.; Tita, V. Effective properties evaluation for smart composite materials. *J. Braz. Soc. Mech. Sci.* **2012**, *34*, 362–370. [CrossRef]
34. Naghdinasab, M.; Farrokhabadi, A.; Madadi, H. A numerical method to evaluate the material properties degradation in composite RVEs due to fiber–matrix debonding and induced matrix cracking. *Finite Elem. Anal. Des.* **2018**, *146*, 84–95. [CrossRef]

35. Rodriguez-Ramos, R.; Medeiros, R.D.; Guinovart-Diaz, R.; Bravo-Castillero, J.; Otero, J.A.; Tita, V. Different approaches for calculating the effective elastic properties in composite materials under imperfect contact adherence. *Compos. Struct.* **2013**, *99*, 264–275. [CrossRef]
36. Montero-Chacon, F.; Zaghi, S.; Rossi, R.; Garcia-Perez, E.; Heras-Perez, I.; Martinez, X.; Oller, S.; Doblare, M. Multiscale thermo–mechanical analysis of multi–layered coatings in solar thermal applications. *Finite Elem. Anal. Des.* **2017**, *127*, 31–43. [CrossRef]
37. Guan, X.; Liu, X.; Jia, X.; Yuan, Y.; Cui, J.; Mang, H.A. A stochastic multiscale model for predicting mechanical properties of fiber reinforced concrete. *Int. J. Solids Struct.* **2015**, *56*, 280–289. [CrossRef]
38. Santana, H.B.; Thiesen, J.L.M.; Medeiros, R.D.; Ferreira, A.; Tita, V. Multiscale analysis for predicting the constitutive tensor effective coefficients of layered composites with micro and macro failures. *Appl. Math. Model.* **2019**, *75*, 250–266. [CrossRef]
39. Nasirov, A.; Gupta, A.; Hasanov, S.; Fidan, I. Three–scale asymptotic homogenization of short fiber reinforced additively manufactured polymer composites. *Compos. Part B Eng.* **2020**, *202*, 108269. [CrossRef]
40. Zhang, L.; Yu, W. Variational asymptotic homogenization of elastoplastic composites. *Compos. Struct.* **2015**, *133*, 947–958. [CrossRef]
41. Han, F.; Cui, J.; Yu, Y. The statistical second–order two–scale method for thermomechanical properties of statistically inhomogeneous materials. *Int. J. Numer. Meth. Eng.* **2009**, *46*, 654–659. [CrossRef]
42. Nasution, M.R.E.; Watanabe, N.; Kondo, A.; Yudhanto, A. Thermomechanical properties and stress analysis of 3–D textile composites by asymptotic expansion homogenization method. *Compos. Part B Eng.* **2014**, *60*, 378–391. [CrossRef]
43. Zhao, J.; Li, H.; Cheng, G.; Cai, Y. On predicting the effective elastic properties of polymer nanocomposites by novel numerical implementation of asymptotic homogenization method. *Compos. Struct.* **2016**, *135*, 297–305. [CrossRef]
44. Yang, Z.; Sun, Y.; Cui, J.; Yang, Z. Thermo–mechanical coupling analysis of statistically inhomogeneous porous materials with surface radiation by second–order two–scale method. *Compos. Struct.* **2017**, *182*, 346–363. [CrossRef]
45. Vega, C.R.; Pina, J.C.; Bosco, E.; Flores, E.S.; Guzman, C.; Yaneza, S. Thermo–mechanical analysis of wood through an asymptotic homogenisation approach. *Constr. Build. Mater.* **2021**, *315*, 125617. [CrossRef]
46. Bosco, E.; Claessens, R.J.M.A.; Suiker, A.S.J. Multi–scale prediction of chemo–mechanical properties of concrete materials through asymptotic homogenization. *Cem. Concr. Res.* **2020**, *128*, 105929. [CrossRef]
47. Hennessy, M.G.; Moyles, I.R. Asymptotic reduction and homogenization of a thermo–electrochemical model for a lithium–ion battery. *Appl. Math. Model.* **2020**, *80*, 724–754. [CrossRef]
48. Zhou, L.; Tang, J.; Tian, W.; Xue, B.; Li, X. A multi–physics coupling cell–based smoothed finite element micromechanical model for the transient response of magneto–electro–elastic structures with the asymptotic homogenization method. *Thin Wall. Struct.* **2021**, *165*, 107991. [CrossRef]
49. Zhang, Y.; Shang, S.; Liu, S. A novel implementation algorithm of asymptotic homogenization for predicting the effective coefficient of thermal expansion of periodic composite materials. *Acta Mech. Sin.* **2017**, *33*, 368–381. [CrossRef]
50. Cheng, G.; Cai, Y.; Xu, L. Novel implementation of homogenization method to predict effective properties of periodic materials. *Acta Mech. Sin.* **2013**, *29*, 550–556. [CrossRef]
51. Cai, Y.; Xu, L.; Cheng, G. Novel numerical implementation of asymptotic homogenization method for periodic plate structures. *Int. J. Solids Struct.* **2014**, *51*, 284–292. [CrossRef]
52. Chen, Z.; Xie, Y.M.; Wang, Z.; Li, Q.; Zhou, S. A comparison of fast fourier transform–based homogenization method to asymptotic homogenization method. *Compos. Struct.* **2020**, *15*, 111979. [CrossRef]
53. Qi, L.; Tian, W.; Zhou, J. Numerical evaluation of effective elastic properties of composites reinforced by spatially randomly distributed short fibers with certain aspect ratio. *Compos. Struct.* **2015**, *13*, 843–851. [CrossRef]
54. Tian, W.; Qi, L.; Zhou, J.; Liang, J.; Ma, Y. Representative volume element for composites reinforced by spatially randomly distributed discontinuous fibers and its applications. *Compos. Struct.* **2015**, *131*, 366–373. [CrossRef]
55. Tian, W.L.; Qi, L.H.; Chao, X.J.; Liang, J.; Fu, M. Periodic boundary condition and its numerical implementation algorithm for the evaluation of effective mechanical properties of the composites with complicated micro–structures. *Compos. Part B Eng.* **2019**, *162*, 1–10. [CrossRef]
56. Rodríguez-Ramos, R.; Otero, J.A.; Cruz-González, O.L.; Guinovart-Díaz, R.; Sevostianov, I. Computation of the relaxation effective moduli for fibrous viscoelastic composites using the asymptotic homogenization method. *Int. J. Solids Struct.* **2020**, *190*, 281–290. [CrossRef]
57. Feng, J.W.; Cen, S.; Li, C.F.; Owen, D.R.J. Statistical reconstruction and Karhunen–Loève expansion for multiphase random media. *Int. J. Numer. Meth. Eng.* **2015**, *105*, 3–32. [CrossRef]
58. Zhang, W.; Song, L.; Li, J. Efficient 3D reconstruction of random heterogeneous media via random process theory and stochastic reconstruction procedure. *Comput. Methods Appl. Mech.* **2019**, *354*, 1–15. [CrossRef]
59. Kuga, Y.; Phu, P. Experimental studies of millimeter–wave scattering in discrete random media and from rough surfaces–summary. *J. Electromagn. Wave* **1996**, *10*, 451–453. [CrossRef]
60. Chu, L.; Dui, G.; Ju, C. Flexoelectric effect on the bending and vibration responses of functionally graded piezoelectric nanobeams based on general modified strain gradient theory. *Compos. Struct.* **2018**, *186*, 39–49. [CrossRef]
61. Chen, L.; Huang, J.; Yi, M.X.; Xing, Y.F. Physical interpretation of asymptotic expansion homogenization method for the thermomechanical problem. *Compos. Struct.* **2019**, *227*, 111200. [CrossRef]

62. Pierson, H.O. *Handbook of Refractory Carbides and Nitrides*, 2nd ed.; William Andrew: Hillegom, The Netherlands, 1996; pp. 8–16.
63. Yao, D.; Li, H.; Wu, H.; Fu, Q.G.; Qiang, X.F. Ablation resistance of ZrC/SiC gradient coating for SiC–coated carbon/carbon composites prepared by supersonic plasma spraying. *J. Eur. Ceram. Soc.* **2016**, *36*, 3739–3746. [CrossRef]
64. Ding, X.; Ma, J.; Zhou, X.; Zhao, X.; Li, G. Fabrication of CeO_2–Nd_2O_3 microspheres by internal gelation process using M(OH)m and [MCit·xH_2O] (M = Ce^{3+}, Ce^{4+}, and Nd^{3+}) as precursors. *J. Sol-Gel Sci. Techn.* **2019**, *92*, 66–74. [CrossRef]
65. Flem, M.L.; Allemand, A.; Urvoy, S.; Cédat, D.; Rey, C. Microstructure and thermal conductivity of Mo–TiC cermets processed by hot isostatic pressing. *J. Nucl. Mater.* **2008**, *380*, 85–92. [CrossRef]

MDPI AG
Grosspeteranlage 5
4052 Basel
Switzerland
Tel.: +41 61 683 77 34

Materials Editorial Office
E-mail: materials@mdpi.com
www.mdpi.com/journal/materials

www.ingramcontent.com/pod-product-compliance
Lightning Source LLC
LaVergne TN
LVHW070455061126
851858LV00011B/240